IE&EM 2019

Chen-Fu Chien · Ershi Qi · Runliang Dou
Editors

IE&EM 2019

Proceedings of the 25th International Conference on Industrial Engineering and Engineering Management 2019

Springer

Editors
Chen-Fu Chien
Department of Industrial Engineering
and Engineering Management
National Tsing Hua University
Hsinchu, Taiwan

Ershi Qi
Department of Industrial Engineering
Tianjin University
Tianjin, Tianjin, China

Runliang Dou
Department of Information Management
and Management Science
Tianjin University
Tianjin, Tianjin, China

ISBN 978-981-15-4529-0 ISBN 978-981-15-4530-6 (eBook)
https://doi.org/10.1007/978-981-15-4530-6

This Springer imprint is published by the registered company Springer Nature Singapore Pte Ltd.
The registered company address is: 152 Beach Road, #21-01/04 Gateway East, Singapore 189721, Singapore

Contents

Simulation and Optimization

A Hybrid Instance Selection Method Based on Convex Hull and Nearest Neighbor

Shanghe Wang and Jin Tian(✉)

College of Management and Economics, Tianjin University, Tianjin, China
jtian@tju.edu.cn

Abstract. With the increasing size of the data, reducing dataset to reduce computational complexity has become an important task. Instance selection is one of the common preprocessing processes in data mining, which can delete the redundant instances and noisy points from dataset. In the past, various instance selection algorithms have been proposed. Most of them are effective for selecting convex instances, but fail to achieve good performances when dealing with concave instances. This paper proposes a hybrid instance selection algorithm based on convex hull and nearest neighbor information. Firstly, he proposed algorithm identifies the nearest enemy for each data point, and then divides the dataset into several subsets by grouping the points, which have the same nearest enemy, into one subset. Finally, the convex hull algorithm is executed on each subset to select the convex hull points. Our algorithm is evaluated on 14 datasets and compared with some traditional instance selection algorithms. The experimental results show that the proposed algorithm performs better than other traditional algorithms.

Keywords: Nearest enemy · Convex hull · Instance selection · SVM

1 Introduction

In recent years, with the rapidly development of the e-commerce and social networks, a large amount of data have been produced, and the data has accumulated at an unprecedented rate. How to identify noise points and redundant samples from massive data, extract the useful instances and obtain the knowledge of the whole data by analyzing the typical instances, which is still a challenge.

Instance selection is a method of data preprocessing [1], whose purpose is to select a subset from the whole dataset and achieve the original goal of the data mining application. In general, instance selection includes a family of methods which are using different rules. The objective of instance selection is to select a subset as small as possible and not lead to the reduction of the performance.

Actually, instance selection can be used to reduce the scale of the original dataset, resulting in the reduction of the computational resources, and select a subset that can perfectly represent the whole dataset. Instance selection algorithms can remove the noises and the redundant instances, remaining the important minority instances before applying learning techniques. So, the accuracy of the learning techniques will increase after using the instance selection methods.

C.-F. Chien et al. (Eds.): *IE&EM 2019*, pp. 3–13, 2020.
https://doi.org/10.1007/978-981-15-4530-6_1

In recent years, lots of instance selection methods have been proposed [2]. Some of them are to select the boundaries points to represent the whole dataset [3–7], although the subset's scale will be small, but there are still some problems. Most of these algorithms typically have high time complexity, which is often not feasible when dealing with large scale datasets. So it is usually inefficient for these algorithms to look for boundary instances by traversing the whole dataset. The convex hull algorithms can select the outer contour of the dataset. However, it's always inefficient while solving the instances on the concave surface. Therefore, this paper proposes an instance selection method based on convex hull and the nearest enemy. Our approach uses the nearest neighbor to divide the original dataset into several small subsets, and selects the boundary points on each small subset. Simultaneously, dividing the whole dataset into subsets also divides the classification decision surface into several small faces, which can reduction the limitation of the concave surface of the convex hull algorithms. First, we use an Edited Nearest Neighbor (ENN) method filter the noises [8]. Second, our algorithm divides the whole dataset into several subsets based on the distance from each point to its nearest enemy. The convex hull algorithm is used on each subset to obtain the convex hull points. Then select some convex hull points and map the remaining convex hull points to the lines of each pair of the consecutive selected points. Finally, find the closest points of the mapping point in the entire dataset. Our approach was evaluated on 14 well-known datasets and its performance was compared with 4 other important algorithms. The SVM classifier is used to verify the effectiveness of the algorithms. The results show that the proposed method performs the best accuracy, comparing with other algorithms.

Section 2 introduces our methods. Section 3 presents and analyzes the results of the experiments. Section 4 sets out the conclusions.

2 Methodology

In this section, we propose a hybrid instance selection algorithm based on convex hull and nearest enemy, called CNEIS.

Some concepts are introduced in advance. In a real vector space, for a given dataset $\boldsymbol{X}$, the intersection $T(\boldsymbol{X})$ of all convex sets containing $\boldsymbol{X}$ is called the convex hull of $\boldsymbol{X}$. $\boldsymbol{x}$ is an instance in $\boldsymbol{X}$ and t is an convex hull points of $\boldsymbol{X}$. $NE(\boldsymbol{x})$ is the nearest enemy of instance $\boldsymbol{x}$. The nearest enemy is an instance which is closest to $\boldsymbol{x}$ and has a different class label. And $c(\boldsymbol{x})$ represent the class of $\boldsymbol{x}$. knx represents the k nearest neighbors of $\boldsymbol{x}$. $l(a, b)$ represents the lines connecting points a and points b.

We firstly use an ENN method to filter the noises. We calculate the k nearest neighbors of each training instance. $\boldsymbol{x}$ is a given instance. If the number of the instances which have the same class with $\boldsymbol{x}$ in knx is smaller than a threshold p, $\boldsymbol{x}$ is considered to be a noise point and should be deleted from training set. The threshold p is an artificial value set by us.

$$p_x = \sum\nolimits_{t=1}^{|k|} \alpha(c(\boldsymbol{x}_t), c(\boldsymbol{x})) \tag{1}$$

where $\boldsymbol{x}_t$ is an instance in *knx* and the function *a* is defined as follows:

$$\alpha(c(\boldsymbol{x}_t), c(\boldsymbol{x})) = \begin{cases} 1 & if \;\; c(\boldsymbol{x}_t) = c(\boldsymbol{x}) \\ 0 & if \;\; c(\boldsymbol{x}_t) \neq c(\boldsymbol{x}) \end{cases} \tag{2}$$

If P_x is smaller than *p*, $\boldsymbol{x}$ is considered to be an noise point and should be deleted.

Then we identify each point's nearest enemy in the dataset and put the points which have the same nearest enemy into one subset. So the original dataset is divided into several subsets. Figure 1 shows a dataset which have two classes and Fig. 2 shows the subsets of points in class1 according to their nearest enemy. Instances of different shape markers represent different subsets and are recorded as follows:

$$SE(\boldsymbol{y}) = \{\boldsymbol{x}_l | NE(\boldsymbol{x}_l) = \boldsymbol{y}\} \tag{3}$$

where *SE(**y**)* is a subset in which the instance's nearest enemy is $\boldsymbol{y}$. Note that $\boldsymbol{y}$ is an instance which has the different class label with $\boldsymbol{x}$.

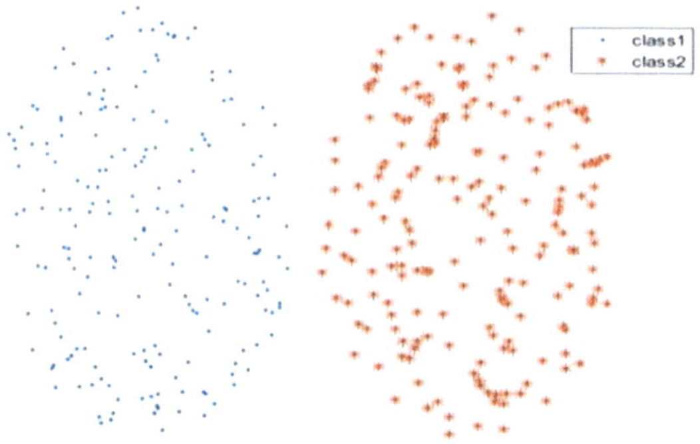

Fig. 1. A two-class dataset

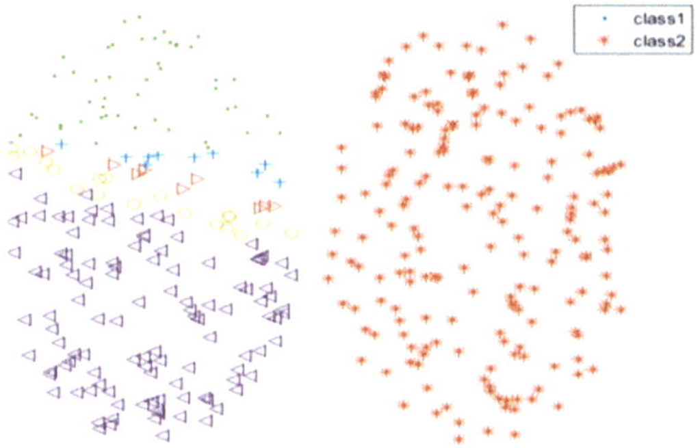

Fig. 2. The subsets of points in class1

Thus the following process can be performed in parallel in each subset. A convex hull algorithm is used on the subset *SE*($\boldsymbol{y}$) to obtain the convex hull points *T*(*SE*($\boldsymbol{y}$)). According to the geometric position of the convex hull points and its nearest enemy, when the line connecting the convex hull points to its nearest enemy is the tangent of the convex hull, the convex hull points between the two tangential points is the points to be selected in this step. These points are put into a new set ***S***. Figure 3 shows the process. Take a subset of Fig. 2 as an example. The hollow o-shaped dots are the convex hull points of the subset, and the triangle-shaped point is the nearest enemy of the subset. The two solid lines represent the two tangent lines from the nearest enemy to the convex hull of the subset. Thus there are two intersection points of the tangent and the convex hull. The yellow o-shaped points represent the final selected points, which are exactly the convex hull points between the two intersection points.

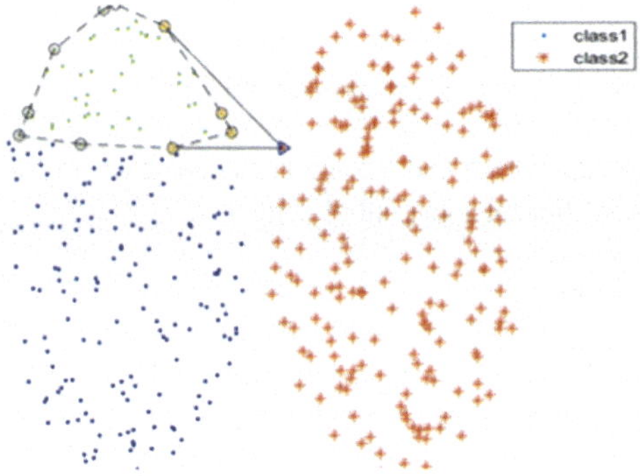

Fig. 3. Instance selection in the convex hull points of a subsets in class1

The points selected in the previous step are respectively connected to its adjacent points in the selected points set, and the remaining convex hull points are mapped to the lines of these selected points. Specifically, the remaining convex hull points ***x*** is connected with its nearest enemy *NEx* and intersects with the line connecting the selected points. Then the intersection points obtained find its nearest points in the entire dataset. These nearest points are the mapping points that we need.

The union of the mapping points and the selected points is the set of points obtained by the proposed method. The points circled by the black circles in Fig. 4 are the final selected points of class1.

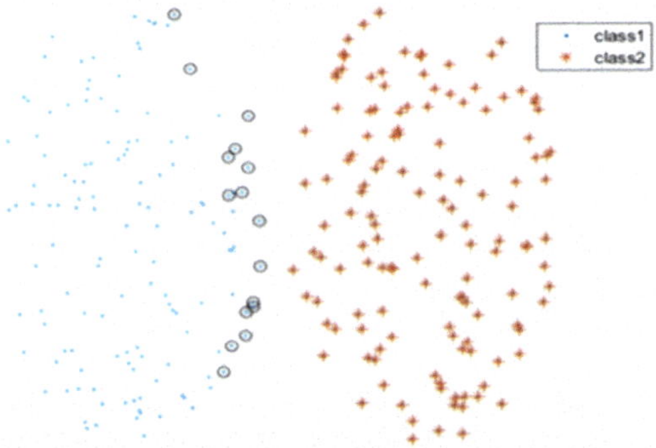

Fig. 4. The final selected points in class1

3 Experimental Results

To testify the performance of the proposed method, we compare the algorithm with four well-known important instance selection algorithms: the Instance Based 3 (IB3) [9], the Iterative Case Filtering (ICF) [10], Kernel Subclass Convex Hull (KSCH) [11] and Fast Condensed Nearest Neighbor (FCNN) [12]. We conducted the experiments on 14 datasets which are briefly described in Table 1.

Table 1. 14 Datasets used in experiments

Dataset	Instances	Attributes	Classes
Flame	240	2	2
15r	600	2	15
Pathbn	670	2	3
Ring3	2783	2	3
Jain	373	2	2
Slabcircle	1000	2	2
Sharpcorner	873	2	2
31d	3100	2	31
44overalcross	3500	2	2
Phoneme	5404	5	2
Monk	432	6	2
Newthyoid	215	5	3
Bupa	345	6	2
Haberman	306	3	2

Our experiments were designed for comparing the classification accuracy.

$$accuracy = \frac{|Success(Test)|}{|Test|} \tag{4}$$

where *Test* is a given set of instances used to testify how the instance selection algorithm performs. *Success*(*Test*) is a set of correctly classified instances in the *Test*. $|\cdot|$ represents the number of the instances in a set.

For evaluating the accuracy of our methods, we applied a SVM classifier. And conducted 30 experiments on each dataset. In each experiment, the dataset was randomly divided into two parts, a 70% training set *X* and a 30% test set *Test*. Each algorithm processes the train set *X*, use the result *S* train the classifier, and classifies the test set *Test* with the classifier to obtain the classification accuracy.

Firstly, we compare the proposed method with four instance selection methods and Table 2 shows the accuracies of the five algorithms.

Shown in Table 2, the proposed CNEIS performs best in most datasets. The KSCH algorithm mainly selects the convex instances and performs well on some datasets in which the distribution of each class of points is roughly circular, such as 15r and 30d. However, KSCH performs poorly on other datasets, such as Phoneme. The FCNN algorithm achieves lower accuracies on some datasets compared with the CNEIS, such as Jain and Ring3. As for ICF algorithm, the accuracies are much lower than those of CNEIS on 12 of 14 datasets. The IB3 algorithm only achieves similar accuracies on phoneme and flame with CNEIS, and performs worse on the other datasets. Thus we can conclude that CNEIS has an advantage on all datasets, especially on monk, haberman and bupa.

Table 2. The accuracy of all algorithms

Accuracy	KSCH	FCNN	IB3	ICF	CNEIS	ICF
Flame	0.8998	0.9157	0.9199	0.9786	0.9816	0.9786
15r	0.9921	0.9707	0.9534	0.9887	0.9958	0.9887
Pathbn	0.9231	0.8796	0.9045	0.9210	0.9959	0.9210
Ring3	0.8015	0.7997	0.8647	0.9337	1	0.9337
Jain	0.7298	0.8767	0.9754	0.9123	0.9867	0.9123
Slabcicle	0.5418	0.6754	0.7864	0.8425	0.9781	0.8425
Sharpcorner	0.5629	0.6923	0.7884	0.8218	0.9685	0.8218
31d	0.9713	0.9167	0.9468	0.9163	0.9727	0.9163
44overalcross	0.8721	0.7825	0.8651	0.8772	0.9786	0.8772
Phoneme	0.5504	0.8551	0.8343	0.7342	0.8719	0.7342
Monk	0.9667	0.6942	0.6995	0.7253	0.9975	0.7253
Newthyoid	0.9348	0.8858	0.8435	0.8497	0.9508	0.8497
Bupa	0.6999	0.5445	0.5459	0.5763	0.7220	0.5763
Haberman	0.5504	0.5074	0.5276	0.5197	0.7326	0.5197

Secondly, in the ENN algorithm used in the proposed method to filter the noises, we need to design a threshold p. For a given instance $\boldsymbol{x}$, if the number of the instances which have the same class to $\boldsymbol{x}$ in the set of k nearest neighbors is lower than p, $\boldsymbol{x}$ is considered to be a noise and should be deleted from the training set. To testify the effect of the threshold p on the classification accuracy, we set p as $\{0.3, 0.4, 0.5, 0.6\}$.

Table 3. The effect of different p values on the accuracy of the algorithm

p	0.3	0.4	0.5	0.6
Flame	0.9530	0.9530	0.9749	0.9763
15r	0.9944	0.9944	0.9944	0.9944
Pathbn	0.9940	0.9944	0.9944	0.9941
Ring3	1	1	1	1
Jain	0.9926	0.9928	0.9930	0.9924
Slabcicle	0.9979	0.9966	0.9966	0.9963
Sharpcorner	0.9532	0.9534	0.9534	0.9526
31d	0.9667	0.9689	0.9718	0.9718
44overalcross	0.9993	0.9993	0.9912	0.9666
Phoneme	0.8723	0.8641	0.8641	0.8533
Monk	0.9936	0.9936	0.9789	0.9680
Newthyoid	0.9179	0.9179	0.9000	0.8856
Bupa	0.6641	0.6641	0.6618	0.6518
Haberman	0.7253	0.7253	0.7272	0.7180

From Table 3, we can conclude that the accuracy of the algorithm varies with the value of p. With the p value is gradually increased, the accuracy of the algorithm increases. Because with the increasing of p value, the denoising intensity is strengthened, and the noises in dataset are deleted so that the accuracy is improving. However, when the p value is increased to a certain extent, with the p value increasing, the accuracy will gradually decrease. This is because that if the p value is too large, the denoising intensity will be too large, so that the effective boundary instance points are regarded as noisy points and deleted, and the accuracy is decreasing. In this paper, we use the appropriate denoising threshold for different datasets, and remove the noisy instances, then select instances from datasets.

We further conduct additional experiments to compare the instance selection performances of CNEIS and KSCH on two 2-dimensional datasets flame and ring3. Figure 5 show the instance selection results of CNEIS and KSCH on flame dataset. Different shapes represent different types of instance points. The points circled by red circles represent the instances finally selected and the points circled by the blue triangle indicates the instances that were misclassified in the test set.

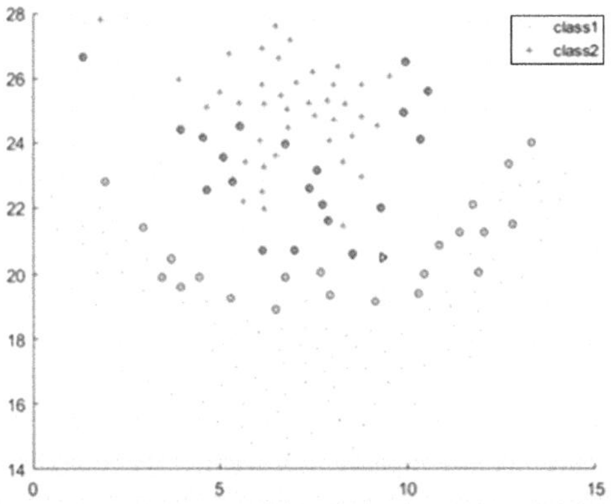

(a) CNEIS

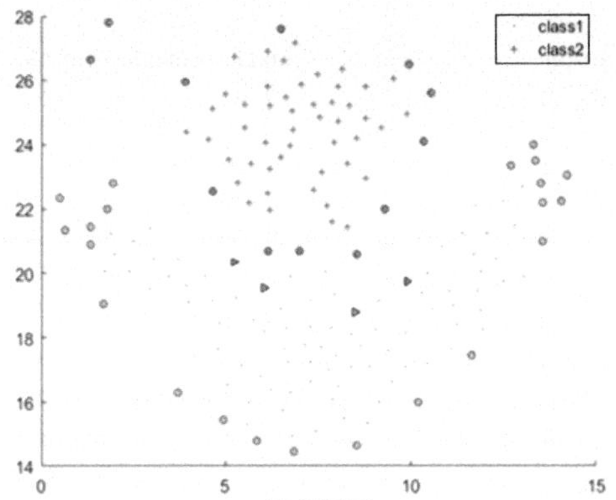

(b) KSCH

Fig. 5. The instance selection result on flame

Shown in Fig. 5, KSCH selects the convex instance points but fails to select the concave instances of class2, which lead KSCH to misclassify both two classes and more than 13% instances of class2 are misclassified. CNEIS only misclassifies a few points in class1.

Table 4 shows the classification accuracies in different classes on the flame dataset. KSCH has a high classification accurate on class1, because KSCH moves the classification hyperplane down which leads to increase the classification accuracy in class1 but damage the classification accuracy in class2.

Table 4. The accuracy of flame in different classes

Predicted			1	2
CNEIS	Real	1	0.9630	0
		2	0.0370	1
KSCH	Real	1	0.9955	0.1387
		2	0.0045	0.8613

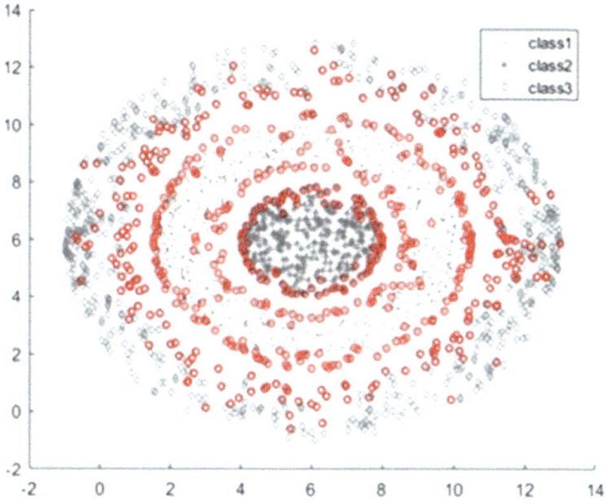

(a) CNEIS

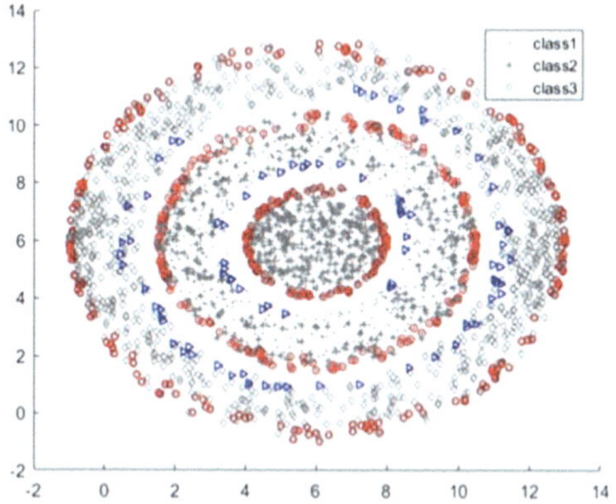

(b) KSCH

Fig. 6. The instance selection result on Ring3

Figure 6 shows the instance selection results of CNEIS and KSCH on Ring3 dataset.

Shown in Fig. 6, we can find that the KSCH algorithm can classify the instances in class2 completely correctly, but for the instances in class1 and class3, since the convex hull algorithm can only select the convex samples, the instances in the inner ring, which are close to the classification decision surface, can't be selected. The classification decision surface moved out, resulting in a large number of instances in class1 and class3 being misclassified. However, CNEIS performs well on all three classes of instances, and the accuracy is close to 100%.

Table 5 shows the classification accuracies in different classes on the ring3 dataset.

Shown in Table 5, we can find that the classifier trained by instances that KSCH algorithm identifies more than 9% of the instances in class2 as class1, and more than 15% of the instances in class3 are classified into class2.

Table 5. The accuracy of ring3 in different classes

Predicted			1	2	3
CNEIS	Real	1	1	0	0
		2	0	1	0
		3	0	0	1
KSCH	Real	1	1	0.0961	0
		2	0	0.9139	0.1647
		3	0	0	0.8453

4 Conclusion

In this paper, we propose an instances selection method based on the convex hull and the nearest enemy, which can select the boundary instance close to the classification decision surface. It can reduce the restriction of the concave instances by dividing the dataset into subsets. At the same time, the method executes in parallel on each subset, which improves the effectiveness of the method. The experimental results show that CNEIS can select powerful instances and yield the best accuracy compared to other instance selection methods.

In the future, we plan to find the points on the contour of the dataset while finding the instances close to the classification decision surface. This can solve the problem that the dataset is too large to be imported and should be processed in batches.

References

1. Carbonera, J.L., Abel, M.: A density-based approach for instance selection. In: 2015 IEEE 27th International Conference on Tools with Artificial Intelligence (ICTAI). IEEE Computer Society (2015)
2. García, S., Derrac, J., Cano, J.R., et al.: Prototype selection for nearest neighbor classification: taxonomy and empirical study. IEEE Trans. Pattern Anal. Mach. Intell. **34**(3), 417 (2012)
3. He, H., Ma, Y.: Class imbalance learning methods for support vector machines, pp. 83–99. Wiley-IEEE Press (2013)
4. Nikolaidis, K., Goulermas, J.Y., Wu, Q.H.: A class boundary preserving algorithm for data condensation. Pattern Recogn. **44**(3), 704–715 (2011)
5. Leyva, E., González, A., Pérez, R.: Three new instance selection methods based on local sets: a comparative study with several approaches from a bi-objective perspective. Pattern Recogn. **48**(4), 1523–1537 (2015)
6. Lin, W.-C., Tsai, C.-F., Ke, S.-W., Hung, C.-W., Eberle, W.: Learning to detect representative data for large scale instance selection. J. Syst. Softw. **106**, 1–8 (2015)
7. Arnaiz-González, Á., Díez-Pastor, J.-F., Rodríguez, J.J., et al.: Instance selection of linear complexity for big data. Knowl.-Based Syst. **107**, 83–95 (2016)
8. Wilson, D.L.: Asymptotic properties of nearest neighbor rules using edited data. IEEE Trans. Syst. Man Cybern. **SMC-2**(3), 408–421 (1972)

9. Aha, D.W., Kibler, D., Albert, M.K.: Instance-based learning algorithms. Mach. Learn. **6**(1), 37–66 (1991)
10. Brighton, H., Mellish, C.: Advances in Instance Selection for Instance-Based Learning Algorithms. Kluwer Academic Publishers (2002)
11. Zhou, X., Jiang, W., Tian, Y., et al.: Kernel subclass convex hull sample selection method for SVM on face recognition. Neurocomputing **73**(10–12), 2234–2246 (2010)
12. Angiulli, F.: Fast nearest neighbor condensation for large data sets classification. IEEE Trans. Knowl. Data Eng. **19**(11), 1450–1464 (2007)

Integrating Explicit Trust and Implicit Trust for Product Recommendation

Ling Ma and Jin Tian(✉)

College of Management and Economics, Tianjin University, Tianjin, China
jtian@tju.edu.cn

Abstract. Recommender systems discovers users' interests through users' historical activities, and provides personalized recommendation for users. With the development of E-commerce, there are more and more users and items, which lead recommender systems to face a lot of challenges, such as data sparsity, cold start, scalability and so on. Adding trust information to recommender system provides a new way to solve the problem of data sparsity and cold start. There are two kinds of trust relationships between users. One is explicit trust, which can get from users' trust list or friends list. The other is implicit trust, which can be obtained through users' historical activities. In this paper, we propose a recommender system based on explicit trust and implicit trust. Each user's predictive ratings consist of two parts, one is from the user's explicit trust friends, and the other is from the user's implicit trust friends. Experimental results on two datasets demonstrate that the proposed approach outperforms other state-of-the-art recommendation algorithms.

Keywords: Recommendation systems · Implicit trust · Explicit trust · Trust propagation

1 Introduction

With the development of the Internet, a lot of data are generated. It is becoming more and more difficult to quickly find the required information in a large amount of data. This is the problem of information overload. Recommender systems are proposed to solve this problem. Recommendation system can analyze user preferences according to historical data, then help users filter information, and display the useful information to users. Recommender systems are widely used in the field of e-commerce. On the one hand, it can make the user experience better and improve the loyalty of users. On the other hand, it can help businesses sell more things and increase sales.

Recommender systems can be divided into content-based recommender systems, collaborative filtering systems and hybrid recommender systems. Collaborative filtering is one of the most widely used recommender systems [1]. Collaborative filtering assumes that two users with similar historical ratings also have similar ratings for other items that have not been rated. Collaborative filtering includes three steps: the first step is to calculate the similarity between users according to their historical ratings; the second step is to find the nearest neighbors who are most similar to the target users as the nearest neighbors of the target users; and the third step is to calculate the weighted

C.-F. Chien et al. (Eds.): *IE&EM 2019*, pp. 14–20, 2020.
https://doi.org/10.1007/978-981-15-4530-6_2

average of the nearest neighbors' ratings on the target items as the predicted rating. Collaborative filtering does not require domain knowledge, and is intuitive and easy to understand. However, traditional collaborative filtering faces the following problems: data sparsity, cold start, scalability and so on [2]. Adding trust relationship between users in recommender systems provides a new way to solve the above problems.

Some scholars have tried to use trust relationship to improve recommender system. Ma et al. propose SoRec which map user-item rating matrix and trust relation matrix to low dimensional space at the same time, and combine the two matrixes by sharing the same user latent feature space [3]. Tidaltrust performs a modified breadth-first search in social networks to predict the target user's rating on the target item [4]. MoleTrust is similar to TidalTrust, the different is that MoleTrust considers all users who have rated the target item up to a maximum-depth [5]. Jamali et al. proposed a random walk method called TrustWalker, which is a combination of trust-based and item-based recommendations [6]. Deng et al. proposed an algorithm called RelevantTrustWalker, which is similar to TrustWalker, but using matrix factorization to calculate the similarity between users and the similarity between items [7]. Jamali and Ester proposed a recommender system which combines the collaborative filtering and trust-based approach to improve Top-N Recommendation [8]. Though these papers have improved recommender systems by consider trust information, most of these paper did not combine the explicit trust and implicit trust. In fact, the implicit trust is also really useful in recommendation.

In this paper, we proposed a trust based recommender system named RS-exp-imp, which combine explicit trust influence and implicit trust influence. Firstly, we mine the implicit trust between users by calculating similarity of historical ratings between users. Then, for every user, we select K explicit trust users and K implicit trust users. The predicted rating is a combination of the rating from explicit trust users and implicit trust users. Besides, we consider trust propagation to use more effective information. In specific, if the target users have not rated the target item, then his rating is predicted ratings from his own explicit trust users and implicit trust users.

2 Methodology

Suppose there are m users and n items. The historical ratings can be expressed as a matrix $\boldsymbol{R} = [r_{ij}]_{m \times n}$. r_{ij} is the rating of the jth item given by the ith user, which is usually an integer number from 1 to 5 with interval of 1. $\boldsymbol{T} = [T_{if}]_{m \times m}$ denotes the explicit trust matrix. If the ith user trust the fth user, $T_{if} = 1$, otherwise $T_{if} = 0$. The problem we study in this paper is as follows: predict the ratings for items given by users using $\boldsymbol{R}$ and $\boldsymbol{T}$.

A. Mining the implicit trust between users

Explicit trust is obtained directly from the user's trust list or friends list and is defined by the user himself. Implicit trust is mined through user's historical ratings. The idea behind implicit trust is that there is a trust relationship between the target user and those users who have similar historical ratings with him. So we calculate the similarities between users. In this paper, we use Pearson Correlation Coefficient (PCC) to measure the similarities between users. The calculation formula of PCC is 1.

$$sim_{ig} = \frac{\sum_{j \in \mathbf{I}_i \cap \mathbf{I}_g} (r_{ij} - \bar{R}_i) \times (r_{gj} - \bar{R}_g)}{\sqrt{\sum_{j \in \mathbf{I}_i \cap \mathbf{I}_g} (r_{ij} - \bar{R}_i)^2} \times \sqrt{\sum_{j \in \mathbf{I}_i \cap \mathbf{I}_g} (r_{gj} - \bar{R}_g)^2}} \tag{1}$$

where sim_{ig} is the similarity of the ith user and the gth user, $\mathbf{I}_i$ is the set of the items that the ith user has rated r_{ij} is the rating that the ith user gave to the jth item $\bar{R}_i$ is the average rating of the ith user. The value of sim_{ig} is between -1 and 1. The larger the value is, more similar the two users are.

For every user, we select top K users who are the most similar to him as his implicit trust users and build implicit trust relationship between users.

B. Trust propagation

For the target user, his predicted rating is a combination of predicted rating from his explicit trust users and his implicit trust users. As shown in Fig. 1.

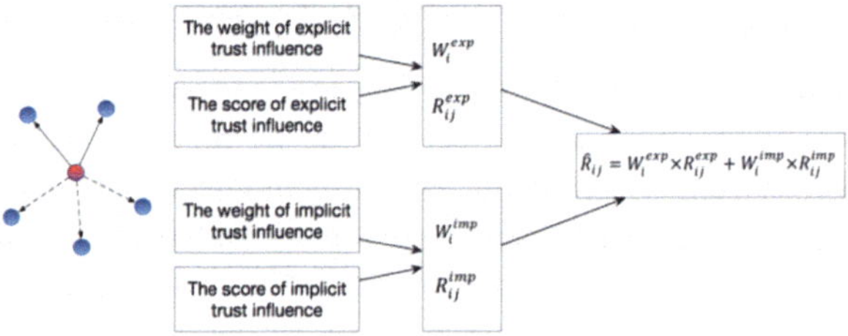

Fig. 1. Recommender systems based on explicit trust and implicit trust

The target user is presented as a red point in Fig. 1. The dotted line represents the implicit trust relationship between users, while the solid lines represent explicit trust relationships between users. Suppose the target user has two explicit users and three implicit users, his predicted rating for the target item consists of two part. One is from his two explicit users, which is defined as R_{ij}^{exp}. Another is from his three implicit users, which is defined as R_{ij}^{imp}. The weights of R_{ij}^{exp} and R_{ij}^{imp} are W_i^{exp} and W_i^{imp}, respectively.

$$W_u^{exp} + W_u^{imp} = 1 \tag{2}$$

R_{ij}^{exp} is the weighted average of the ratings from all explicit trusted users. The formula of R_{ij}^{exp} is 3.

$$R_{ij}^{exp} = \frac{\bar{R}_i \sum_{m \in T_i^{exp}} Trust_{im}^{exp} \times (R_{mj} - \bar{R}_m)}{\sum_{m \in T_i^{exp}} Trust_{im}^{exp}} \tag{3}$$

where $\bar{R}_i$ is the average rating of the ith user. T_i^{exp} is the set users who are explicit users of the ith user. $Trust_{im}^{exp}$ is the trust degree of the ith user to the mth user. The more the target user trusts the mth user, the large the weight is and the greater the influence from the mth user, $Trust_{im}^{exp}$ is calculated by 4.

$$Trust_{im}^{exp} = \frac{1}{1 + \exp(-\frac{|I(i) \cap I(m)|}{2})} \times sim_{im} \quad (4)$$

where $I(i)$ is the item set the ith user has rated. sim_{im} is calculated using 1.

Similar to explicit trust, the formula of R_{ij}^{imp} is defined in 5

$$R_{ij}^{imp} = \frac{\bar{R}_i \sum_{n \in T_i^{imp}} Trust_{in}^{imp} \times (R_{nj} - \bar{R}_n)}{\sum_{n \in T_i^{imp}} Trust_{in}^{imp}} \quad (5)$$

where T_i^{imp} is the set users who are implicit users of the ith user. $Trust_{in}^{imp}$ is the trust degree of the ith user to the nth user, which is calculated by 6.

$$Trust_{in}^{imp} = \frac{1}{1 + \exp(-\frac{|I(i) \cap I(n)|}{2})} \times sim_{in} \quad (6)$$

In this paper, we set $W_u^{exp} = W_u^{imp} = 0.5$.

If a user who is trusted by the target user has not rated the target item, we predict a rating for from his own explicit trust users and implicit trust users, and the predicted rating are seen as his own rating and is fed back to the target user. In this way, propagation of both explicit trust and implicit trust is considered into the recommender system.

3 Results

A. Datasets

In this section, we select two well-known datasets FilmTrust and Ciao to do experiments to compare the recommender system proposed in this paper and other recommender systems.

FilmTrust is a movie website. Users in this website are able to rate the movies in the range of 0.5 (min) to 4.0 (max) with step 0.5. Moreover, we can get the explicit trust relationship between users. The Ciao dataset contains users' ratings on an online-shopping website Ciao.com. The values of the ratings are in the range of 1 (min) to 5

Table 1. Statistics of FilmTrust and Ciao

Statistics	FilmTrust	Ciao
Users	1642	4770
Items	2071	5079
Social relations	1853	12883
Ratings	35494	44716
Maximum number of ratings per user	244	679
Maximum number of social relations per user	59	73

(max) with step 1. We can also get the explicit trust relationship between users. The general statistics of the two datasets are shown in Table 1.

B. Metrics

We use the Mean Absolute Error (MAE) and the Root Mean Square Error (RMSE) to measure the accuracy of predicted ratings.

The MAE is defined as:

$$MAE = \frac{\sum_{(i,j)\in \mathrm{Te}} \left| r_{ij} - \hat{r}_{ij} \right|}{|\mathbf{Te}|} \tag{7}$$

where $\hat{r}_{ij}$ is the predicted value of the rating r_{ij} and $\boldsymbol{Te}$ is the testing rating set.

The RMSE is defined as

$$RMSE = \sqrt{\frac{\sum_{(i,j)\in \mathrm{Te}} \left(r_{ij} - \hat{r}_{ij} \right)^2}{|\mathbf{Te}|}} \tag{8}$$

The smaller the value of MAE or RMSE is, the better the recommendation performance is.

C. Comparison

We compare our proposed recommender system and the following recommender systems.

(1) UBCF: user-based CF approach, which predicts the preference of a target user by collecting the ratings from other similar users.
(2) TrustWalker: proposed by Jamali and Ester, which is a random walk method combining the trust-based and item-based recommendations [6].
(3) RelevantTrustWalker: proposed by Deng et al. [7]. The target of each walk is selected according to the trust relevancy among users instead of being selected randomly.
(4) Trust-CF: proposed by Jamali and Ester [8]. It combines the collaborative filtering and trust-based approach.

K is the number of implicit trusted users. We do 5-fold cross validation six times. A total of 30 experiments were conducted and T-test was carried out. The results are shown in Tables 2 and 3.

Table 2. Accuracy comparisons of rating prediction on FilmTrust

Recommender systems	FlimTrust	
	RMSE	MAE
UBCF	0.8551***	0.6531***
Trustwalker	0.8975***	0.6646***
RelevantTrustWalker	0.8958***	0.6683***
Trust-CF	1.0709***	0.8077***
RS-exp-imp (K = 5)	0.8383	0.6369
RS-exp-imp (K = 10)	0.8383	0.6305

*p < 0.05; **p < 0.01; ***p < 0.001

Table 3. Accuracy comparisons of rating prediction on Ciao

Recommender systems	Ciao	
	RMSE	MAE
UBCF	1.0817***	0.7937***
Trustwalker	1.2128***	0.8550***
RelevantTrustWalker	1.1795***	0.8409***
Trust-CF	1.5302***	1.1141***
RS-exp-imp (K = 5)	1.0059	0.7473
RS-exp-imp (K = 10)	1.0036	0.7477

*p < 0.05; **p < 0.01; ***p < 0.001

From Tables 2 and 3 we can see that our proposed recommender system can predict ratings more accurate than other 4 recommender systems and T test are significant. When $K = 10$, RS-exp-imp performs better than $K = 5$, but the results is not significant. So the parameter K can affect the results, but the effect is not obvious.

D. The Advantage of Explicit Trust and Implicit Trust

In this section, we do experiments to explore the advantages of combining explicit trust and implicit trust. We compare the performance of only consider explicit trust, only consider implicit trust and consider both of explicit trust and implicit trust. The results are shown in Table 4.

It can be seen that considering both explicit trust and implicit trust perform best, which confirm the advantage of explicit trust and implicit trust.

Table 4. The advantage of explicite trust and implicite trust

K = 10	Ciao		FilmTrust	
	RMSE	MAE	RMSE	MAE
RS-exp	1.0896	0.7916	0.8507	0.6388
RS-Imp	1.0052	0.7499	0.8310	0.6307
RS-exp-imp	1.0036	0.7477	0.8303	0.6305

4 Conclusion

In this paper, we proposed a recommender system RS-exp-imp, which combines both explicit trust and implicit trust. We calculate the similarity between users and mine the implicit trust between users. Then predicted rating for the target user is the combination of predicted rating from users who are explicit trusted by him and predicted rating from users who are implicit trusted by him. The degree of trust determines the weight of rating. If the direct trusted users have not rated the target items, we predict a rating for him in the same way as his rating. In this way, trust propagation is considered into RS-exp-imp. Experimental results on two datasets demonstrate that the proposed RS-exp-imp outperforms other state-of-the-art recommender systems and confirms the advantage of combine the explicit trust and implicit trust.

Acknowledgment. This work was supported by the General Program of the National Science Foundation of China (Grant No. 71871156, 71471127).

References

1. Goldberg, D., Nichols, D., Oki, B.M., Terry, D.: Using collaborative filtering to weave an information tapestry. Commun. ACM **35**(12), 61–70 (1992)
2. Aggarwal, C.C.: Recommender Systems: The Textbook. Springer, Cham (2016)
3. Ma, H.Y., Lyu, M.R., King, I.: SoRec: social recommendation using probabilistic matrix factorization. In: ACM Conference on Information and Knowledge Management, pp. 931–940 (2008)
4. Golbeck, J.A.: Computing and applying trust in web-based social networks. University of Maryland at College Park (2005)
5. Massa, P., Avesani, P.: Trust-aware recommender systems. In: ACM Conferencc on Recommender Systems, pp. 17–24 (2007)
6. Jamali, M., Ester, M.: TrustWalker: a random walk model for combining trust-based and item-based recommendation. In: The 15th ACM SIGKDD International Conference on Knowledge Discovery and Data Mining, pp. 397–405 (2009)
7. Deng, S., Huang, L., Xu, G.: Social network-based service recommendation with trust enhancement. Expert Syst. Appl. **41**(18), 8075–8084 (2014)
8. Jamali, M., Ester, M.: Using a trust network to improve top-N recommendation. In: ACM Conference on Recommender Systems (2009)

Prediction of Engineering Parameters Based on Improved Artificial Neural Network

Jiang Shen[1], Ling-ling Li[1], and Dong-ming Lin[2(✉)]

[1] Department of Management and Economics, Tianjin University, Tianjin, China
[2] Logistics Center of Shenzhen Power Supply Bureau, China Southern Power Grid, Shenzhen, China
13728662879@163.com, chengzhi0821@163.com

Abstract. The prediction of engineering parameters is crucial for engineering management. This study proposes an improved integrated artificial neural network (ANN) model, combining ANN with AdaBoost algorithm and cost sensitive method to predict engineering parameters. Through the integration of ANN and the set of prediction error threshold and cost value, it not only improves the generalization ability of ANN, but also ensures the scientificity of engineering prediction. The proposed model is applied to predict concrete compressive strength, and compared with artificial neural network, linear support vector machine, support vector regression, multiple linear regression, classification and regression tree. The results show that the mean absolute percentage errors are respectively 14.23%, 16.54%, 30.07%, 17.91%, 31.85%, and 26.08%. Compared with the other five models, the proposed model shows best prediction accuracy. The study shows that the proposed model is superior in engineering prediction.

Keywords: ANN · AdaBoost algorithm · Engineering prediction

1 Introduction

In recent years, Building Information Modeling (BIM) has been widely used in engineering projects. BIM contains a large amount of project information, such as geometric and physical data of components, functional features, etc. As a platform for data integration and sharing, BIM can provide reliable data and result visualization for algorithms [1]. Many researchers have studied the combination of BIM and various algorithms for cost prediction, construction site layout, etc. The capability of Artificial Neural Network (ANN) to perform nonlinear mapping of output from multiple inputs makes them suitable for handling estimation problems in construction that typically involve complex input-output relations [2–4]. ANN demonstrated excellent predictive ability [5–7]. Reference [8] used BP neural network to predict construction cost based on BIM.

In the process of machine learning, the prediction results of ANN depend on its network topology, weights, thresholds and other parameters. Therefore, the results are often unstable. Researchers have optimized the parameters of ANN to improve the

C.-F. Chien et al. (Eds.): *IE&EM 2019*, pp. 21–27, 2020.
https://doi.org/10.1007/978-981-15-4530-6_3

prediction accuracy. Lahiri and Khalfe [9] developed a new hybrid procedure to find the optimum ANN architecture and tunes the ANN parameters. This method incorporated hybrid ANN and differential evolution technique (ANN-DE) for efficient tuning of ANN meta parameters. Jo *et al.* [10] introduced three optimization algorithms to search for the optimal values of the ANN's hyper-parameters. They were random search (RS), tree-structured Parzen estimator (TPE), and hyper-parameter optimization via radial basis function and dynamic coordinate search (FIORD). Lo *et al.* [11] used a parameter automatic calibration (PAC) approach to adjust the training parameters and found the results yielded by the ANN-PAC model were more reliable. Lu *et al.* [12] pointed out that the genetic algorithm (GA) performs better in determining the weighting and threshold of ANN models and is therefore combined with BP to develop this study's ANN model. Although hyperparametric optimization of ANN can improve its prediction accuracy, the optimized ANN model has poor generalization ability. Therefore, when the model is applied to new data sets, and robustness of the model is weak. In order to overcome this problem, Sun and Gao [13] combined AdaBoost algorithm with neural network to overcome the instability of single neural network and provide more accurate and stable prediction for new data sets. AdaBoost algorithm is an iterative algorithm, which is mainly applied to the classification problem [14]. After several years of development, it has been extended to the prediction field [15]. AdaBoost algorithm can improve the prediction accuracy of any given weak predictor.

In this study, AdaBoost algorithm is used to achieve predictors ensemble. ANN is regarded as weak predictor, and the sample weight is adjusted according to the prediction error of ANN. The updated sample weight is used again to train the next new weak predictor, and finally a series of weak predictors are combined, getting a strong predictor and the final output. In order to test the accuracy of the proposed algorithm in engineering prediction, it is applied to the prediction of concrete compressive strength. In order to avoid engineering risks, the prediction error needs to be controlled within a certain range. Therefore, the proposed algorithm is combined with cost sensitive method, threshold value is set for prediction errors.

2 Methodology

In order to improve the accuracy of prediction, the AdaBoost algorithm is used to build a hybrid prediction framework containing multiple ANNs. The prediction results of all ANNs are summarized to obtain the final prediction results. The weight of each ANN is controlled by the AdaBoost algorithm. In order to distinguish the importance of different training samples, the weight of the training samples is jointly controlled by the AdaBoost algorithm and the cost sensitive method, and the cost sensitive method is used to set an appropriate threshold for the prediction error. When the error exceeds the threshold, the sample weight needs to be increased on the basis of high cost value, that is, the sample whose prediction error falls outside the threshold has higher prediction cost. When the prediction error is within the set threshold, the sample weight is decreased based on the lower cost value.

In order to ensure the validity and scientificity of the engineering prediction results, the prediction whose error exceeds the set threshold is regarded as a wrong prediction, that is, the prediction result that is too high or too low is unacceptable, therefore, the prediction problem can be converted into classification problem. The prediction results are divided into two categories, one is the correct prediction, the error of which is within the threshold, and the prediction is regarded as +1, and the other is the wrong prediction, the error of which exceeds the threshold, and the prediction is regarded as −1. The prediction result is recorded as $(h_i(x_i), y_i)$, y_i is the actual result, $(+1, +1)$ represents that the prediction is correct, and the cost value is C_1; $(+1, -1)$ represents that the prediction is wrong, and the cost value is C_2. The cost of $(+1, -1)$ is higher than $(+1, +1)$, that is, $C_2 > C_1$. By combining the AdaBoost algorithm with the cost-sensitive method, the prediction accuracy of ANN is improved.

ANN is applied for prediction. AdaBoost algorithm and the cost-sensitive method are used to adjust the sample weights, and then several ANNs (weak predictors) are combined by AdaBoost algorithm to achieve the high-precision prediction. The calculation steps of the algorithm are as follows:

Step 1. Select the training data set $S_n = \{X_i, Y_i\} = \{(x_1, y_1), (x_2, y_2), \ldots, (x_n, y_n)\}$ $(i = 1, 2, \ldots, n)$. The weight of the sample S_n is $\{W_t(i)\}$ $(t = 1, 2, \ldots, T)$. The initial weight of the sample is $1/n$.

$$\{W_1(i)\} = 1/n \quad (i = 1, 2, \ldots, n) \tag{1}$$

n is the sample size. T is the number of the predictors.

Step 2. Use the predictor H_t to predict the output variable $\{Y_i\}$, and calculate the prediction error $\{E_t\}$.

$$E_t = \sum\nolimits_{i:f_t(x_i) \neq y_i} w_t^i \tag{2}$$

Step 3. Calculate the weight of the predictor.

$$\alpha_t = \frac{1}{2}\ln\left(\frac{1 - E_t}{E_t}\right) \tag{3}$$

Step 4. Update the sample weight $\{W_t(i)\}$. The error threshold value is set to θ.

$$\begin{cases} W_t(i) = \frac{C_i W_{t-1}(i)\exp(-\alpha_t C_i l_t^i)}{Z_t} \\ Z_t = \sum\limits_i C_i W_{t-1}(i)\exp(-\alpha_t C_i l_t^i) \end{cases} \tag{4}$$

$$l_t^i = \begin{cases} 1, f_t(x_i) = y_i \\ -1, f_t(x_i) \neq y_i \end{cases} \tag{5}$$

$$C_i = \begin{cases} C_1, E_t \leq \theta \\ C_2, E_t > \theta \end{cases} \tag{6}$$

Calculating according to the sample weight update formula above can minimize the prediction error with high cost [16].

Step 5. Repeat step 1–4 until all predictors have been executed.

Step 6. Summarize all the predictors in the AdaBoost framework to form the final strong predictor.

$$H = \sum_{t=1}^{T} \alpha_t H_t \tag{7}$$

The engineering parameters are used to predict the concrete compressive strength. The data set used in the experiment is from the UCI database, which include1030 samples. The sample attributes include cement (m^3/kg), blast furnace slag (m^3/kg), fly ash (m^3/kg), water (m^3/kg), superplasticizer (m^3/kg), coarse aggregate (m^3/kg), fine aggregate (m^3/kg) and age (day).

In practical engineering projects, the data of these basic parameter can be directly read from BIM through the API interface using algorithm software, and then analyzed by the algorithm proposed in this paper. In order to test the superiority of the improved artificial neural network, the data set is obtained from the engineering database.

The three-layer ANN model is applied, the number of input neurons is 8, the number of output neurons is 1, the number of hidden layer neurons is 32, the learning rate is 0.2, the number of iterations is 500, and the number of experiments is 25. The error threshold is set to 0.15. The AdaBoost algorithm integrates 10 ANN models. 75% of the total samples are randomly selected as the training set, and the remaining 25% as the test set.

3 Results

Four evaluation indicators are selected to evaluate the performance of the model, namely mean absolute error (MAE), mean square error (MSE), root mean square error (RMSE) and mean absolute percentage error (MAPE).

To obtain comparable experimental results, the same problem is solved by five different methods. AdaBoost-ANN algorithm is compared with artificial neural network (ANN), linear support vector machine (linear SVM), support vector regression (SVR), multiple linear regression (MLR), classification and regression tree (CART) by calculating the values of the prediction error. The experimental results are shown in Table 1.

Table 1. Experimental results of six models

	AdaBoost-ANN	ANN	Linear SVM	SVR	MLR	CART
MAE (MPa)	4.20 ± 0.28	4.85 ± 1.01	8.15 ± 0.35	4.95 ± 0.22	8.24 ± 0.31	7.22 ± 0.32
MSE (MPa2)	33.35 ± 5.36	42.23 ± 16.89	118.96 ± 14.71	45.10 ± 5.01	108.02 ± 7.76	85.97 ± 7.03
RMSE (MPa)	5.75 ± 0.46	6.40 ± 1.11	10.89 ± 0.68	6.71 ± 0.37	10.39 ± 0.38	9.26 ± 0.38
MAPE (%)	14.23 ± 1.16	16.54 ± 3.27	30.07 ± 2.25	17.91 ± 1.75	31.85 ± 2.23	26.08 ± 1.65

It can be seen from Table 1 that the mean absolute error of the AdaBoost-ANN algorithm is reduced by 0.65–4.04 compared with the other five algorithms. In addition, among the other five algorithms, the mean absolute error of the ANN model is the lowest. Similar results are also shown in other indicators. Compared with the other five algorithms the mean square error, root mean square error, and average absolute percentage error of the proposed algorithm are reduced by 8.88–85.61, 0.65–5.14 and 2.31%–17.62%, respectively. Among the other five algorithms, the ANN model outperforms other algorithms. The mean square error is reduced by 8.88–85.61. The root mean square error is reduced by 0.65–5.14. The mean absolute percentage error is reduced by 2.31%–17.62%. Therefore, compared with the other five algorithms, the proposed algorithm has higher prediction accuracy, which indicates that it is superior to the other five algorithms in engineering prediction.

In addition, For SVR, the variance of each evaluation indicator is the lowest among the other five algorithms. For the proposed algorithm, the variance of MAE and MSE is only higher than that of SVR, and the variance of MAPE is the lowest. In addition, the variance of every indicator of the ANN model is the highest.

4 Discussion

Table 1 shows that compared with the other five algorithms, the proposed algorithm has the highest prediction accuracy, and each evaluation indicator of the method is lower, indicating that the proposed algorithm outperforms the other five algorithms and the prediction stability is better. Among the other five algorithms, the ANN model has the best performance in prediction accuracy, but it has the worst performance in prediction stability, which indicates that the ANN model can accurately predict the compressive strength of concrete, but its poor stability can sometimes lead to excessive differences in prediction accuracy, which limits the practical application of the ANN model. Compared with the ANN model, the AdaBoost-ANN algorithm further improves the accuracy of the prediction, and more importantly, it significantly improves the stability of the prediction, and can avoid the situation, to some extent, where the prediction error is large, thus having better generalization ability. The proposed algorithm has higher engineering applicability in practical engineering.

5 Conclusion

From the perspective of engineering application, in order to reduce the risk of engineering practice and improve the prediction accuracy of engineering parameters, the cost-sensitive method is combined with ANN to set a threshold for the prediction error, and the samples whose prediction errors are beyond the threshold are set to higher cost values, thus ensuring higher prediction accuracy and improving the safety and scientificity of engineering prediction. In order to overcome the dependence of single ANN prediction on hyperparameters and samples, improve the generalization ability of the model and the accuracy of prediction, the AdaBoost algorithm is used to integrate multiple ANNs so as to overcome the dependence of single ANN on hyperparameters and samples and improve the generalization ability of the model. By applying the proposed model to concrete compressive strength prediction and comparing with the experimental results of five other algorithms, it is proved that the proposed model has high precision in engineering prediction.

Combining BIM with ANN provides a new idea for engineering data analysis of BIM project. With the promotion and application of BIM, it will become a new direction for the informatization development of engineering management to combine BIM with algorithms to make full use of the massive data in BIM. By applying the results of data analysis to engineering practice, technical support can be provided for engineering decisions.

There are a large number of engineering parameters in the construction project. This study only uses the compressive strength of concrete as an example to verify the applicability of the improved artificial neural network in engineering prediction. Therefore, in order to meet the actual needs of engineering practice, it is necessary to conduct further research in other engineering parameters in BIM.

Acknowledgment. The research is supported by Shenzhen Power Supply Bureau of China Southern Power Grid commission project: Material Storage and Distribution Management Method Based on Digital Portrait and Intelligent Analysis Technology (grant no. 0918002019 030304WLCP00003).

References

1. Gentili, G.B., Tesi, V., Linari, M., Marsili, M.: A review of BIM-based artificial intelligence methods. J. Graph. **39**(5), 797–805 (2018)
2. Chao, L.C., Skibniewski, M.J.: Neural network method of estimating construction technology acceptability. Constr. Eng. Manage. **121**(1), 130–142 (1995)
3. Boussabaine, A.H.: The use of artificial neural networks in construction management: a review. Constr. Manage. Econ. **14**(5), 427–436 (1996)
4. Boussabaine, A.H., Kaka, A.P.: A neural networks approach for cost flow forecasting. Constr. Manage. Econ. **16**(4), 471–479 (1998)
5. Hossain, K.M.A.: Regression and artificial neural network models for strength properties of engineered cementitious composites. Neural Comput. Appl. **29**(9), 631–645 (2018)
6. Erzin, Y., Cetin, T.: The prediction of the critical factor of safety of homogeneous finite slopes using neural networks and multiple regressions. Comput. Geosci. **51**, 305–313 (2013)
7. Leu, S.-S., Lo, H.-C.: Neural-network-based regression model of ground surface settlement induced by deep excavation. Autom. Constr. **13**(3), 279–289 (2004)
8. Tang, H.-Y., Liu, R.-G., Han, Y., Wang, J.: The construction of project cost prediction system based on 5D-BIM. J. Eng. Manage. **29**(4), 107–112 (2015)
9. Lahiri, S.K., Khalfe, N.: Modeling of commercial ethylene oxide reactor: a hybrid approach by artificial neural network & differential evolution. Int. J. Chem. Reactor Eng. **8**(1) (2010)
10. Jo, Y., Min, K.: Comparative study of the artificial neural network with three hyper-parameter optimization methods for the precise LP-EGR estimation using in-cylinder pressure in a turbocharged GDI engine. Appl. Therm. Eng. **149**(25), 1324–1334 (2019)
11. Lo, D.-C., Wei, C.-C.: Parameter automatic calibration approach for neural-network-based cyclonic precipitation forecast models. Water **7**(7), 3963–3977 (2015)
12. Lu, W., Peng, Y., Chen, X., Skitmore, M., Zhang, X.: The S-curve for forecasting waste generation in construction projects. Waste Manage. **56**, 23–34 (2016)

13. Sun, W., Gao, Q.: Exploration of energy saving potential in China power industry based on AdaBoost back propagation neural network. J. Clean. Prod. **217**(20), 257–266 (2019)
14. Zhong, X.Y., Ling, J.: AdaBoost detector based on multiple thresholds for weak classifier. Comput. Eng. Appl. **45**(19), 160–162 (2009)
15. Ge, Q.F., Feng, X.: Classification and prediction of rockburst using AdaBoost combination learning method. Rock Soil Mech. **29**(4), 943–948 (2008)
16. Sun, Y.: Cost-sensitive boosting for classification of imbalanced data. Pattern Recogn. **40**, 3358–3378 (2007)

Research on BP Neural Network Optimization Based on Improved Global Artificial Fish Swarm Algorithms

Shun-tang Zhang and Xue-meng Wei(✉)

School of Management Science and Engineering,
Shandong Technology and Business University, Yantai, China
974460394@qq.com

Abstract. This paper uses the improved global artificial fish swarm algorithm to optimize BP neural network, which reduces the training error of the neural network. Firstly, the structure of BP neural network is established to determine the number of network layers and neurons. Secondly, the improved global artificial fish swarm algorithm is used to optimize the algorithm, which mainly searches for the initial weights and thresholds of the global optimum according to the convergence and search ability of the algorithm. Finally, the optimized BP neural network is used to detect, the weights and thresholds are assigned to the BP neural network. Then the weights and thresholds are trained to minimize the training error. Through experimental analysis, the improved global artificial fish swarm algorithm can optimize BP neural network better, reduce errors and training times, improve the training speed of the network, and enrich the research on the improvement of the neural network.

Keywords: BP neural network · Global artificial fish swarm algorithm · Optimization · Simulation

1 Introduction

The error function of BP neural network is a non-linear function with Sigmoid function as its independent variable. There inevitably exists a local minimum with zero gradient on the error surface [1]. The training process is mainly accomplished by two stages: the forward propagation of signals and the reverse propagation of errors. The error of each connection weight of the neural network can be reduced continuously [2]. BP network has strong ability of non-linear mapping [3], but because the weights and thresholds of the neural network are adjusted according to the direction of negative gradient, the training results will have errors with the actual situation. Therefore, many experts and scholars are studying how to reduce network errors, and put forward many improvement methods, such as: BP Neural Network Constraints [4]. Optimizing Neural

Foundation projects: National Natural Science Foundation of China (41601593, 71373148); Shandong Soft Science Project (2014 RKB01021); Shandong High Level Applied Industrial Engineering Professional Group Construction Project; Shandong Natural Fund Project (ZR2011GL001).

C.-F. Chien et al. (Eds.): *IE&EM 2019*, pp. 28–35, 2020.
https://doi.org/10.1007/978-981-15-4530-6_4

Networks by Genetic Algorithms [5, 6]. Neural Network Based on Bee Colony Optimization [7]. Optimizing BP Neural Network by Improved Differential Evolution Algorithms [8]. Optimizing Neural Networks by Particle Swarm Optimization [9]. In the aspect of application, the analytic hierarchy process and neural network are combined and applied to the actual situation [10]. And more research results have been achieved. Guo [11] proposed to improve BP neural network, and then reduce the error. Yu and Xidongmin [12] proposed an artificial fish swarm algorithm to optimize the initial weights of the PID neural network, aiming at the characteristics of decoupling control for multivariable systems, which accelerated the convergence speed of the PID neural network. Liu [13] introduced chaotic initialization strategy and immune regulation mechanism to overcome the shortcomings of basic AFSA, and used the improved AFSA to optimize the wavelet neural network. In BP neural network, the initial weights and thresholds of the network are set randomly, not artificially controlled. As a result, in the process of training the network, BP neural network is easy to fall into local extremes and increase training errors.

In order to improve the prediction accuracy of BP neural network, this paper improves the global artificial fish swarm algorithm, and constantly adjusts the initial weights and thresholds of BP neural network, combines the improved global artificial fish swarm algorithm with BP neural network to optimize the network, so that the prediction results have higher accuracy and applicability. To improve the training speed and error recognition rate.

2 Brief Introduction of Global Artificial Fish Swarm Algorithms

Li [14] first proposed artificial fish swarm algorithm (AFSA) in 2002. This algorithm is a new top-down autonomous search model. It mainly studies the behavior of fish swarm and imitates the behavior of fish such as foraging, clustering, tailing and stochastic to achieve global optimization. In the basic fish swarm algorithm, the position updating formula of artificial fish has only local optimal information, but not global optimal information [15]. Therefore, adding global optimal fish information can effectively improve the global search ability of the algorithm, and improve the accuracy and accuracy of the algorithm.

The mathematical description of the global artificial fish swarm algorithm is as follows:

$X = (x_1, x_2, \ldots, x_n)$: indicates the state of an individual of an artificial fish. The variables to be sought are $x_i (i = 1, 2, \ldots, n)$; Steps: step of artificial fish; visual: field of artificial fish; try_number: number of attempts; delta: crowding factor; distance between two artificial fish: $d_{ij} = |X_i - X_j|$; food concentration of location: $Y = f(X)$; T_value: threshold of swallowing behavior in global artificial fish swarm algorithm.

(1) foraging behavior

Assuming that the current state of artificial fish i is X_i, a state X_j is selected randomly, and the artificial fish moves randomly in one direction. If a larger food concentration is found, the artificial fish moves further in that direction. On the contrary, the state X_j is selected randomly again. If the try-number times are repeated, and the conditions of advance are not satisfied, the artificial fish will move randomly.

Mathematical representation is:

$$\begin{cases} X_j = X_i^t + \frac{(X_j - X_i^t) + (X_{best\_of} - X_i^t)}{\|(X_j - X_i^t) + (X_{best\_of} - X_i^t)\|} \bullet Step \bullet Rand(), \; Y_i > Y_j \\ X_j = X_i^t = Visual \bullet Rnd(), \; else \end{cases}$$

Rand () is a random number between 0 and 1.

(2) Cluster Behavior

Assuming the current status of artificial fish is X_i, the number of partners NF and the central location Xc in the region are also found. If *Yc/nf*⟩ Y_i, the cluster behavior occurs, X_i moves one step in the direction, otherwise the previous behavior is executed.

Mathematical representation is:

$$\begin{cases} X_j = X_i^t + \frac{(X_c - X_i^t) + (X_{best\_af} - X_i^t)}{\|(X_c - X_i^t) + (X_{best\_af} - X_i^t)\|} \bullet Step \bullet Rand(), \frac{Y_c}{n_f} > \delta Y_i \\ foraging, \; else \end{cases}$$

(3) Tail-chasing behavior

When an artificial fish X_i finds a higher food concentration, the nearby artificial fish will automatically gather here, that is to say, each artificial fish will move towards a higher food concentration.

Mathematical representation is:

$$\begin{cases} X_j = X_i^t + \frac{(X_j - X_i^t) + (X_{best\_af} - X_i^t)}{\|(X_j - X_i^t) + (X_{best\_af} - X_i^t)\|} \bullet Step \bullet Rand(), \frac{Y_j}{n_f} > \delta Y_i \\ foraging, \; else \end{cases}$$

(4) Jumping behavior

Jumping behavior is mainly to improve the global search ability. When jumping behavior is introduced, artificial fish can be effectively avoided from falling into local optimum by forcibly changing the parameters of artificial fish and then changing the behavior of artificial fish.

The expression is expressed as:

$$X_{some}(t+1) = X_{some}(t) + \beta \bullet Visual \bullet Rand()$$

β is a parameter or a function that changes the parameters of artificial fish.

(5) Swallowing behavior

In order to reduce the complexity of the algorithm, swallowing behavior is introduced on the basis of artificial fish swarm algorithm. When the number of artificial fish increases, it will occupy more space and increase the complexity of the algorithm. The introduction of swallowing behavior can effectively solve this problem. When the artificial fish swarm algorithm passes through n iterations, if the objective function value of an artificial fish does not satisfy the condition, in order to reduce the space, the artificial fish must be discarded. The total number of artificial fish: Af_total = Af_total-1.

3 Improved Global Artificial Fish Swarm Algorithm

Compared with other algorithms, the global artificial fish swarm algorithm has many advantages, such as good convergence and robustness, but it also has some disadvantages, such as low optimization accuracy and easy to fall into local optimum. In this paper, the global artificial fish swarm algorithm is improved, so as to achieve higher optimization effect, jump out of local extremum, and reduce the optimization error.

(1) The adaptive step size is added to the global artificial fish swarm algorithm. In the process of executing each behavior, the Step parameter will be involved in each iteration. In the past, the Step is an invariant value, which easily makes the artificial fish fall into local optimum. In this paper, the step size is set to adaptive step size.

$$Step_{t+1} = \frac{\beta \bullet N - t}{N} Step, \beta \in (1.1, 1.5)$$

Among them, N represents the total number of iterations.

(2) Attenuation factor was introduced into foraging behavior. With the increase of iteration times of artificial fish, the visual field of artificial fish is gradually reduced, and then the optimization effect is improved. That is, $Visual_{k+1} = \alpha + Visual_k$, and $\alpha \in [0, 1]$
At the same time, in order to improve the research accuracy, this paper removes the congestion factor.
(3) Tabu tables are introduced into the global artificial fish swarm algorithm to record the routes each artificial fish has traveled and the best advantages of its arrival. These information are updated in the tabu table in every time to avoid repetitive routes for artificial fish, and to search the points that have not yet arrived, expand the search scope, enhance the global search ability and search efficiency.

4 Improved Global Artificial Fish Swarm Algorithms to Optimize BP Neural Network

Global artificial fish swarm optimization of BP neural network is mainly to set the artificial fish state as the initial weights and thresholds of the neural network. the neural network is represented by artificial fish, and the reciprocal of the training error of BP neural network is expressed by the food concentration of artificial fish. To improve the global artificial fish swarm algorithm is to continuously adjust the initial weights and thresholds of BP neural network. The improved global artificial fish swarm algorithm can optimize the initial weights and thresholds of the neural network to the global optimal value, which can minimize the training error. The point with the greatest food concentration is the point with the smallest error in the BP neural network.

Improving the implementation steps of global artificial fish swarm algorithm to optimize BP neural network: Firstly, determining the basic structure of BP neural network; Secondly, using the improved global artificial fish swarm algorithm to optimize BP neural network; Finally, through testing the optimized BP neural network, the running speed and error of the algorithm are improved. The network structure is determined by the number of layers and neurons in the network. The improved global artificial fish swarm algorithm is mainly searches for the initial weights and thresholds of the global optimum according to the convergence and search ability of the algorithm. Finally, the optimized BP neural network is used to detect, and the weights and thresholds are assigned to the BP neural network. Then the weights and thresholds are trained to minimize the training error. The specific algorithm flow is as follows:

Step1, Determine the structure of BP neural network, that is, the number of layers of neural network and the number of neurons.
Step2, Parameter initialization, including the number of artificial fish, step size, visual field, crowding factor, number of attempts, initial position.
Step3, the algorithm is initialized, and the state of the global optimal artificial fish is recorded and counted into the bulletin board.
Step4, In the initialization of the fish flock, artificial fish begin to perform foraging and tail-chasing behavior. Finally, the results of the implementation are evaluated and the target location is determined.
Step5, compared the result with that of the bulletin board. If a better function value is obtained, it is moved to the bulletin board, otherwise, turned step (4);
Step6, Perform the artificial fish selection and update the position information of each artificial fish.
Step7, the artificial fish state in the bulletin board is assigned to the neural network as initial weights and thresholds.
Step8, simulate the BP neural network and test the simulation error.

5 Analysis of Experimental Results

In order to verify the effectiveness of the improved global artificial fish swarm algorithm to optimize the BP neural network, the experimental errors before and after the improvement of the neural network were tested. 500 groups of training samples were randomly selected and 150 groups of data were tested.

(1) parameter setting. According to the establishment of BP neural network, the hidden layer contains seven neurons, the transfer function is logsig, the middle layer contains four neurons, the transfer function is purelin, the training function is

traingdx, the training times in BP neural network are 500, the total number of artificial fish is 20, the maximum number of iterations is 20, and the maximum number of probes is 20. The number of artificial fish was set to 5, the visual field of artificial fish was set to 0.1, the crowding factor was set to 0.618, and the step size was set to 0.2.

(2) The experimental results. By comparing the experimental results before and after the improvement, it is found that after the improvement of the global artificial fish swarm algorithm, the optimization of the neural network can better reduce the error. Table 1 and Fig. 1 show the simulation error and training error requirement of the improved BP neural network. According to Table 1, it can be seen that with the increase of iteration times, the error is gradually decreasing. When the number of iterations reaches 12, the error will not change. At the same time, according to Fig. 1, when the number of training reaches 497 times, it meets the expected error requirement.

Table 1. Global artificial fish swarm algorithm optimization BP neural network simulation error before improvement

Number of iterations	1	2	3	4	5	6	7	8	9	10
Error	0.5733	0.5733	0.5733	0.5733	0.5733	0.56	0.56	0.56	0.56	0.56
Number of iterations	11	12	13	14	15	16	17	18	19	20
Error	0.56	0.5467	0.5467	0.5467	0.5467	0.5467	0.5467	0.5467	0.5467	0.5467

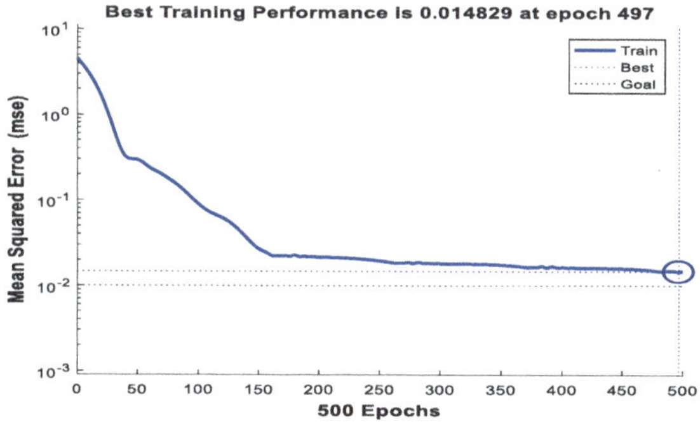

Fig. 1. Training error curve

Through the improvement of BP neural network optimized by global artificial fish swarm algorithm, the improved simulation error results and training error are obtained. As shown in Table 2 and Fig. 2.

Table 2. Improvement of global artificial fish swarm algorithms to optimize BP neural network simulation errors

Number of iterations	1	2	3	4	5	6	7	8	9	10
Error	0.373	0.373	0.373	0.33	0.33	0.33	0.33	0.33	0.33	0.33
Number of iterations	11	12	13	14	15	16	17	18	19	20
Error	0.33	0.33	0.33	0.33	0.33	0.33	0.33	0.33	0.33	

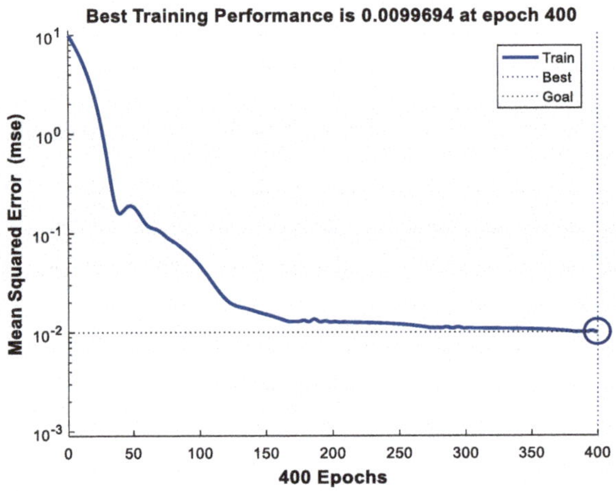

Fig. 2. Training error curve

According to Table 2, the weight and threshold of BP neural network are optimized by using the improved global artificial fish swarm algorithm. When the number of iterations reaches 4 times, the error tends to be stable, and the recognition rate of the error is 97%. According to the training results, as shown in Fig. 2, after 400 training sessions, the training error of BP neural network meets the expected training error requirements.

By comparing the results before and after the improvement, it can be concluded that the weight and threshold of BP neural network can be optimized by improving the global artificial fish swarm algorithm. The improved BP neural network can better reduce errors, get higher accuracy, and improve the number of network training. It shows that the improved global artificial fish swarm algorithm can better optimize the initial weights and thresholds of BP neural network, avoid falling into local extremes, reduce errors and improve the error recognition rate.

6 Conclusion

Standard BP neural network is easy to fall into local extremum, resulting in increased error. In order to reduce the error of the neural network, this paper improves the global artificial fish swarm algorithm by continuously adjusting the initial weights and thresholds, and assigns the adjusted weights and thresholds to the neural network. The simulation results show that this method effectively reduces the training error and simulation error of BP neural network, and improves the recognition rate of the error. Through the improvement of the neural network, the feasibility of the network is enriched, the scope of application of the network is expanded, and the improvement of the neural network is further studied.

References

1. Bo, G., Feiyan, Z.: An improved artificial fish swarm algorithm for the optimization of BP neural network. J. Hunan Univ. Sci. Technol. (Nat. Sci. Ed.) **31**(01), 86–90 (2016)
2. Feng, Y., Lu, H., Yinhong, C.L.: Research on fault diagnosis model based on BP neural network. Comput. Eng. Appl. **55**(06), 24–30 (2019)
3. Wei, L., Zhang, J., Liu, Q.: Neural network optimization based on improved artificial fish swarm algorithm. Control Eng. **21**(01), 84–87, 93 (2014)
4. Zhang, L., Wang, F., Sun, T., Xu, B.: A constrained optimization method based on BP neural network. Neural Comput. Appl. **29**(2), 413–421 (2018)
5. Wen, C., Ma, W., Liu, P.: Structural optimization of RBF neural network based on improved genetic algorithm. Comput. Eng. Sci. **41**(05), 917–923 (2019)
6. Wang, L., Qiao, L., Wei, L.: An optimal convolutional neural network learning method based on genetic algorithms. Comput. Eng. Des. **38**(07), 1945–1950 (2017)
7. Liu, D., Pu, G., Wang, G.: Environmental air quality prediction based on bee colony optimization neural network. Comput. Digit. Eng. **46**(04), 639–643 (2018)
8. Chen, C., Cao, X.: Improved differential evolution algorithms to optimize BP neural network for intrusion detection. Comput. Appl. Softw. **35**(04), 310–316, 324 (2018)
9. Zou, H., Li, X., Shang, X., Miao, C., Huang, C., Lu, J.: Hyperspectral estimation of soil organic matter based on particle swarm optimization neural network. Mapp. Sci. **44**(05), 146–150, 170 (2019)
10. Liu, Y.: An improved AHP and BP neural network method for service quality evaluation of city bus. Int. J. Comput. Appl. Technol. **58**(1), 37–44 (2018)
11. Guo, Y.F., Zhao, Z.S., Huang, L.M.: SoC estimation of lithium battery based on improved BP neural network. In: 8th International Conference on Applied Energy (ICAE 2016) (2017)
12. Yu, K., Liu, X.: PID neural network decoupling control optimized by artificial fish swarm algorithms. Comput. Simul. **31**(10), 350–353 (2014)
13. Liu, J., Gan, X., Dai, Z., Wei, Y.: Wavelet neural network model for improved artificial fish swarm algorithms. Firepower Command Control **40**(12), 115–118 (2015)
14. Li, Xiaolei, Shao, Z., Qian, J.: An optimization model based on animal autonomy: fish swarm algorithm. Theor. Pract. Syst. Eng. **11**, 32–38 (2002)
15. Huang, W., Tao, Z.: Research on VRPSPDTW based on improved global artificial fish swarm algorithm. Comput. Eng. Appl. **52**(21), 21–29 (2016)

Cloud Service Pricing Strategy Based on Service Level Agreement with Default Compensation

Xing Wan, Fu-zan Chen(✉), and Min-qiang Li

College of Management and Economics, Tianjin University, Tianjin, China
fzchen@tju.edu.cn

Abstract. The emergence of cloud computing has completely changed the entire IT world. There is no doubt that making a profit is a cloud service provider (CSP)'s ultimate goal. To make a profit, a CSP needs a reasonable pricing strategy. At the same time, in order to increase the attractiveness of cloud service, CSPs tend to write default clauses in the service level agreement (SLA) and the default cost will affect the profit of CSPs. To determine an optimal pricing strategy, this paper builds some bi-level programming models of CSPs and users. The basic model does not consider default compensation and the default cost is included in the extended model. Finally, the comparison results of the two models prove that it is significant for the CSP to set default compensation clauses.

Keywords: Bi-level programming · Cloud service · Default compensation · Pricing strategy

1 Introduction

In the 21st century, cloud services have developed rapidly in terms of technical services and market size. The global scale has increased from $68.3 billion in 2010 to $305.8 billion in 2018. According to Gartner's forecast, by 2020 this figure will exceed $400 billion.

Hence, it is crucial for a cloud computing providers (CSP) to make a reasonable pricing strategy for cloud computing services. The configuration of cloud service resources is an important factor of pricing [1]. It is because resource allocation directly affects the operating cost. In addition, relevant research indicates that it is more reasonable to adopt a dynamic and tiered pricing strategy [2].

To better coordinate the transaction cooperation between cloud service providers, CSPs usually sign service level agreement (SLA) with users. Generally, SLA includes an agreement on the price of the service, the level of service, and the amount of compensation for breach of contract. The latest version of the Alibaba Cloud Elastic Cloud Service (ECS) Service Level Agreement, which came into force on February 1st, 2018, specifically defines service availability level indicators and compensation plans. One of the clauses is "For a single ECS instance, if the service availability is less than 99.95% but equal to or higher than 99%, the user is compensated for 10% of the monthly service fee."

C.-F. Chien et al. (Eds.): *IE&EM 2019*, pp. 36–46, 2020.
https://doi.org/10.1007/978-981-15-4530-6_5

The default compensation item will inevitably lead to certain default costs, which will thus affect the profit of the CSP. Obviously, this impact is worthy of deep study. The default issue can be treated as whether a transaction is generated [3]. However, using quantitative methods to analyze the cost of default is more realistic.

The main contribution of this paper consists of three parts:

- This paper focuses on the analysis of the default compensation for cloud service pricing. Therefore, this study established two comparative bi-level programming models.
- This paper uses quantitative methods to analyze the cost of default. Furthermore, the response time is defined as the service quality indicator and could be calculated by the queuing theory model.
- This paper clearly provides the optimal pricing formula and optimal resource allocation formula for the basic model. Moreover, the evaluation between the basic model and the extended model shows it is valuable for CSP to set default clauses.

The rest of this paper is organized as follows. This paper discusses related work on cloud service pricing in Sect. 2. The problem description and two comparative bi-level programming models are presented in Sect. 3. The cost function and the user utility function are defined in Sect. 4. In Sect. 4, this paper lists the optimal pricing formula and the optimal resource allocation formula for the basic model and shows the comparison results of the two models. The conclusion is given in Sect. 5.

2 Related Work

In the previous research on cloud service pricing, scholars have established pricing models from different perspectives. Jin *et al.* propose the cloud service pricing optimization scheme on a fine-grained scale [4]. And they calculate the maximum value of the social welfare so that both parties are satisfied. Wang *et al.* established a key parameter determination model including price and service quality index based on multiple game relationships among big data reporting organizations, cloud service organizations, and customers [5]. It can be seen that cloud service pricing is an essential game process between CSPs and users.

For the game problem of cloud service pricing, Cardellini *et al.* propose the optimal pricing strategy based on the relationship between SaaS and IaaS [6]. They studied two different IaaS provider pricing strategies: the first assumes that the IaaS provider sets a unique price; the second assumes that the IaaS provider can set different prices for different customers. For each pricing strategy, the literature confirms the existence of game equilibrium.

Most of the previous related literature use simple linear functions to calculate the cloud service cost. Chiang *et al.* consider the cost of hardware resource, energy consumption, buffer, user opportunity and congestion in the cost efficiency analysis of cloud services [7]. The total cost is a linear combination function for each cost. When studying the cost model of cloud services, Liu *et al.* even consider energy consumption factors, involving physical voltage, capacitance, and frequency. But the total cost is also a linear function of the decision variable of the number of servers [8]. What utility function should be used depends on the definition of "quality of service" in the research.

There are a variety of existing user utility functions. Sumanta *et al.* construct a linear user utility function in the study of cloud service pricing, which is a linear combination of a vector composed of utility positive factors and a vector composed of negative factors [9]. Tang *et al.* establish a logarithmic user utility function for service quality in their research, based on the law of diminishing marginal utility in economics [10].

Given the problem of cloud service default, most of the previous literature chose response time as a service level indicator to investigate default and adopted a variety of processing methods. Zhang *et al.* use the default condition as a constraint on the model to ensure that the response time of the service could not exceed the time specified in the agreement [3]. Sen *et al.* construct the resource request model of the service system based on the first-come-first-served MMC model [11]. They propose that the probability of completing the request within the time limit should be greater than the probability set in SLA. Macías *et al.* propose that users can pay a certain tip to allow cloud service providers to guarantee the quality of service [12].

3 Bi-level Programming Model for Pricing Cloud Service

This paper studies the optimal pricing strategy for real-time cloud services considering SLA in a monopolistic environment, that is, the game relation between one CSP and multiple users.

CSP takes profit maximization as the decision objective, and its decision variable is price and resource allocation. Users take the utility maximization as the decision objective, and the decision variable is to choose the service or not.

The detailed game process is as follows:

1. In the beginning, the CSP provides the users with service information, including price, promised response time and default compensation clauses.
2. The users decide whether to select the service according to the service information provided and their demand for the urgency of the service. Then, they issue purchase instructions to the supplier.
3. The CSP assigns service queues to users based on their purchase instruction, and the users start to wait in line for the service.
4. The users accept the provider's cloud service and feedback on the status of the service to the CSP.
5. The CSP judges whether there is a breach of contract on the basis of the users' feedback. If there is a default, the users will be compensated according to the clauses of default.

According to the problem description above, this paper takes the decision of CSP as the upper decision and the decision of the user as the lower decision.

To strongly prove the value of the compensation clauses set in the SLA, this paper establishes two comparative models. The basic model does not consider default compensation and the default cost is included in the extended model.

In the basic model, CSP aims at maximizing its profit π, taking the price p and the number of configured servers NS as decision variables. When there are n users purchasing the services, the CSP's profit is the part of the revenue from the users minus the operating cost C of the cloud service. The CSP's programming is expressed as (1).

$$\max \pi(NS, p) = pn - C \tag{1}$$

Assuming there are M users in the cloud service market, the utility of user j (j = 1, 2, …, M) is a function U of the price and promised response time PRT in the SLA. Each user's goal is to maximize their utility u_j. This study assumes that the CSP only provides one service, and users are free to choose whether to purchase the service or not. The users' programming is expressed as (2) and (3).

$$\max u_j(x_j) = U_j(PRT, p)x_j \tag{2}$$

$$\text{s.t.} \quad x_j = 0\, or\, 1 \tag{3}$$

Equation (3) means that when x_j is equal to 0, user j does not purchase the service, and when x_j is equal to 1, user j purchases the service. It can be inferred that user j would purchase the service only if the utility function value is greater than zero.

In the basic model of this study, the promised response time PRT is not directly treated as a decisive variable. Since there is no default clause, the CSP should consciously make the promised response time PRT as equal as possible to the actual response time RT of each user. It should be noted that RT is a random variable.

Because it is a real-time cloud service, this paper uses the queuing theory model to analyze the response time of the system. Furthermore, the system can be viewed as an MM1 model in which multiple service stations are connected in parallel [4]. This paper assumes that each server is homogeneous. The service rate μ of the system represents the number of users each server can serve in a unit of time. The arrival rate of the system represents the number of users entering the system per unit time. In this study, the arrival rate can be directly reflected by the occupied market size n. Thus, the promised response time PRT can take the average value of the actual response time RT, which is expressed as (4).

$$PRT = E(RT) = \frac{1}{\mu - \frac{n}{NS}} \tag{4}$$

In the extended model, the CSP needs to consider the issue of default costs DC. The CSP's profit is equal to the revenue minus the operating cost and the default cost. Meanwhile, the promised response time PRT is a decisive variable, not necessarily equal to the average value of the actual response time RT. Anyhow, the value of PRT should not be exaggerated. This paper assumes that the minimum value of PRT is the reciprocal of the service rate μ. So, the CSP's programming in the extended model is expressed as (5) and (6).

$$\max \pi(NS, PRT, p) = pn - C - DC \tag{5}$$

$$\text{s.t. } PRT \geq \frac{1}{\mu} \tag{6}$$

The users' programming in the extended model is the same as that in the basic model, which is expressed as (2) and (3).

In the extended model, whether a default occurs or not depends on whether the actual response time *RT* exceeds the promised response time *PRT*. If the actual response time of user *j* exceeds the promised response time, the CSP needs to provide the user with default compensation and bear the cost.

However, according to the fact and relevant literature, the compensation provided by CSP cannot be infinite. In other words, there should be an upper limit α for compensation. When the actual response time increases to a certain extent, the default cost paid by the CSP does not increase.

Thus, the default function is defined as (7) and (8). And the relationship between default cost of user *j* and the response time is shown in Fig. 1.

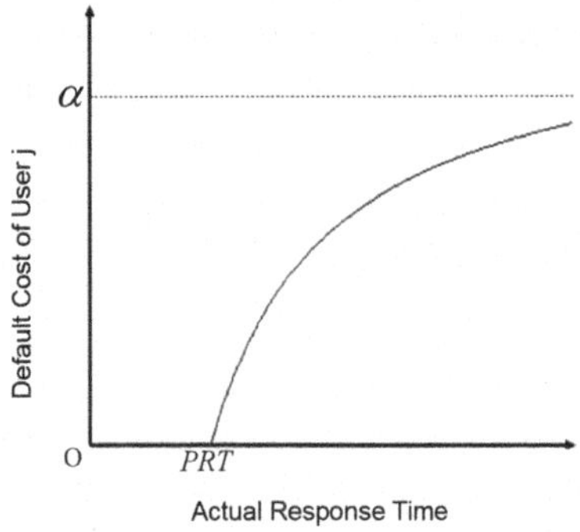

Fig. 1. The relationship between default cost of user *j* and the response time

$$DC = \sum DC_j \tag{7}$$

$$DC_j(RT) = \begin{cases} 0, & RT \leq PRT \\ \alpha(1 - \frac{PRT}{RT}), & RT > PRT \end{cases} \tag{8}$$

Both in the basic model and the extended model, the CSP needs to consume a lot of power resources, maintain the servers regularly, and pay for relevant personnel. These would make the CSP bear operating costs.

This paper assumes that each cost in the operation is proportional to the number of servers *NS*. For each server, the operating cost is *a*. The operating cost function is defined as (9).

$$C = aNS \tag{9}$$

For the lower decision of the models, cloud services with lower prices and better quality are more popular with the user community. In the eyes of users, the smaller the promised response time is, the better quality service would be. So, the utility function is defined as (10).

$$U = v_0 - \delta_j PRT - p \tag{10}$$

In (10), v_0 indicates users' basic utility, which is the same for each user. δ_j represents the sensitivity of user j to the response time. The higher δ_j means the higher time requirements of user j. This paper assumes that δ_j obeys a uniform distribution of zero to a positive number b.

4 Results and Analysis

4.1 Optimal Decision of the Basic Model

User j would purchase the service only if the utility function value is greater than zero and δ_j obeys a uniform distribution of zero to a positive number. Therefore, according to (10), the occupied market size n could be expressed as (11).

$$n = \frac{v_0 - p}{PRT} \tag{11}$$

Then, according to (4) and (11), n is solved as (12).

$$n = \frac{\mu NS(v_0 - p)}{NS + v_0 - p} \tag{12}$$

If the optimal price p^* and the optimal number NS^* of servers exist, the partial derivatives of the profit function with p and NS are both equal to 0.

The second derivative test proves that the optimal price and the optimal must exist. Finally, the optimal number NS^* of servers and the optimal price p^* of the basic model are obtained as (13) and (14).

$$NS^* = (\sqrt{1 + v_0} - 1)(\sqrt{a\mu\sqrt{1 + v_0}(\sqrt{1 + v_0} - 1)} - 1) \tag{13}$$

$$p^* = NS^* + v_0 - \sqrt{NS^*(NS^* + v_0)} \tag{14}$$

4.2 Comparative Evaluation

Based on the queuing theory, the probability density function of the customer's stay time is expressed as (15).

$$f_{RT}(t) = \begin{cases} (\mu - \frac{n}{NS})e^{-(\mu-\frac{n}{NS})t}, & t \geq 0 \\ 0, & t < 0 \end{cases} \quad (15)$$

In terms of probability theory, the calculation method of the default cost expectation of user j in the extended model is expressed as (16).

$$E(DC_j) = \int_{-\infty}^{+\infty} f_{RT}(t) \cdot DC_j(t)dt \quad (16)$$

Then, the profit function of the extended model is transformed into (17).

$$\pi = pn - aNS - \alpha n \int_{PRT}^{\infty} (\mu - \frac{n}{NS})e^{-(\mu-\frac{n}{NS})t}(1 - \frac{PRT}{t})dt \quad (17)$$

Equation (17) is not an elementary function. Therefore, this paper has to use the numerical method to solve the extended model. The parameter settings are listed in Table 1.

Table 1. Parameter settings

Parameter	Value
μ	2
a	1
v_0	3

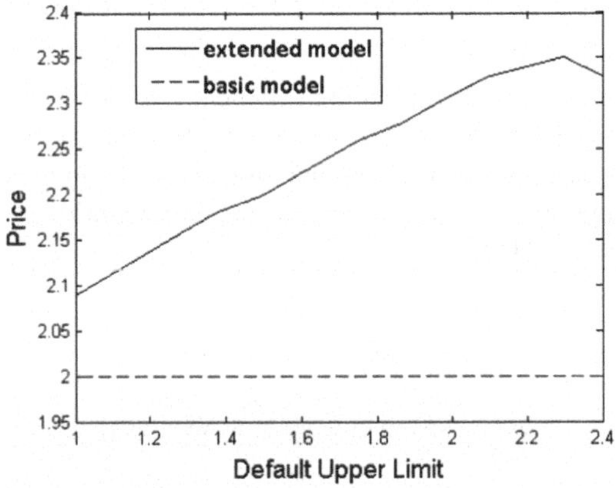

Fig. 2. The optimal price in the cases of different values of the default upper limit α

Next, the enumeration method is used to search and evaluate the optimal solution of the extended model with the default upper limit α taking multiple values.

Finally, the parameters are substituted into (13) and (14) to evaluate the basic model, and the results of the two models are compared and analyzed.

Figures 2, 3 and 4 show the optimal solution of the basic model and the extended model when α takes different values. Figures 5 and 6 present the occupied market size, cost and profit when the decisive variables are optimal.

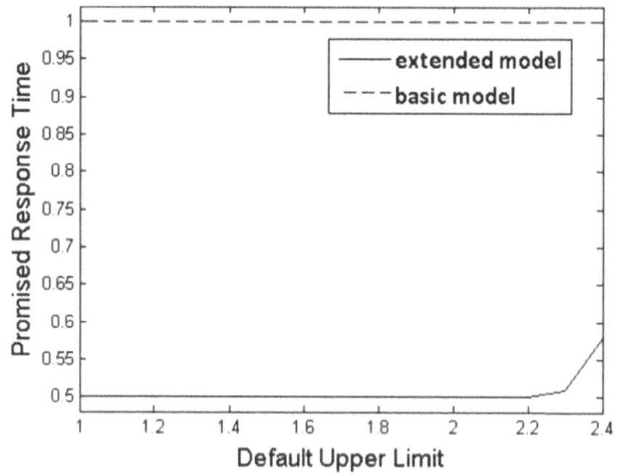

Fig. 3. The promised response time in the cases of different values of the default upper limit α

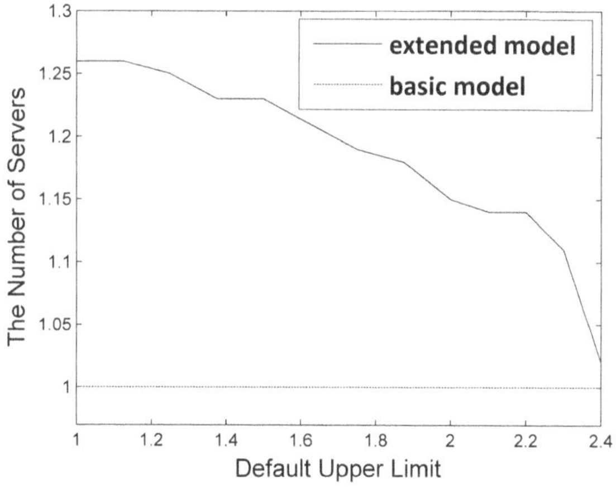

Fig. 4. The number of serves in the cases of different values of the default upper limit α

When α is lower than 1.9 (i.e. less than 82% of the corresponding price), the profit in the extended model is higher than the profit of the base model. In this case, the price and the occupied market size in the extended model are both greater than those in the basic model, though the number of servers (the operating cost) and the default cost are also more.

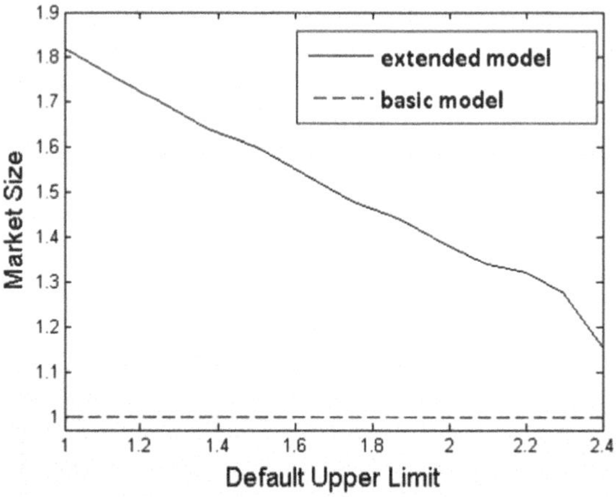

Fig. 5. The occupied market size when the decisive variables are optimal in the cases of different values of the default upper limit α

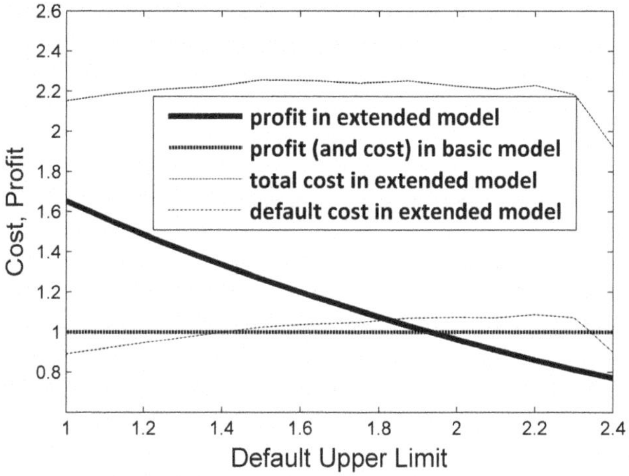

Fig. 6. The profit when the decisive variables are optimal in the cases of different values of the default upper limit α

When α is higher than 1.9 (i.e. more than 82% of the corresponding price), the profit in the extended model is less. However, due to the very short *PRT*, the CSP still occupies a greater market size in the extended model. At worst, the market size in the extended model has increased by 15%, from 1 to 1.15.

According to the results, it is significant for the CSP to set default compensation clauses in SLA. Such a strategy for cloud services is beneficial for the CSP to expand its market size, and this strategy would even give the CSP a chance to increase its profit. When the default upper limit is low, if the CSP adopts the default compensation strategy, it would gain more profit. It is because the price and the occupied market size get improved obviously, whereas the costs only increase slightly. When the default upper limit is high, the default compensation strategy still makes the CSP occupy a larger market size thanks to the adjustable promised response time.

5 Conclusion and Future Work

This article studies the optimal pricing strategy for real-time cloud services considering SLA with default compensation in a monopolistic environment.

In the model section, the article establishes two comparative bi-level programming models (the basic model and the extended model) based on the game relation between one CSP and multiple users. The basic model does not consider default compensation and the default cost is included in the extended model. Simultaneously, this paper provides the default cost function, the operating cost function, and the utility function for the models. For the default cost function, this study chooses the response time as the service quality index, which is calculated by the queuing theory model.

In the analysis section, the article first analyzes the basic model by using analytical methods to figure out the optimal pricing formula and optimal resource allocation formula. Then, the article compares the evaluation between the basic model and the extended model. The results show that it is better for the CSP to set default compensation clauses. The default compensation strategy gives CSP a great opportunity to increase its profit. Even though the profit level does not rise, the market size would expand.

In the future, we plan to study the tiered pricing strategy for cloud services. Also, we are considering adding the expectation of default cost to the utility function.

Acknowledgment. The work was supported by the general program of national natural science foundation of China (No. 71771169) and the key program of national natural science foundation of China (No. 71631003).

References

1. Tanaka, M., Murakami, Y.: Strategy-proof pricing for cloud service composition. IEEE Trans. Cloud Comput. **4**(3), 363–375 (2016)
2. Pan, W., Yu, L., Wang, S.: Dynamic pricing strategy of provider with different QoS levels in web service. J. Netw. **4**(4), 228–235 (2009)
3. Zhang, Z., Tan, Y., Dey, D.: Price competition with service level guarantee in web services. Decision Support Systems **47**(2), 93–104 (2009)
4. Jin, H., Wang, X., Wu, S., Shi, X.: Towards optimized fine-grained pricing of IaaS cloud platform. IEEE Trans. Cloud Comput. **3**(4), 436–448 (2015)
5. Wang, J., Liu, A., Zhang, S.: Key parameters decision for cloud computing: insights from a multiple game model. Concurrency Comput. Pract. Experience **29**(7), e4200 (2017)
6. Cardellini, V., Valerio, V.D., Presti, F.L.: Game-theoretic resource pricing and provisioning strategies in cloud systems. IEEE Trans. Services Comput. **99**, 1 (2016)

7. Chiang, Y., Ouyang, Y., Hsu, C.: Performance and cost-effectiveness analyses for cloud services based on rejected and impatient users. IEEE Transactions on Services Computing **9** (3), 446–455 (2016)
8. Liu, C., Li, K., Li, K., Buyya, R.: A new cloud service mechanism for profit optimizations of a cloud provider and its users. IEEE Trans. Cloud Comput. **99**, 1 (2017)
9. Sumanta, B., Soumyakanti, C., Megha, S.: Pricing cloud services—the impact of broadband quality. Omega **50**, 96–114 (2015)
10. Tang, L., Chen, H.: Joint pricing and capacity planning in the IaaS cloud market. IEEE Trans. Cloud Comput. **99**, 57–70 (2017)
11. Sen, S., Raghu, T.S., Vinze, A.: Demand heterogeneity in it infrastructure services: modeling and evaluation of a dynamic approach to defining service levels. Inf. Syst. Res. **20**(2), 258–276 (2009)
12. Macías, M., Guitart, J.: SLA negotiation and enforcement policies for revenue maximization and client classification in cloud providers. Future Gener. Comput. Syst. **41**, 19–31 (2014)

Research on Unequal Area Facility Layout Problem of Packaging Test Workshop Manufacturing Units in Semiconductor Company

Hao Wu[1,2], Zhen-hua Gao[2(✉)], and Ming Li[2]

[1] School of Artificial Intelligence, Ma'anshan University, Maanshan 243100, China
[2] School of Management Science and Engineering, Anhui University of Technology, Maanshan 243032, China
seugzh@163.com

Abstract. In order to solve the phenomenon that the layout of packaging test workshop facilities in semiconductor company is complex and not suitable for production requirements, this paper puts forward an improved bat algorithm for re-layout planning based on the double-layer structure of the workshop, the unequal area of the workshop and the production status of multi-variety products, with the goal of minimizing material handling cost and maximizing the utilization rate of workshop area. The results after the calculation of the algorithm obviously improve the layout of the packaging test workshop, reduce the cost of material handling, and increase the area utilization rate.

Keywords: Unequal area facility layout · Improved hybrid bat algorithm · Semiconductor company packaging test workshop

1 Introduction

For manufacturing industry, to enhance the viability of industry, reduce the cost of production processes and enhance the production capacity of multi-variety products has become the main goal in its production improvement. According to a study by the Market Intelligence & Consulting Institute (MIC) of the Taiwan Policy on China, the global semiconductor industry market will increase by more than 10% in 2018, and the growth of this market is mainly due to the product demand of communication terminals and the further development of artificial intelligence technology. But World Semiconductor Trade Statistics (WSTS) in its latest 2019 annual forecast that the tide of higher prices for semiconductor products in 2019 is likely to fade, and that the world semiconductor market is expected to enter a steady pace of development. Because of the inherent characteristics of semiconductor products, there are many characteristics in the process of processing, such as large quantity, variety, complex production process, many auxiliary materials and high time constraints. Therefore, it is possible to reduce the manufacturing cost and improve the production efficiency through the adjustment of the facility layout.

C.-F. Chien et al. (Eds.): *IE&EM 2019*, pp. 47–57, 2020.
https://doi.org/10.1007/978-981-15-4530-6_6

Because of the importance of facility layout to manufacturing enterprises, many scholars have carried out in-depth research on this problem. Cho and Bai applied the research results to the enterprise, solved the existing problems [1, 2]. To solve the bi-objective facility layout problem, Sahin used simulated annealing (SA) and it performed better than the previous works [3]. Abdulateef applied the Ant Colony technique to make the Ant colony system which can solve high flexibility facility layout problems [4]. Hasda proposed an iterative based two steps genetic algorithm [5]. Ning and Li studied the single row layout problem and proposed CE-based algorithm [6]. Zhang and Murray, they developed the mixed integer programming model to solving the double-row layout problem by a corrected formulation [7]. With the continuous study of the layout of the facility, some scholars began to consider the impact of other objective factors on it. There are scholars in the layout process to consider the security and other factors [8, 9].

As the pace of market change continues to improve, the needs of enterprises for the layout of facilities also have a change. In this case, many scholars began to study the dynamic layout problem [10–13] and the robust layout problem [14].

In the above research, most of the areas to be laid out are assumed to be rectangular, to a certain extent ignored the objective limitations of the area to be arranged, in this case, some scholars have also studied this aspect. Some scholars also use algorithms for the optimization of unequal area facility layout problems [15, 16].

In the follow-up research of this paper, according to the actual layout of the packaging test workshop of semiconductor manufacturing company, in order to minimize the material handling cost and maximize area utilization as the optimization goal, an improved bat algorithm is proposed to optimize the layout of the workshop.

2 Mathematic Modeling

In this paper, the objective function of the package workshop aiming at the minimum logistics handling cost and the maximum used area utilization as the target is as follows:

$$\begin{aligned} T &= \alpha * rate_U + \beta * rate_C = \alpha * \left(\left(\frac{(P_u - P_n)}{P_u} \right) \right) \\ &+ \beta * \left(\sum_{t=1}^{t} \sum_{i=1}^{m} \sum_{j=1}^{m} D_{ij} * F * S_{ij}^{n} \right) \end{aligned} \tag{1}$$

Subjected to

$$rate_c = \frac{currentC}{C_*}, \; rate_u = \frac{currentU}{U_*} \tag{2}$$

$$P_u = [(\max x_i + \frac{l_j}{2}) - (\min x_j - \frac{l_j}{2}) * (\max y_k + \frac{h_k}{2} - \min y_l - \frac{h_l}{2})] \quad (3)$$

In order cater for real production condition and avoid overlapping between facilities, the constraints should be considered as follows:

$$x_{ir} - x_{jr} \geq \frac{1}{2} * (L_i + L_j) + d_{ij} \quad (4)$$

$$y_{ir} - y_{jr} \geq \frac{1}{2} * (H_i + H_j) + H_{ij} \quad (5)$$

$$\sum_{i=1}^{m} n_{ir} + \sum_{j=1}^{m} \sum_{i=1}^{m} d_{ij} * R_{ir} * R_{jr} < L \quad (6)$$

$$R_{ir} = \begin{cases} 1, & \textit{unit I is placed on line R} \\ 0, & \textit{unit I is not placed on line R} \end{cases}, \sum_{i=1}^{m} R_{ir} = 1 \quad (7)$$

As shown in (1), the objective function value T of this mathematical model is composed of the maximum area usage and the minimum logistics handling cost. α, β as the target weight, take the value of 0.7, 0.3. In order to ensure the consistency of the units between the two optimization targets, the (2) is normalized.

The meanings represented by each letter are as follows:

P_u represents the area used in the current layout mode; P_n is the sum of the area required for the job unit; $\max x_i$, $\min y_l$ correspond to the largest and smallest working units of transverse and longitudinal coordinates respectively;

C represents the cost of material handling under a certain layout; m is the sum of all operating units in the workshop; t is the type of product quantity processed in the workshop; i, j are integers, and $i \neq j \in [1, m]$; S_{ij}^t refers to the number of parts t required to be transported between the operating unit i and the operating unit j, and F is the unit cost of material handling; D_{ij} represents the manhattan distance between the operating unit i and the operating unit j.

3 The Improved Bat Algorithm for Unequal Area Facility Layout

In order to solve the problem that the basic bat algorithm is prone to fall into local optimal solution and solve the problem of insufficient precision, an improved bat algorithm is proposed and applied to the layout of unequal area.

3.1 Adding Parallel Grouping Schemas

To enhance the global search ability of bat algorithm and enrich the population diversity in the iterative process, the population structure in the algorithm is adjusted. When the initial population is produced, the population is randomly divided into N groups, and the best solution between the groups is *Group_j_best* (j = 1, 2, …, N) is recorded separately during the iteration, and the relevant data is updated after each iteration for the next step.

3.2 Improvement of Speed Update Equations

Bat algorithm in the speed update of global search, because only the impact of the current optimal solution is considered, the diversity of the population will be lost in the later stage of the algorithm. Therefore, this paper designs three kinds of speed update equations, combined with the parallel grouping architecture proposed in the previous step, the improved Speed update pseudo code is as follows:

if $rand \le k1$;

$$v_i^c = v_i^{c-1} + f_i(x_i^{c-1} - Group\_j\_best)\,) \quad (8)$$

elseif $rand>k1\&rand<k2$;

$$v_i^c = \omega * v_i^{c-1} + c1 * f_i(x_i^{c-1} - Group\_j\_best) + c2 * f_i(x_i^{c-1} - x_*) \quad (9)$$

elseif $rand \ge k2$;

$$v_i^c = v_i^{c-1} + f_i(x_i^{c-1} - x_*) \quad (10)$$

end

In the above equations, $k1$ and $k2$ are the fixed thresholds initially set, and the speed update equation of individual i in the c iteration is judged by comparing the threshold value with the numerical size of the random number rand. The speed update equation of particle swarm algorithm, which is used to calculate the common influence of the optimal solution *Group_j_best* and the current global optimal solution x_* on individual speed renewal, was calculated using (9). $c1, c2$ is the position learning factor, and the size of the learning factor will affect the direction of speed renewal; Eq. (8) based on the improvement of the speed update equation in the basic bat algorithm, the current global optimal solution x_* is replaced with the optimal solution *Group_j_best* between groups to calculate the effect of the optimal solution between groups on the individual, while the (10) is the original speed update equation that is retained. According to the pseudo-code mentioned above, it can be seen that when the speed of the bat individual is updated, it is no longer limited to the influence of the optimal solution of the current population, considering the influence of the suboptimal solution (that is, the optimal solution between groups) on the speed update and the speed update equation in the particle swarm algorithm, which expands the global search scope To a certain extent, the situation of falling into local optimal solution is avoided, and the ability to deal with high dimensional multi-peak problem is enhanced.

3.3 Improvement of Domain Search Equations

(1) Improvements to Solution set updates
In the face of high dimensional multi-peak problem and population falling into local optimal solution, it is difficult to jump out of local optimal solution and cause the search ability of the algorithm to decrease and the population diversity to be lost. For this purpose, a multi-layer dynamic domain search model is designed, that is, during the iterative process, when $rand > r_i^0$, each bat in the population flies around the optimal bat in the group in which it is located, creating a new individual, as shown in the (11):

$$X_{new} = \delta A^c + Group\_j\_best \tag{11}$$

After iterating through the Eq. (11), it is possible that X_{new} is not a viable solution, so X_{new} is compared with x_i^{c-1} to find different solution fragments and assign them to $mm(i)$, while assigning different fragments in X_{new} to 0 and $mm(i)$ in order to X_{new}. The new value is 0 in the position.
(2) Interchange operator
In order to strengthen the search ability of the algorithm for better solution, this paper uses two kinds of domain search methods to use the optimal solution *Group_j_best* and the Global optimal solution x_* respectively, and the specific domain search method diagram is shown in Fig. 1a and b.

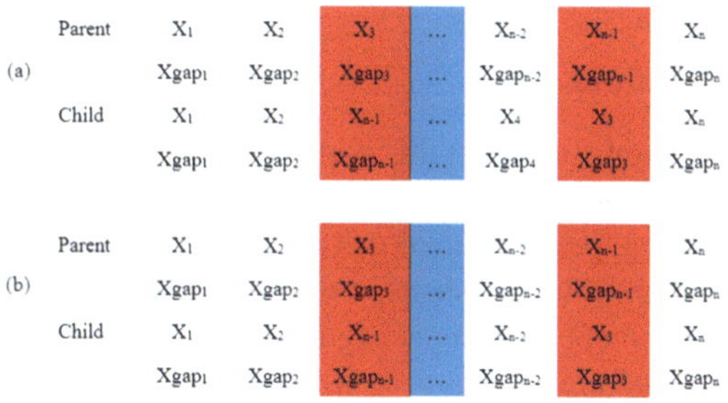

Fig. 1. (a) Domain search method of group optimal solutions. (b) Domain search method of global optimal solution

3.4 Longitudinal Coordinate Dynamic Adjustment

When the line between the length and spacing of each operating unit in the horizontal direction is determined by the constraint between the sum of the work units and the length of the shop floor, the coordinates of the second line first are the baseline with the highest vertical distance in the preceding line, plus the minimum spacing in the vertical direction ygap and half of the width in the vertical direction of the After all the job unit locations have been determined, the longitudinal coordinates are adjusted to the position difference in the vertical direction of the previous line of work, but the vertical spacing between the two operating units must be no less than ygap. This step is adjusted to make the job Unit layout compact.

4 Case Study

4.1 Present Situation of Workshop

The package test workshop studied in this paper contains a double-decker building can be arranged area is 10596.6 m^2 per layer, workshop size data for 144 m * 60.5 m (the upper part of the workshop) and 139.6 m * 13.5 m, the workshop layer is 4 m high, the workshop contains more than 20 processing processes and more than 10 auxiliary processes or equipment.

In the existing layout of the packaging test workshop, according to the processing process for the dust-free requirements of the different, the workshop is divided into the front section of the workshop and the post-processing segment. Through the analysis of the existing layout, we can see that beyond the processing area, more than 30% of the workshops are used as additional office areas, and the space utilization in the area is low. At the same time, it is precisely because of the existence of this office area, so that the subsequent increase in equipment can only be placed on the second floor. The plane workshop has been cut and divided horizontally, in the premise of only one elevator for cross-floor transport, there has been a cross-transport situation, which has a negative impact on the degree of product cleanliness.

4.2 Process Flow Analysis

The package test workshop studied in this paper is mainly based on QFN, SOP, QFP, BGA in the classification of these four types of packaging technology, there are more than 15 kinds of product types. The total daily production batch is about 130 times, a total of 112,680 Semiconductor finished products. After the actual investigation and operation manual requirements in the packaging test workshop, the main processes in the production process and area data are shown in Table 1, and the handling batches is shown in Table 2, respectively. In Table 2, A, B, C, D, E, F represent the removal batches, with values of 120, 100, 80, 60, 40, 20 and each batch containing 864 semiconductor components.

Due to the small weight of the semiconductor products themselves, hundreds of semiconductor processing after the completion of the finished product weight to reach more than two grams, so this article in the calculation of logistics handling costs to set the unit weight of 1, unit distance removal costs set to 1. Based on these two hypotheses and the data in Table 2, the current workshop handling cost can be calculated, the handling cost is 107240, in which the packaging process results in a handling cost of 82890.5, the test process resulting in a handling cost of 24349.5.

Table 1. The main processes in production process

Num	Unit Name	Size (m*m)	Num	Unit Name	Size (m*m)
1	Wafer bank	17*20	13	LM	21*18
2	Taping	4.8*2.5	14	DD	10*7
3	B/G	6.5*17	15	Dipping	10*6.5
4	WS	10*16	16	Deflash	6*6.5
5	DB	47*16	17	Tin leaching	50*20
6	EC	8*14	18	FS	10*20
7	PC	3.2*7.4	19	BM	5*14
8	WB	85*20	20	SS	16*20
9	Auto Mapping	6*5	21	TT	9.2*6
10	Frame Tapping	6*6	22	FT	60*42
11	MD	28*14	23	Packing	20*15
12	PMC	4*12			

Table 2. Logistics from packaging and testing workshop to table (batch)

From	To	Number	From	To	Number
1	2	A	12	19	F
2	3	D	13	12	F
2	5	D	13	14	E
3	4	D	13	20	E
4	5	D	14	17	D
5	6	A	15	13	E
6	7	C	16	17	E
6	10	E	17	13	E
7	8	C	17	18	D
9	11	C	18	21	D
10	11	E	19	20	F
11	12	E	20	21	D
11	13	F	21	22	B
12	14	F	21	23	F
12	15	E	22	23	B
12	16	E			

4.3 Related Constraints

In order to reduce the cost of equipment movement and downtime due to the process of equipment movement, TT, FT and packing operations will be placed on the second floor of the workshop according to the requirements of R's Industrial engineering department. Therefore, the package, test two processes separate layout, that is, the first floor of the workshop packaging process layout design and office area placement, the second floor of the workshop to carry out the layout design of the test process. Therefore, all semiconductor products that need to be sent to TT, FT, packing for further processing need to be transported across the floor through the elevator, where the elevator coordinates are determined to be (96.1, 232.4).

4.4 Analysis of Optimization Results

In order to make the codec process in the algorithm process more concise and intuitive, select the two-stage encoding method here is: $\{m_1, m_2, \ldots, m_n | \Delta_1, \Delta_2, \ldots, \Delta_n\}$, in this encoding mode, the real number sequence of the first n items represents the order of the placement of n job units, each of which appears only once; The latter n item represents the spacing between the various operating units in order to ensure the smooth transportation of the transport equipment during the production process, and its value range is [0.8,1.8].

In order to verify the performance of improved bat algorithm, the basic bat algorithm is added for comparison, and sets the remaining relevant parameters in the algorithm as follows: Frequency range $f_{\max} = 2, f_{\min} = 0$, loudness attenuation coefficient q = 0.9, pulse increase rate e = 0.9, threshold $k1 = 0.3, k2 = 0.7$, minimum spacing ygap = 0.8 in vertical direction, maximum iteration number g = 500, population size nchr = 40.

Through the algorithm calculation, the order of the operation units in the optimal solution is {5, 10, 2, 8, 20, 15, 14, 16, 9, 7, 11, 12, 19, 6, 3, 4, 21, 13, 7, 18, 23, 22, 21}, and the spacing between the working units is {1.19, 1.51, 1.61, 1.41, 1.60, 1.54, 1.11, 1.00, 1.76, 0.85, 1.61, 1.39, 1.43, 1.37, 0.91, 1.22, 0.90, 1.42, 1.77, 1.02, 1.33, 1.41, 1.02}.

The optimal solution is 44291.8 and the area of operation area is $P_u = 9170\,\text{m}^2$ The algorithm iteration is shown in Fig. 2.

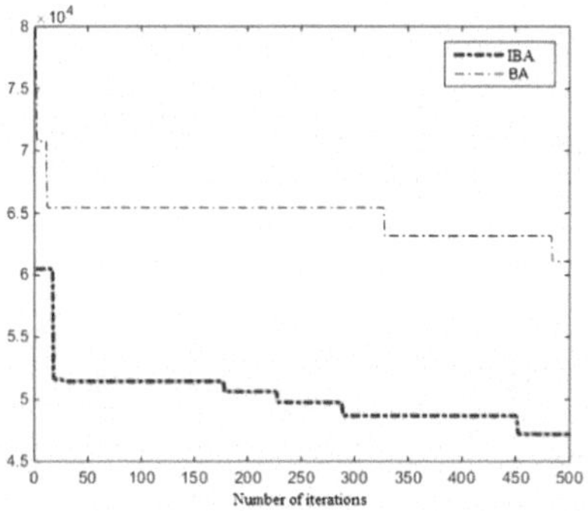

Fig. 2. Algorithm iteration diagram

After the arrangement of the operation unit inside the packaging test workshop, according to the relevant requirements of the company, the relevant auxiliary units in the workshop are placed in place. The requirements are as follows:

(1) The placement of auxiliary units shall not have an impact on the semiconductor manufacturing process;
(2) Auxiliary unit placement should be coherent, as far as possible to avoid the phenomenon of scattered placement;
(3) The specific specifications of the auxiliary unit can be adjusted according to the actual layout, but it is necessary to keep the footprint of the corresponding units unchanged;

Under the above three requirements, combined with the relevant requirements of the process manual and the existing layout, the packaging test workshop in the remaining units to be placed, after the adjustment of the package test workshop layout diagram as shown in Figs. 3 and 4.

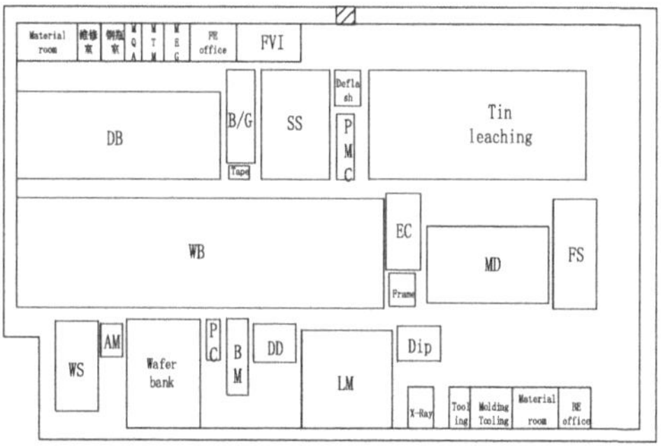

Fig. 3. Packaging test workshop first floor final layout diagram

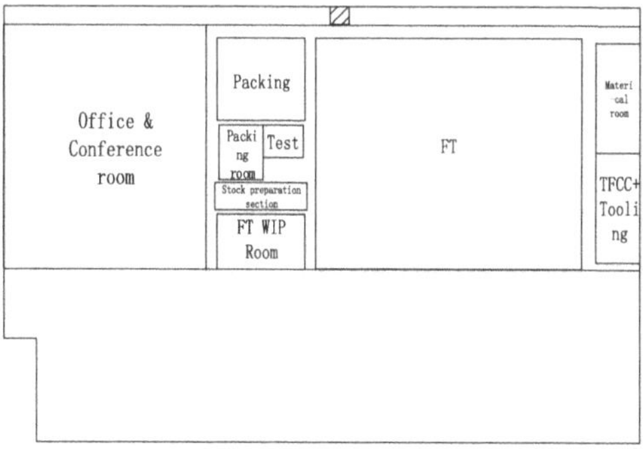

Fig. 4. Packaging test workshop second floor final layout diagram

The removal cost of the test process calculated by the algorithm is 44291.8401, and compared with the existing layout conditions, the test process reduces the handling cost by 38598.7 and the cost reduction reaches 46.6%, and the total cost of handling in the adjusted Package Test workshop is 63587.4, The removal cost is 40.7% lower than under the current layout conditions, the utilization rate of the area used in the first floor of the workshop is 91.5%, and the utilization rate of the area used in the second floor of the workshop is 91.4%.

5 Conclusion

Aiming at the unequal area condition and actual layout of the packaging test workshop, this paper puts forward a multi-objective optimization model, and optimizes the design with the improved bat algorithm. First of all, through the actual investigation and data collection of the packaging test workshop, through the process analysis of four main production products in the workshop, summed up the current packaging test workshop problems and other data. Secondly, in view of the unequal area of the workshop to be arranged, an improved bat algorithm is proposed and used in the layout design of the packaging test workshop. Through the iterative calculation of the algorithm, the handling cost of the package test workshop is reduced to 63587.4, the handling cost is reduced by 40.7% compared with the current layout condition, the utilization rate of the first floor of the workshop is 91.5%, and the utilization rate of the second floor of the workshop is 91.4%, which saves the area used in the workshop. Finally, after the correlation calculation of the algorithm, according to the actual situation and specific needs of the company, the algorithm results are improved two times, so that the improved layout mode is more in line with the actual situation of the workshop, more practical significance.

References

1. Cho, K., Moon, I., Yun, W.: System analysis of a multi-product, small-lot-sized production by simulation: a Korean motor factory case. Comput. Ind. Eng. **30**(3), 347–356 (1996)
2. Bai, J., Pan, W., Sun, L.: The layout optimization of the motor wheelset's assembly workshop. In: International Conference on Information Management. IEEE (2011)
3. Sahin, R.: A simulated annealing algorithm for solving the bi-objective facility layout problem. Expert Syst. Appl. **38**(4), 4460–4465 (2011)
4. Abdulateef, O.F., Moosa, A., Kadim, D.: Enhancing facility layout via ant colony technique (act). Innov. Syst. Des. Eng. **4**(12), 1–12 (2013)
5. Hasda, R.K., Bhattacharjya, R.K., Bennis, F.: Modified genetic algorithms for solving facility layout problems. Int. J. Interact. Des. Manuf. **11**, 713–725 (2016)
6. Ning, X., Li, P.: A cross-entropy approach to the single row facility layout problem. Int. J. Prod. Res. **45**(11), 1–14 (2018)
7. Zhang, Z.Q., Murray, C.: A corrected formulation for the double row layout problem. Int. J. Prod. Res. **50**(15), 4220–4223 (2012)
8. Azadeh, A., Moradi, B.: Simulation optimization of facility layout design problem with safety and ergonomics factors. Int. J. Ind. Eng. **21**(4), 209–230 (2014)

9. Latifi, S.E., Mohammadi, E., Khakzad, N.: Process plant layout optimization with uncertainty and considering risk. Comput. Chem. Eng. **106**, 224–242 (2017)
10. Ulutas, B., Islier, A.A.: Dynamic facility layout problem in footwear industry. J. Manuf. Syst. **36**, 55–61 (2015)
11. Azevedo, M.M., Crispim, J.A., de Sousa, J.P.: A dynamic multi-objective approach for the reconfigurable multi-facility layout problem. J. Manuf. Syst. **42**, 140–152 (2017)
12. Li, J.Y., Tan, X., Li, J.C.: Research on dynamic facility layout problem of manufacturing unit considering human factors. Math. Probl. Eng. **2018**, 1–13 (2018)
13. Tayal, A., Gunasekaran, A., Singh, P.S., et al.: Formulating and solving sustainable stochastic dynamic facility layout problem: a key to sustainable operation. Ann. Oper. Res. **253**, 1–35 (2016)
14. Moslemipour, G., Lee, T.S., Loong, Y.T.: Performance analysis of intelligent robust facility layout design. Chin. J. Mech. Eng. **30**, 407–418 (2017)
15. Kulturel-Konak, S., Konak, A.: A new relaxed flexible bay structure representation and particle swarm optimization for the unequal area facility layout problem. Eng. Optim. **43** (12), 1263–1287 (2011)
16. Kulturel-Konak, S., Konak, A.: Linear programming based genetic algorithm for the unequal area facility layout problem. Int. J. Prod. Res. **51**(14), 4302–4324 (2013)

Optimum Design of Ticketing Service System of Guangzhou-Zhuhai Intercity Railway Zhuhai Station

Da-wei Zhou[1(✉)], Jin-jin Huang[1], Lin-lin Gan[1], Dao-zhi Zhao[2], Shao-luan Wang[1], Kai-xuan Hu[1], and Xin-feng Zou[1]

[1] School of Industrial Automation, Beijing Institute of Technology, Zhuhai, Zhuhai, China
450570503@qq.com
[2] College of Management and Economics, Tianjin University, Tianjin, China

Abstract. Aiming at the problems existing in the ticketing service system of Guangzhou-Zhuhai Intercity Railway Zhuahai Station, the paper used the basic principles of IE and queuing theory to model and simulate the system using Flexsim software, and put forward the optimized scheme of the system of Zhuhai Station, which can alleviate the serious situations that the ticketing service time was so long, meanwhile, reduced the number of ticket sellers and made passengers travel more convenient. And this provided new ideas for the optimization of the service platform of Zhuhai Station.

Keywords: Facilities layout · Modeling and simulation · Queuing theory · Service efficiency

1 Introduction

How to scientifically and reasonably set up the number of service windows and layout of facilities in railway service system has always been the research content in the railway passenger station construction.

Queuing theory in operational research is a commonly used method to improve and optimize the efficiency of existing systems in IE research. It takes the number of queues, queue length and arrival rate in service system as the research object, and obtains the optimal service by observing, recording and calculating the field data in detail [1].

Facility planning is a new, improved and expanded manufacture or service system. It comprehensively considers various factors and makes analysis, plans and designs so that the resources can be rationally allocated and the system can be effectively operated after completion to achieve various expected goals [2].

System modeling and simulation technology [3] has been successfully applied in service industries such as banks and highway passenger stations, and achieved good results, such as Flexsim-based highway passenger station simulation system whose

Sponsored by Advantage Discipline of Mechanical Engineering in Zhuhai (Zhuhai Key Project).

C.-F. Chien et al. (Eds.): *IE&EM 2019*, pp. 58–66, 2020.
https://doi.org/10.1007/978-981-15-4530-6_7

author is Hu, and Xi [4]. It provides many ideas for the use of simulation technology in relevant passenger service areas.

Industrial Engineering is mostly used in production systems, while service systems are relatively few. This paper takes the ticketing service system of Guangzhou-Zhuhai intercity railway Zhuhai Station as the research object. Based on queuing theory, Flexsim Simulation and Facility Planning, it explores a simple, effective and practical improvement plan in the paper.

2 Basic Information

Zhuhai Station is the terminal of Guangzhou-Zhuhai Intercity Railway, at the same time, it is also the starting station of Zhuhai-Jinwan Airport intercity railway under construction, which is close to Gongbei Port, the largest land crossing in China. There are more than 140 trains are sent to Guangzhou South, Guilin, Chengdu, Nanning, Zhengzhou, Kunming, Beijing and Shanghai every day. Since February 2013, the average daily passenger flow of Zhuhai Station has reached more than 15,000 people, and which on the peak day during National Day in 2017 has reached more than 60,000 people.

The main building of the station has four floors. The first and second floors of the underground are taxi station and social parking space, the first floor is ticket selling and waiting hall, and the second floor is train arrival and departure platform. On the first floor, there are 10 manual windows, 29 self-service ticket machines and 10 self-service ticketing machines. Figure 1 shows the layout of the ticket office and the entrance.

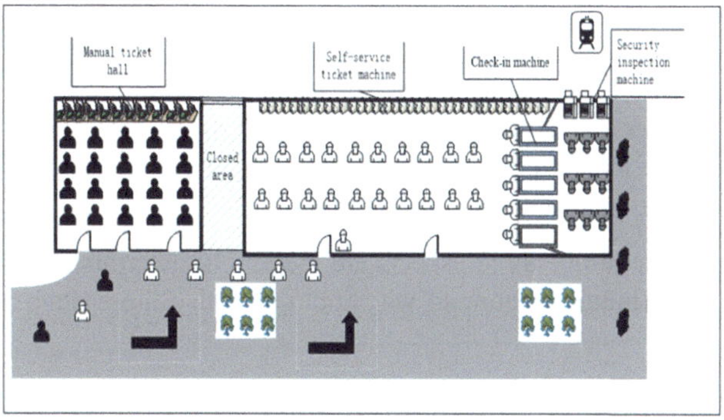

Fig. 1. Plane layout of ticket office and entrance of Zhuhai Station

3 Situation Analysis and Improvement

The status of Zhuhai Station is analyzed by using the system thinking method of Industrial Engineering, System modeling and simulation, questionnaire survey, 5W1H questioning method and ECRS improvement principle [5].

Flexsim software is used to model and simulate [6] the process from ticket selling to ticket checking in Zhuhai Station. According to the data collected from field investigation and questionnaire, the Flexsim is imported to set and run the relevant parameters. The results are shown in Fig. 2.

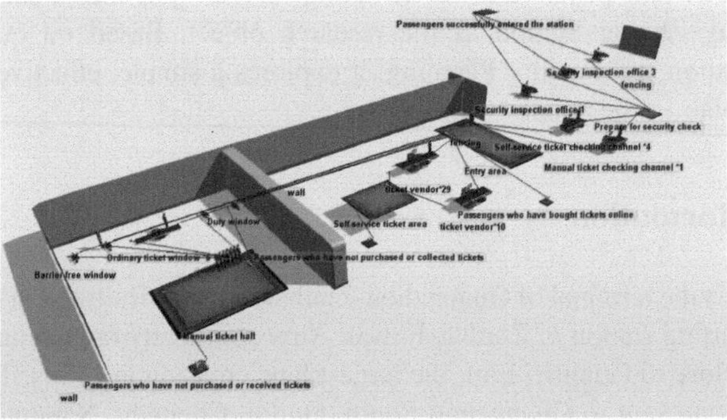

Fig. 2. Flexsim simulation diagram of ticketing service system (before)

The simulation results show that the flow of passenger is not balanced in the manual and the self-service ticketing areas, the manual ticketing window is congested, that the waiting time is long, and some self-service ticket machines are idle.

Using 5W1H method, combined with the results of simulation, field investigation and questionnaire, the main reasons are found as follows:

a. The manual ticketing window business distribution is inappropriate and the processing efficiency is low.
b. The improper placement of self-service ticketing equipment leads most people to go to the manual ticketing hall, and some machines are idle. It takes a long time for passengers to enter the ticketing hall until the tickets are purchased.
c. The unreasonable location planning of manual ticketing hall and self-service ticketing area leads to crowding in manual ticketing hall.

After analysis, the above three aspects are the key to determine the inefficiency of ticketing service at Zhuhai Station, so we should optimize these three aspects.

3.1 Artificial Ticketing Hall

Statistical analysis [7] of field observation data of effective days shows that the ticketing service system of Zhuhai Station is busy at 11:00–12:00 and 16:00–17:00. The total number of arrivals is 2424, of which 90% are for ordinary business. Therefore, the arrival rate of passengers handling business at ordinary windows:

$$\lambda = 2424 \times 90\%/120 = 18.18 \ \text{(person/min)}$$

At the same time, the time of 100 passengers in the manual service windows were observed in these two periods. According to the statistics of relevant data, after calculation and test, it was determined that the service processing time obeyed Poisson

Distribution at a certain significant level, and the average time of each passenger's service processing in a window was obtained (v value was taken as the median value of the interval). Assuming that each window has the same business capability, meanwhile, six windows have the same service rate:

$$v = \frac{\sum vf_v}{100} = 118.42 \ \text{(sec/person)}$$

Average number of business processes completed (service rate):

$$\mu = 60/118.42 \times 6 = 3.05 \ \text{(person/min)}$$

According to: c = 6, λ = 18.18 (person/min),

$$\mu = 3.05 \ \text{(person/min)},$$

We derive the following calculations:
Busy rate of six ordinary windows

$$(\rho) = \frac{\lambda}{c\mu} = \frac{18.18}{6 \times 3.05} = 99\%$$

$$P_0 = 1/[\sum_{n=0}^{c-1} \frac{(\lambda/\mu)^n}{n!} + \frac{(\lambda/\mu)^c}{c!} \cdot \frac{1}{1-(\lambda/Cu)}] = 2.28 \times 10^{-5}$$

According to the system idle probability P_o, the following data are calculated respectively:

Length of queue: $L_q = \frac{P_0(\lambda/\mu)^c\rho}{c!(1-\rho)^2} = 14.1$ (person/min)

Number of people in queuing: $L_s = L_q + \frac{\lambda}{\mu} = 20.06$ (person/min)

Average waiting time: $W_q = \frac{L_q}{\lambda} = 0.78$ (min/person)

Average length of stay: $W_s = \frac{L_s}{\lambda} = 1.10$ (min/person)

Passenger handling per unit time: $Z = 60/W_s = 60/1.10 = 55$ (person)

As can be seen above, when the number of the manual ticketing window is 6, the manual busy rate is as high as 99%. In this queuing system, the best waiting time calculated is 1.1 min per person. However, from the field observation time, it can be seen that in reality, the queuing waiting time is far more than 1.1 min per person, so the number of service windows in this queuing system is seriously insufficient.

After further analysis, it is due to the improper distribution of window services, resulting in low service efficiency of each service window. In this regard, based on the actual situation of the manual ticketing hall, without changing the total number of business processing windows, after ESCRI analysis, the refund window, duty room and Hong Kong, Macao and Taiwan business were merged. The two additional windows were changed to ordinary windows, alleviating the queuing situation of ordinary windows and retaining barrier-free windows. The improved manual ticketing system is calculated.

Arrival rate: $\lambda = 18.18$ (person/min)
By adding two windows, $\mu = \frac{60}{118.42} \times 8 = 4.05$ (person/min)
Reception c = 8, $\lambda = 18.18$ person/min, $\mu = 4.05$ person/min
From that, we get this calculation:
Business Rate: $\rho = \frac{\lambda}{c\mu} = \frac{18.18}{8 \times 4.05} = 56\%$

$$P_0 = 1/[\sum_{n=0}^{c-1} \frac{(\lambda/\mu)^n}{n!} + \frac{(\lambda/\mu)^c}{c!} \cdot \frac{1}{1-(\lambda/Cu)}] = 0.13$$

Length of queue: $L_q = \frac{P_0(\lambda/\mu)^c \rho}{c!(1-\rho)^2} = 1.54$ (person/min)
Number of people in queuing system: $L_s = L_q + \frac{\lambda}{\mu} = 6.03$ (person/min)
Average waiting time: $W_q = \frac{L_q}{\lambda} = 0.08$ (min/person)
Average length of stay: $W_s = \frac{L_s}{\lambda} = 0.33$ (min/person)
Passenger handling per unit time: $Z = 60/W_s = 60/0.33 = 181$ (person)

After adding two windows, the artificial busy rate is reduced to 56%, and the queuing time is reduced to 0.33 min per person.

3.2 Self-service Ticket Area

According to the same observation and analysis method, it was found that the improper placement of equipment in the self-service ticketing area cause most people to go to the manual ticket hall, some machines were idle, and the time for passengers to enter the area to purchase tickets was longer. In this regard, under the condition that the overall layout of the self-service ticketing area remains unchanged, the number of self-service equipment is reasonably arranged according to the results calculated by queuing theory,

Table 1. Data computation of self-service ticket area

Calculated Amount	Formula	Before Improvement	After Improvement
Station Numbers of ordinary business C (a)	——	39	17
Arrival rate λ (person/min)	The number of arrivals per unit of time / time	16.0	16.0
Average transaction time (sec/person)	$\frac{\sum v f_v}{100}$	57.57	57.57
Service rate μ (person/min)	$\frac{60}{t} \times C$	1.04	——
Business rate ρ	$\frac{\lambda}{c\mu}$	39%	95%
System idle probability P_0	$1/[\sum_{n=0}^{c-1} \frac{(\lambda/\mu)^n}{n!} + \frac{(\lambda/\mu)^c}{c!} \cdot \frac{1}{1-(\lambda/Cu)}]$	2.31×10^{-7}	1.8×10^{-8}
The length of the queue L_q (person/min)	$\frac{P_0(\lambda/\mu)^c \rho}{C!(1-\rho)^2}$	2.38×10^{-7}	2.90
Number of people in the queue system L_s (person/min)	$L_q + \frac{\lambda}{\mu}$	15.38	18.28
Average waiting time W_q (min/person)	$\frac{L_q}{\lambda}$	1.4875×10^{-8}	0.18
Average staying time W_s (min/person)	$\frac{L_s}{\lambda}$	0.96	1.14

and the number of self-service ticketing machines is reduced from 39 to 17. The calculation method is the same as the above which based on queuing theory, and the process is slight [8]. Data comparison before and after improvement is shown in Table 1.

3.3 Self-service Ticketing and Ticket-Checking Area

After on-the-spot investigation, passengers need to walk out of the gate of the manual ticketing hall and than enter the ticket-checking area from another gate. The distance they need to walk is 68 m. Between the self-service ticketing and ticket-checking area, 29 ticket-selling and ticket-collecting machines and 10 student ticket-collecting machines account for 60% of the total area, and there are intersections between the ticket-selling queue and the ticket-checking queue. The layout before improvement is shown in Fig. 3.

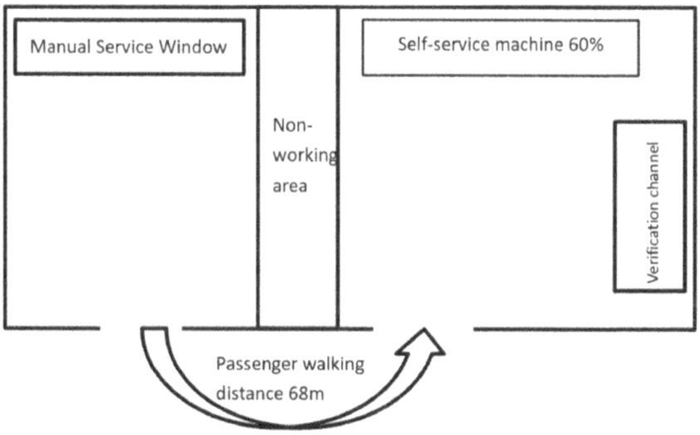

Fig. 3. Self-service ticketing and ticket-checking area (before)

In these two closed areas, there is a delay in the transmission of information, in other words, when passengers queue in the manual ticketing area, they can not receive broadcasting reminders in the check-in area, and the queuers in the manual ticketing area can not see the use of the self-service ticket machine.

According to the principle of space minimization in facility planning [9], we get these improvements:

a. To transform two separate closed ticket-selling areas, manual and self-service, into an open area;
b. The proportion of self-service ticket machines and ticket-checking machines in the self-service ticketing and ticket-checking areas will be adjusted from 6:4 to 4:6.

The improved layout is shown in Fig. 4.

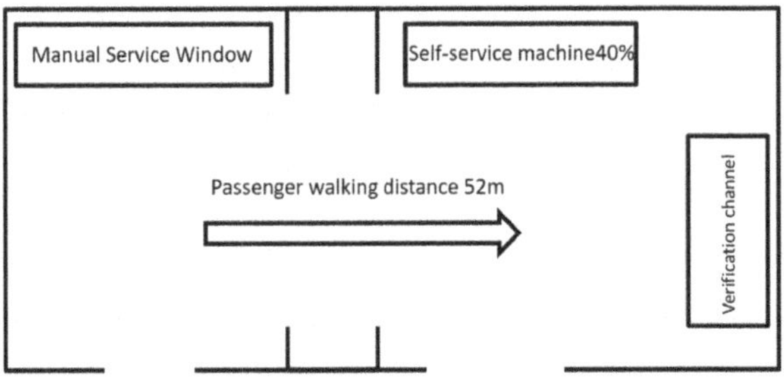

Fig. 4. Self-service ticketing and ticket-checking area (after)

4 Analysis of Improvement Effect

In the manual ticketing hall, three windows (the refund window, Hong Kong, Macao and Taiwan resident window and duty room) are merged, and the busy rate of the whole ticketing hall system is taken as a comparative reference. Under the condition that the passenger arrival rate remains unchanged, the number of people served per minute is increased by increasing the window, the busy rate is reduced from 99% to 73%, and the busy rate of staff is reduced by 26 percentage points, which is handled in unit time. The number of passengers increased from 55 to 181, which effectively alleviated the original queuing phenomenon.

In the aspect of self-service ticketing machines, 39 self-service ticketing machines were reduced to 17 because of the low utilization rate of the machines. Under the condition that the arrival rate of passengers and the service rate of the machines remain unchanged, the busy rate increased to 95%, effectively improving the utilization rate of the machines.

In terms of facility layout, separate closed manual and self-service ticket selling areas were transformed into an open area, and the distance of customers' walking was reduced from 68 m to 52 m, which improved the service level of the system. The proportion of the self-service ticket machines and ticket checking machines in the self-service ticket selling and checking area was adjusted from 6:4 to 4:6, which improved the utilization rate of the station.

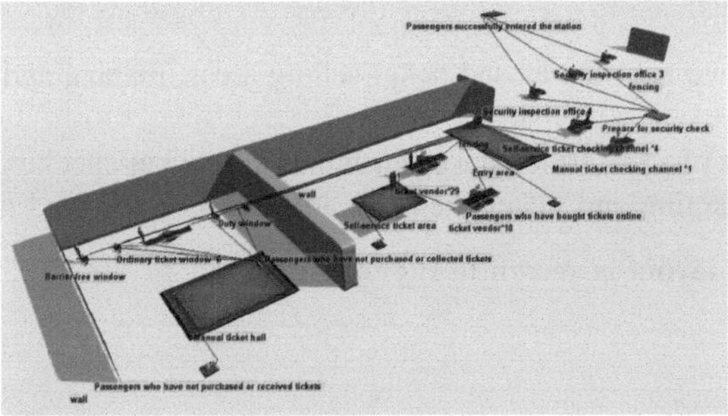

Fig. 5. Flexsim simulation diagram of ticketing service system (after)

Flexsim simulation of Zhuhai Railway Station after improvement is shown in Fig. 5 [10]. The simulation shows that the above improvement measures effectively alleviate the phenomenon of passenger queuing for ticket purchasing.

It is known that the average time spent by passengers in the manual ticketing hall decreases from 1.10 min/person to 0.33 min/person, the straight line distance from the gate of the self-service ticket office to the ticket checking office is 71 m. Before improvement, the distance from the manual ticketing hall to the self-service ticketing area is 68 m while the distance is 52 m after. We assume that the walking speed is 0.75 m/s, and the time spent by passengers in the whole ticketing service is calculated:

Before improvement: $T_I = 1.1 \times 60 + 68/0.75 + 71/0.75 = 251.0$ (s)

After improvement: $T_2 = 0.33 \times 60 + 52/0.75 + 71/0.75 = 183.8$ (s)

Average saving time: $T = T_1 - T_2 = 67.2$ (s)

After the adjustment of ticket windows and areas, the average time spent by passengers decreased from 251.0 s to 183.8 s, and the average time saved by each passenger was 67.2 s.

5 Conclusion

a. Flexsim modeling and simulation software is used to simulate the current situation, find the key problems, and then improve and optimize the system.
b. The application of queuing theory to the calculation and analysis of manual and self-service ticketing areas can alleviate the situation of queuing and idle machines, which also improve the service efficiency and the utilization rate of the self-service ticketing machines per unit time, and reduce the busy rate of artificial windows.
c. Through the spatial layout analysis of Facility Planning, the pedestrian distance is shortened, concurrently the information is more convenient and the situation of cross-flow chaos is eliminated.

Aiming at the problem of system modeling and simulation in the ticketing service system of Zhuhai Station, this optimization design uses queuing theory and Facility Planning to optimize the system, and explores the simple mode of service system optimization design. This method has a good reference value for other service systems.

Clarification. The data is anonymized. The reason why anonymous is chosen in this questionnaire survey is that the questions involved do not need the respondents to provide their invalid information, and the respondents can directly express their true inner thoughts and situations. I promise that the questionnaire survey of this paper is true, the data source and data processing are true, and I am responsible for the authenticity of the questionnaire survey. If there is any falsehood, I am willing to bear all the consequences.

Declarant: Da-wei Zhou

Jan. 9, 2020

References

1. Hu, Y.-Q.: Basis and Application of Operational Research, pp. 286–295. Higher Education Press, Beijing (2014)
2. Ji, S.-F.: Modern Facilities Planning and Logistics Analysis, pp. 94–99. Machinery Industry Press, Norwalk (2016)
3. Luo, Y.-B.: Modeling and Simulation of production system, pp. 6–20. Huazhong University of Science and Technology Press, Wuhan (2014)
4. Hu, Da-Wei, Zeng, J.-G.: Flexsim-based simulation system for highway passenger station. J. Changan Univ. (Science edition) **30**(2), 89–95 (2013)
5. Yi, S.-P., Guo, F.: Second Edition of Basic Industrial Engineering, pp. 30–40. Machinery Industry Press, Beijing (2017)
6. Fang, Y.-D., Shan, Y.: Flexsim-based simulation optimization of industrial engineering integrated laboratory. Comput. Prod. Circ. **6**(1), 87 (2019)
7. Li, J., Ye, Y.-Y.: Management Operational Research Experimental Course, pp. 199–203. South China University of Technology Press, Guangzhou (2015)
8. Gong, M.: Application of operational research in hospital office. Electron. J. Clin. Med. Lit. **5**(57), 180–181 (2018)
9. Han, D.-M., Chen, B., Lu, F.-X.: Discussion on the planning of public service facilities in urban residential areas: taking Jining city as an example. Residential Real Estate **30**(3), 61 (2018)
10. Shi, D.: Simulative research on sorting of express centralized sites. Mod. Bus. Ind. **40**(18), 20 (2019)

A Redesign of Improving the Medical Effect of Hyperbaric Oxygen and Reducing Product Cost of Oxygen Mask Based on Ergonomics

Da-wei Zhou[1(✉)], Jian-hai Ou[1], Jia-qi Wu[1], Dao-zhi Zhao[2], and Xin-feng Zou[1]

[1] School of Industrial Automation, Beijing Institute of Technology, Zhuhai, China
450570503@qq.com
[2] College of Management and Economics, Tianjin University, Tianjin, China

Abstract. Based on the basic principle of ergonomic anthropometry, the structure and size of oxygen mask in hyperbaric oxygen chamber were redesigned by using pro/E software, which greatly increased the patients' pure oxygen inhalation and blood oxygen concentration, enhanced the effectiveness of hyperbaric oxygen treatment, made user more conveniently, and reduced the production cost of oxygen mask.

Keywords: Cost · Facial size · Mask inner volume · Modeling

1 Introduction

Hyperbaric oxygen can increase the oxygen content in human blood and tissues, and increase the oxygen diffusion distance. Hyperbaric oxygen therapy is widely used in a series of diseases caused by hypoxia and ischemia. Increasing blood oxygen concentration is an effective way to treat related diseases [1, 2].

At present, the oxygen inhalation masks in the hyperbaric oxygen chamber are mostly half-masks, wrapped around nose and mouth, forming a sealed space, connected with intake and exhaust pipes, providing patients with a certain amount of pure oxygen for medical treatment. We know that the smaller the space inside the oxygen mask, the less exhaust the patient breathes in the mask, the more oxygen he inhales in the body, and the better the therapeutic effect.

Human factor engineering studies various factors of human anatomy, physiology and psychology in a certain environment, which makes human, machine and environment more harmonious [3, 4]. This design takes the large oxygen inhalation mask of hyperbaric oxygen chamber produced by Wujin Huaxing Medical Equipment Co., Ltd. as the basic research object, fully considers the measurement of the human facial size, and uses pro/E software to improve the design of oxygen inhalation mask modeling. On the

Sponsored by Zhuhai Key Laboratory of Mechanical Manufacturing System Engineering (Zhuhai Key Project).

C.-F. Chien et al. (Eds.): *IE&EM 2019*, pp. 67–75, 2020.
https://doi.org/10.1007/978-981-15-4530-6_8

premise of guaranteeing the application function of oxygen inhalation mask, the capacity of its internal space is reduced as far as possible. The purpose of improving medical effect is to reduce the residual amount of exhaust gas and increase the oxygen inhalation.

2 Current Situation Analysis and Improvement Ideas

2.1 Physiological Analysis of Human Respiratory System

According to the experimental data, the tidal volume (the amount of gas inhaled or exhaled at a time) of normal adults during peaceful breathing is about 500 ml, and the ventilation volume is about 7.5 L/min according to 15 breaths per minute [5].

2.2 Improving Ideas

Setting the tidal volume of human body as V_1, the volume of oxygen mask as V_2, then the oxygen inhaled by patients once per breath is $V = V_1 - V_2$. From the perspective of medical efficacy, the more oxygen inhaled, the more conducive to the treatment of diseases [1, 2]. V_1 is a constant for normal adults, generally ranged from 400 ml to 600 ml [5]. As a result, in order to increase V, we must try to reduce V_2.

2.3 Analysis of Oxygen Mask in Traditional Hyperbaric Oxygen Chamber

The oxygen-absorbing mask of the hyperbaric oxygen chamber is composed of a mask shell, a three-way joint of the valve body and an inlet and outlet bellows. A one-way valve is installed in the end joint of the three-way inlet and outlet pipe. The function of the valve piece is that when the air is inhaled, the helium valve flap is opened, the exhaust valve flap is closed, and the exhalation is reversed [6]. The following studies do not involve inlet and outlet bellows. The cover case and the tee joint are shown in Fig. 1.

Pro/E software was applied to the following calculations (The wall thickness of the mask shell in this design is measured by 2 mm). Figure 2 shows the volume of the oxygen mask model, that is, $V_{21} = 329631\ mm^3$.

Figure 3 shows the volume of the actual oxygen mask model, which means $V_{22} = 45423\ mm^3$.

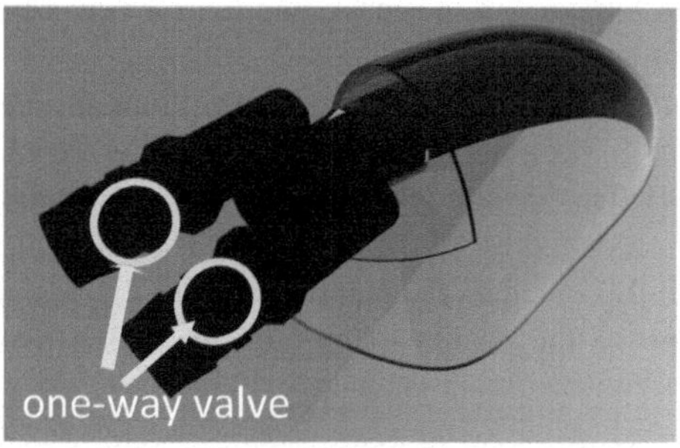

Fig. 1. Model of oxygen mask

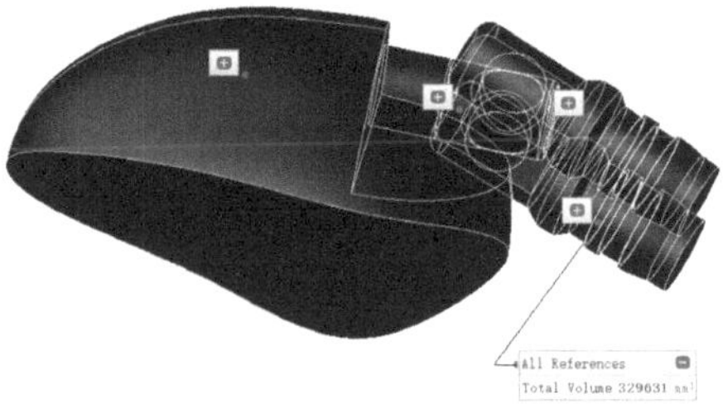

Fig. 2. Volume of physical model of traditional oxygen mask

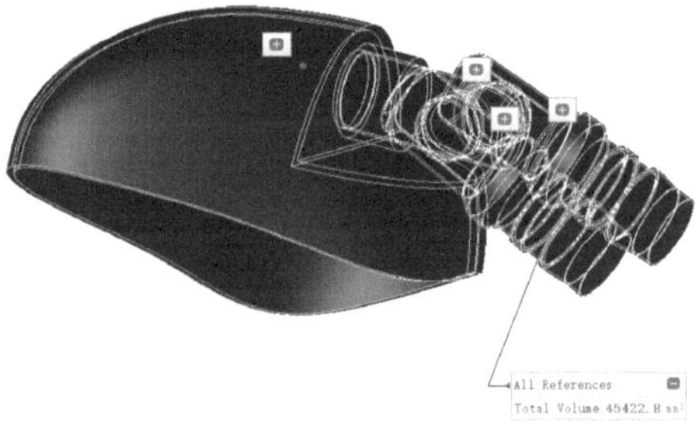

Fig. 3. The actual volume of oxygen mask model

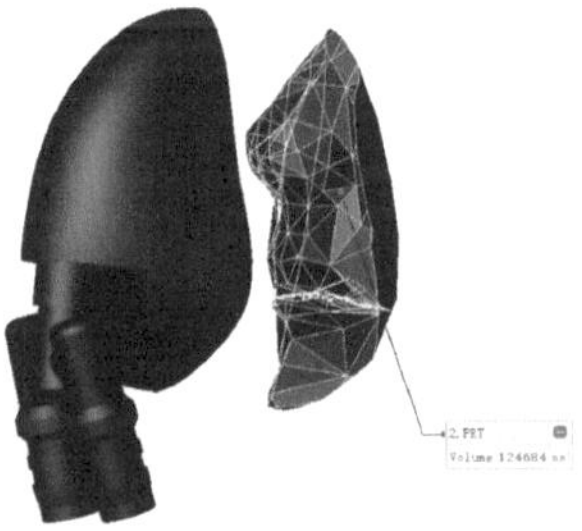

Fig. 4. The volume of the patient's face, mouth and nose entering the oxygen mask model (before)

Figure 4 shows the actual volume of the patient's face, mouth and nose entering the oxygen mask, which means $V_{23} = 124684\ mm^3$.

Subtracting the volume V_{21} of the oxygen mask model, the volume V_{22} of the actual oxygen mask model, and the volume V_{23} of the patient's face, mouth and nose into the oxygen mask, we can conclude that the space volume V_2 of the traditional mask is $159524\ mm^3$.

That is, the space volume of the mask $V_2 = V_{21} - V_{22} - V_{23} = 329631 - 45423 - 124684 = 159524\ (mm^3)$

The position of the T-joint face mask shell of the traditional oxygen mask is about the middle of the person below the nose, and the joint bellows can be adjusted inward and outward.

The total length, width and height of the mask shell are 128 mm, 74 mm and 68 mm. As shown in Fig. 5, the nose depth of 95 percentiles of adult men is 22 mm (18 mm for women) [7]. The corresponding size of the mask is 68 mm, which is far larger than the actual size required. It does not meet the requirements of human engineering, resulting in the excessive volume of the mask and less oxygen uptake of patients.

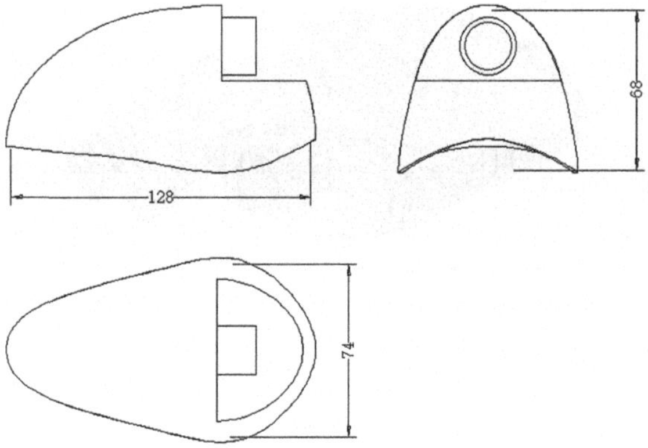

Fig. 5. Size of traditional mask case

3 Improvement Design of Oxygen Mask

The design of oxygen mask involves anatomy, physiology and psychology. The most basic problem in the design is the size of human body [8, 9]. The new design reduces the area of the mask, changes the structure of the mask, adjusts the outlet from the middle and the lower part of the front to the middle part of the left and right oblique sides, adjusts the connector from a tee joint to two straight-through joints, and adjusts the installation position of the one-way valve from the tee joint to the bellows end of the straight-through joint to the shell end of the mask, so as to reduce the total height of the mask. It is as low as the size that conforms to the principle of Ergonomics. Figure 6 shows the overall model of the improved oxygen mask. According to the data of human face size in the national standard of human body size, the main dimensions of the mask shell are set as below [7].

The total length of the mask shell = nose height + distance between the nasal and submental points = 55 + 77 = 132 mm (male 95 percentile size)

The total width of the mask shell = mouth width + function correction + psychological correction = 51 (female 95 percentile size) + 15 + 8 = 74 mm

Total height in the mask case = nose depth + functional correction + psychological correction = 22 (male 95 percentile size) + 18 + 8 = 48 mm

The newly set dimensions of the mask case are shown in Fig. 7.

The pro/E software is used to calculate the volume V_3 of the improved oxygen mask.

Figure 8 shows that the volume V_{31} of the improved oxygen mask model is 208745 mm^3.

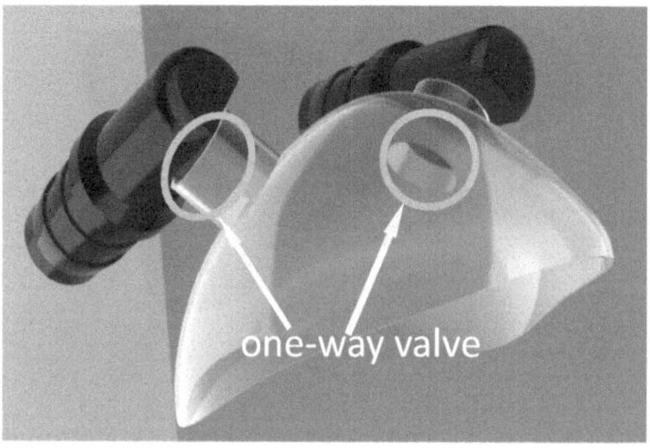

Fig. 6. Modeling of improved oxygen mask

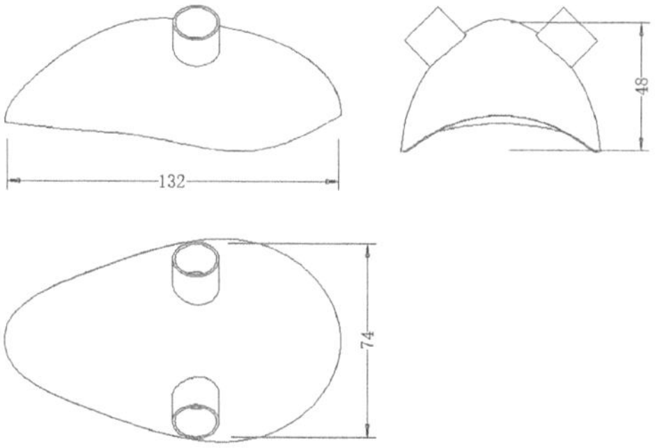

Fig. 7. Improves the main dimensions of the rear housing

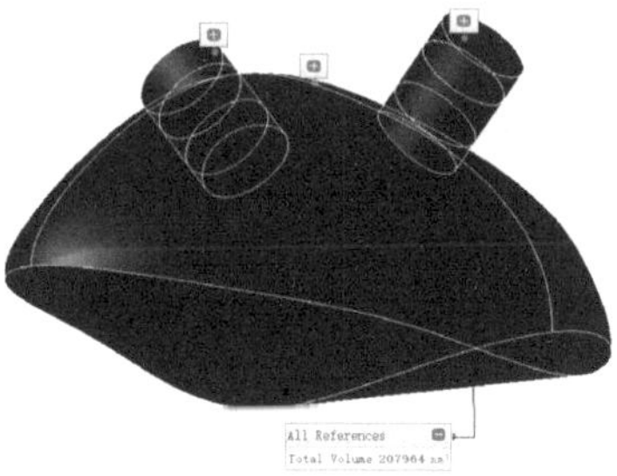

Fig. 8. Entity model Volume of oxygen mask after improvement

The volume V_{32} of the actual oxygen mask model can be improved from Fig. 9 to 31053 mm^3.

Figure 10 shows that the volume of the actual oxygen mask (V_{33}) is 124684 mm^3 when the patient's face, mouth and nose enter the improved mask.

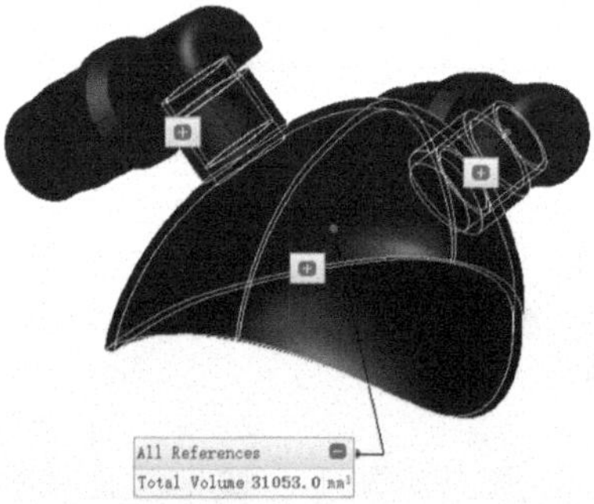

Fig. 9. The actual model volume of oxygen mask after improvement

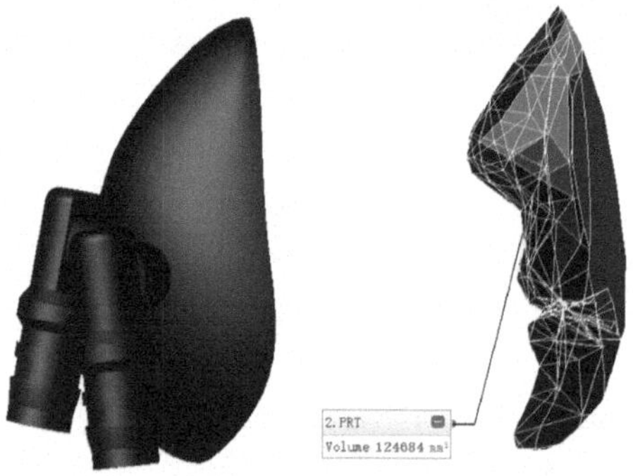

Fig. 10. The volume of the patient's face, mouth and nose entering the oxygen mask model (after)

The volume V_{31} of the improved oxygen mask model, the volume V_{32} of the actual oxygen mask model and the volume V_{33} of the improved oxygen mask can be subtracted. It can be concluded that the volume V_3 of the improved oxygen mask is 53008 mm^3.

That is, the space volume of the mask $V_3 = V_{31} - V_{32} - V_{33} = 208745 - 31053 - 124684 = 53008$ (mm^3).

4 Effect Analysis

Content area reduction before and after improvement of oxygen mask is $\Delta V = V_2 - V_3 = 159524 - 53008 = 106516$ (mm^3).

The volume reduction of shell V_4 and the volume reduction of ventilation joint V_5 of oxygen mask before and after improvement were calculated by pro/E software.

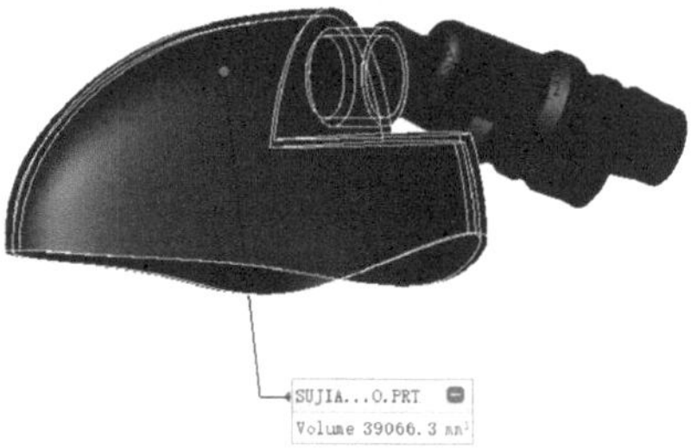

Fig. 11. Volume of mask shell (before)

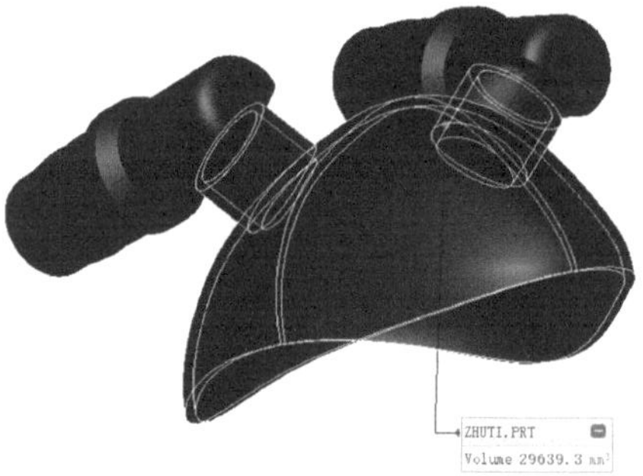

Fig. 12. Volume of mask shell (after)

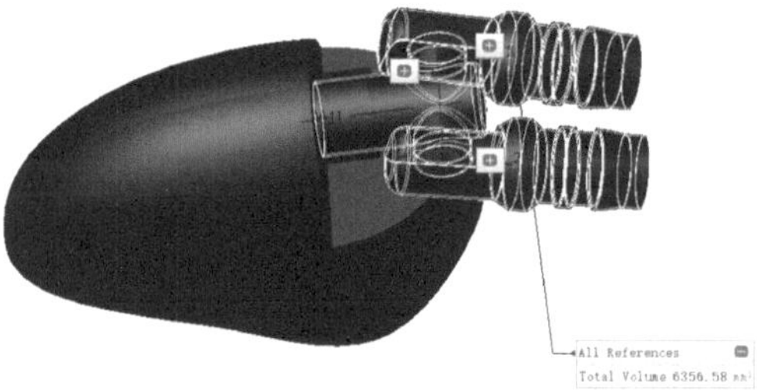

Fig. 13. Volume of joint (before)

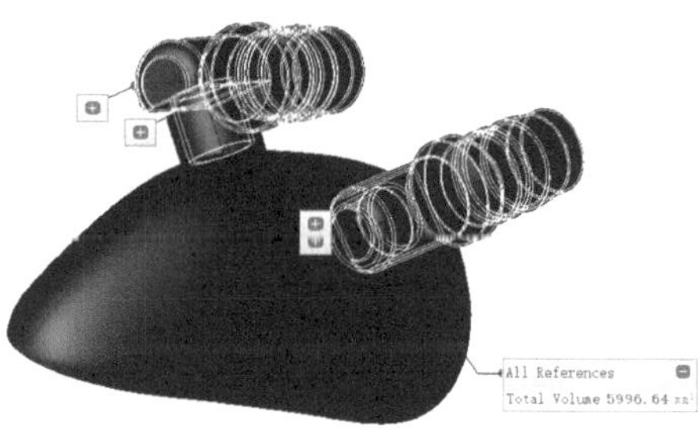

Fig. 14. Volume of joint (after)

From Figs. 11, 12, 13 and 14, we can see that the volume of mask shell before improvement (V_{41}) is 39066 mm^3, the volume of mask shell after improvement (V_{42}) is 29639 mm^3, the volume of joint before improvement (V_{51}) is 6357 mm^3, and the volume of joint after improvement (V_{52}) is 5997 mm^3. Therefore, the volume reduction of the rear mask shell ($V_4 = V_{41} - V_{42} = 39066 - 29639 = 9427$ mm^3) and the volume reduction of the ventilation joint ($V_5 = V_{51} - V_{52} = 6357 - 5997 = 360$ mm^3) are improved.

Table 1 compares the data before and after the improvement of oxygen mask. From this we can see that:

a. After improvement, the volume of oxygen mask decreased by 106516 mm^3, 106.5 ml in total, which means that the amount of pure oxygen inhaled per breath increased by 106.5 ml. According to 60 min of routine treatment in the hyperbaric oxygen chamber, more pure oxygen could be inhaled $106.5 \times 15 \times 60 = 95850$ ml = 95.85 L, and the increase rate of pure oxygen inhalation was $(106.5/500) \times 100\% = 21.3\%$.

It is precisely the 106.5 ml pure oxygen inhaled per breath that can make the peripheral blood vessels and nerves of patients get better oxygen, increase the concentration of blood oxygen, increase the dispersion distance of blood oxygen, and improve the effect of hyperbaric oxygen therapy.

Table 1. Comparison of data before and after improvement of oxygen mask

Project	Before	After	Reduce
Total length/mm	128	132	−4
Total width/mm	74	74	0
Total height/mm	68	48	20
Mask content/mm^3	159524	53008	106516
Mask material/mm^3	39066	29639	9428
Pipeline material/mm^3	6357	5997	360

b. After the improvement, the volume of the cover shell is reduced by 9428 mm^3 and the volume of the valve body joint is reduced by 360 mm^3. Polyvinyl chloride (PVC) material is generally used in the face mask case with a density of 1.38 g/cm^3, and ABS material is generally used in the valve body joint with a density of 1.12 g/cm^3. Each oxygen mask can save $1.38 \times 9428/1000 = 13.01$ g, ABS plastic $1.12 \times 360/1000 = 0.4$ g [10].

Based on the price of 6,900 yuan per ton of PVC (quoted by Tianyuan Group on June 6, 2019), the cost of each oxygen mask can be reduced by $13.01 \times 6900/1000000 = 0.09$ (yuan). If one million sets are produced annually, the cost of materials can be reduced by 90,000 yuan per year.

In addition, when the tee joint connected with bellows is replaced by two directly connected joints with the mask, the bellows can be adjusted in more directions and angles, which is more convenient for patients to use.

5 Conclusion

a. This design is based on the principle of human factor engineering. By reducing the volume of oxygen mask, the patient's pure oxygen inhalation volume can be greatly increased, the blood oxygen concentration and the oxygen diffusion distance can be increased, so as to improve the therapeutic effect. The improvement of curative effect needs further study.
b. The improved design can effectively reduce the material usage of the mask shell, reduce the material cost of the product and improve the economic benefits of the enterprise.
c. The improved mask makes the patient more comfortable and convenient to use.
d. The improvement design of this mask provides some design ideas and methods for other products.

References

1. Tao, H.-Y., Jiang, G.-D., Lin, F.: Clinical Application of Hyperbaric Oxygen. Second Military Medical University Press, Shanghai (2015)
2. Cui, D.-J.: Hypoxia and oxygen therapy. Chin. J. Multi. Org. Dis. Elderly **3**(2), 153–156 (2004)
3. Guo, F., Qian, X.-S.: Human Factor Engineering, 2nd edn, pp. 1–17. Machinery Industry Press, Beijing (2018)
4. Ruan, B.-X., Liu, Y.-X., Dong, M.-M.: Ergonomics of Industrial Design, 3rd edn, pp. 2–4. Machinery Industry Press, Beijing (2016)
5. Wang, T.-H.: Physiology, 9th edn. People's Medical Publishing House, Beijing (2018)
6. Liu, Y., Ma, W.: Difficulty analysis and measures of oxygen inhalation for patients in hyperbaric oxygen chamber. Information of Medical Equipment **7**(10), 112–113 (2007)
7. National Standard of the People's Republic of China GB/T 2428–1998, Head-face Dimensions of Adults, pp. 3–5. China Standard Press, Beijing (1998)
8. Zhang, X., Xie, W., Zhang, K., Hu, Y., Liang, Z.: Design of a new oxygen inhalation device. China Med. Dev. Inf. **24**(13), 25–26, 146 (2018)
9. Ding, Y.-I., Cheng, G.-P.: Ergonomics, pp. 16–32. Beijing Institute of Technology Press, Beijing (2013)
10. Sang, Y.: Plastic Materials and Formulas, 3rd Edition, pp. 9–20, 55–59. Chemical Industry Press, Beijing (2014)

Evolutionary Path of Group Safety Behavior of Grass-Roots Employees

Su-xia Liu, Xu Yang(✉), Yu-qing Zhu, and Qiang Mei

School of Management, Jiangsu University,
Zhenjiang, Jiangsu, People's Republic of China
Aprilyang157@foxmail.com

Abstract. To discover the evolution rules of strategy choice for the group safety behavior of grass-roots employees, the payment matrix for the behavioral interaction among employees is constructed by virtue of the evolutionary economics. Through analysis of evolution model for the interactive process of behavior, the influence of strategy choice for the employees' individual behavior on the evolution of group behavior is revealed. Through numerical simulation, the study on the influence of variation in the different initial proportion and different parameters of group on the evolutionary trend. The result shows that it is conducive for grass-roots employees to have their behavior evolve towards the expected direction if business executives strengthen the safety check, fair in meting out rewards or punishments, improve employees' awareness to the cost of safety noncompliance behavior, lower the satisfaction from noncompliance operation, optimize the operation procedure and pre-job safety education and training.

Keywords: Evolutionary path · Grass-roots employees · Group game · Numerical experiment · Safety behavior

1 Introduction

Grass-roots employees' safety behavior exerts a great influence on the healthy and sustainable development of an enterprise as they directly participate in the production and engage in the element tasks. With the perfection of facilities and equipment and increasingly mature technology, the hard environment for the work safety of employees has witnessed a certain improvement, but the support for the soft environment such as encouragement of workmates to supervise others and communicate with each other about safety is ignored, as a result, the noncompliance production of employees still continues despite repeated prohibition and the risk of accident is increased [1]. It is pointed in the social learning theory put forward by Albert Bandura, an American psychologist that human behavior can be acquired through their own direct experience and model behavior of others, and people can form their own behavior through observing that of people around them [2]. As the grass-roots employees and their surrounding workmates are in the same work environment and their living environment and cultural level are similar, one person's behavior will inevitably have an impact on the others' behavior through exchange and communication about the safety and transfer of all information, and the group behavior of work safety or compliance production can

C.-F. Chien et al. (Eds.): *IE&EM 2019*, pp. 76–93, 2020.
https://doi.org/10.1007/978-981-15-4530-6_9

also be formed [3], which will exert an influence on the employees' choice for their own attitude and behavior. Moreover, employees can judge their workmates' behavior and compare their own behavior with them, once a certain behavior can bring gains to them, other employees will learn such behavior [4].

Grass-roots employees of an enterprise are always in a group consisting of themselves and their workmates during the production process instead of being isolated. It is very easy for grass-roots employees to show the safety attitude and behavior [5] consistent with that of their workmates due to influence of group pressure as most of them have low education level, high labor intensity, weak safety awareness, yearn for integration into the group and being accepted by workmates. The group behavior of grass-roots employees is a stable status which is finally formed from the employees' interaction in the group and constant adjustment of their own strategies, during which employees and their workmates interact with each other, and workmates' safety behavior is a major factor affecting the safety atmosphere of the group. In the literature review by [6] with respect to the safety atmosphere of construction site, it is pointed the workmates' role and influence is an important dimension for the safety atmosphere. Also, there are scholars verifying the workmates' influence and support behavior are the major parts constituting the safety atmosphere of a group [7, 8] through factor analysis. Zohar's (2010) study shows the group atmosphere and workmates' behavior exert a significant influence on the employees' safety behavior, and positive friendship among workmates can reduce the adverse influence of low-level safety atmosphere on the employees' attitude towards safety and safety behavior from the network perspective of employees' friendship [9]. Fugas, Silva and Meliá (2012) discovers the workmates' behavior plays a role of regulating the employees' cognition of standard for safety and attitude towards safety in the study on relations between the safety atmosphere and safety behavior in the transformational leadership organization [10]. Besides, Zhou, Fang and Wang (2008) draw a conclusion from the study that the degree of influence of safety atmosphere (colleague's influence and management commitment) on the safety behavior is greater than that of personal experience thereon according to Bayesian Theory [11].

Thus it can be seen that the grass-roots employees' safety behavior is closely related to their workmates' behavior [12], however, there is few study in the domestic and foreign literatures based on the interaction of their employee' safety behavior. Therefore, this article provides theoretical support and policy suggestion for normalizing the group behavior of grass-roots employees through building game model between employees and their workmates with evolutionary economics theory and discussion about the factor affecting the evolutionary results through analysis of evolution model in the interactive process.

2 Methodology

A. *Model assumption*

In the same group (such as a team), the grass-roots employees' and their workmates' choice for safety behavior is interactive and is a dynamic adjustment process. It is assumed that the employees' attitude towards safety and safety behavior in the group are different, some have strong safety consciousness, while some have weak safety consciousness, therefore, employees in the group are divided into Group A and Group B, two game players with bounded rationality, from which one employee is randomly selected respectively each time to play the game. Behavior abiding by the provisions for work safety is called safety compliance behavior, while the behavior inconsistent with the provisions for work safety is called safety noncompliance behavior [13].

The second it is assumed that employees need to contribute extra labor L in case of choice for work safety behavior, coefficient of gains from work safety behavior is α_1, coefficient of gains from production is α_2, the potential loss to be borne by employees in case of choice for safety noncompliance behavior is S_Z; as the safety compliance behavior may cause injury and death to workmates, the potential loss to be borne by workmates is λS_Z ($\lambda > 0$). The psychological satisfaction obtained by employees who choose safety compliance behavior due to puppyism or follow of their workmates' behavior is called puppyish gains V, and such behavior can bring negative demonstration effect (D) to the workmates. The probability that employees choose safety noncompliance behavior are discovered by enterprise executives is P_{C1}, once being discovered, those will be imposed upon a punishment B (fine and deduction of bonus), and all the workmates in the team getting involved will be also imposed upon a punishment εB ($\varepsilon > 0$) and will complain, as a result, the workmates' words and deeds will bring a certain psychological blow E to the noncompliance employees. If all the selected employees choose the safety noncompliance behavior, the probability for the discovery of noncompliance behavior will rise to P_{C2}, which, together with P_{C1}, reflects the enterprise's strength for supervision over its employees' work safety. Table 1 shows the payment matrix built for the game of grass-roots employees' choice for work safety behavior model.

Table 1. Payment matrix

	Employee A	Employee B
Safety compliance	$a = \alpha_1 L$, $e = \alpha_1 L$	$c = \alpha_1 L - \lambda S_Z - P_{C1}\varepsilon B - D$, $g = \alpha_2 L - S_Z - P_{C1}B + V - P_{C1}E$
Safety noncompliance	$b = \alpha_2 L - S_Z - P_{C1}B + V - P_{C1}E$, $f = \alpha_1 L - \lambda S_Z - P_{C1}\varepsilon B - D$	$d = \alpha_2 L - S_Z - P_{C2}B + V$, $h = \alpha_2 L - S_Z - P_{C2}B + V$

B. *Model evolution*

To build the replicated dynamic equation for the grass-roots employees' choice for work safety behavior model according to the game relationship above, and simulate the replicated game process of these two groups with bounded rationality with such replicated dynamic equation. It is assumed that, at the initial time of the game, proportion of employees in the Group A that choose safety compliance behavior strategy is p, and that choose safety noncompliance behavior strategy is $1 - p$; while proportion of employees in the Group B that choose safety compliance behavior strategy is q, and that choose safety noncompliance behavior strategy is $1 - q$.

Gains U_1 and U_2 respectively from employees' choice for safety compliance behavior and safety noncompliance behavior strategy in Group A is:

$$U_1 = qa + (1-q)c \tag{1}$$

$$U_2 = qb + (1-q)d \tag{2}$$

The replicated dynamic equation for employees' choice for safety compliance behavior in Group A is as follows:

$$F(p) = \frac{dp}{dt} = p(1-p)(U_1 - U_2) = p(1-p)[q(a-b-c+d)+c-d] \tag{3}$$

Gains V_1 and V_2 respectively from employees' choice for safety compliance behavior and safety noncompliance behavior strategy in Group B is:

$$V_1 = pe + (1-p)f \tag{4}$$

$$V_2 = pg + (1-p)h \tag{5}$$

The replicated dynamic equation for employees' choice for safety compliance behavior in Group B is as follows:

$$F(q) = \frac{dq}{dt} = q(1-q)(V_1 - V_2) = q(1-q)[p(e-g-f+h)+f-h] \tag{6}$$

Jacobian Matrix of the equation obtained therefrom is:

$$J = \begin{bmatrix} \frac{dF(p)}{dp} & \frac{dF(p)}{dq} \\ \frac{dF(q)}{dp} & \frac{dF(q)}{dq} \end{bmatrix} = \begin{bmatrix} (1-2p)[q(a-b-c+d)+c-d] & p(1-p)(a-b-c+d) \\ q(1-q)(e-g-f+h) & (1-2q)[p(e-g-f+h)+f-h] \end{bmatrix} \tag{7}$$

By recording the determinant of Jacobian Matrix as Det J and its trace as Tr, we can get:

$$\begin{aligned} \mathrm{Det\,J} = {} & (1-2p)(1-2q)[q(a-b-c+d)+c-d][p(e-g-f+h)+f-h] \\ & - pq(1-p)(1-q)(a-b-c+d)(e-g-f+h) \end{aligned} \tag{8}$$

$$Tr = (1-2p)[q(a-b-c+d)+c-d] + (1-2q)[p(e-g-f+h)+f-h] \tag{9}$$

3 Analysis

The symbols p and q respectively represent the proportion of individuals in Group A and B that choose the compliance behavior strategy in such group, so $0 \leq p \leq 1$ and $0 \leq q \leq 1$, and equilibrium and stability of the system will be discussed about on the plane $M^* = \{(p, q)|0 \leq p, q \leq 1\}$.

Case 1: When $a - b > 0$ (or $e - g > 0$) and $c - d > 0$ (or $f - h > 0$), the system has four equilibriums, among which (0,0) is source point, (1,1) is stable point, while (0,1) and (1,0) are saddle point, and see Fig. 1 for the dynamic interactive process. No matter what kind of strategy the other player chooses, gains from safety compliance behavior strategy are greater than those from safety noncompliance behavior strategy. Therefore, increase the employees' understanding of the importance of work safety, have them fully realize the safety gains and the potential loss from noncompliance behavior; reinforce the strength for the inspection of and punishment upon employees' safety noncompliance behavior; strengthen the construction of safety atmosphere and reduce puppyish gains from employees' safety noncompliance behavior and cut off their adverse demonstration effect to make the system evolve into the mode of (compliance, compliance).

Case 2: When $a - b<0$ (or $e - g<0$) and $c - d<0$ (or $f - h<0$), the system has four equilibriums, among which (1,1) is source point, (0,0) is stable point, while (0,1) and (1,0) are saddle point, and see Fig. 2 for the dynamic interactive process. In this case, the system will evolve into the mode of (noncompliance, noncompliance), and no matter what kind of strategy one player chooses, gains from safety compliance behavior strategy by the other player are less than those from safety noncompliance behavior strategy. The accident is caused by many factors including safety noncompliance behavior, therefore, employees' noncompliance may not cause accident, in this state, their workmates may even imitate their safety noncompliance behavior due to gains-seeking and fluke mind instead of stopping the same, as a result, the safety noncompliance behavior exerts a further influence on the workmates around and causes the whole system to fall into an "adverse" locking mode.

Case 3: When $a - b<0$ (or $e - g<0$) and $c - d > 0$ (or $f - h > 0$), the system has five equilibriums, among which (0,0) and (1,1) are source point, (0,1) and (1,0) are stable point, while (p^*,q^*) is saddle point, and see Fig. 3 for the dynamic interactive process. In this case, the system will evolve into the mode of (compliance, noncompliance) or (noncompliance, compliance).

In such case, the game player can obtain higher gains as long as he or she chooses the strategy different from that of the other players, which means when employees of one player choose compliance behavior, those of the other that choose the safety noncompliance behavior will obtain higher gains and vice versa. Strategy chosen employees will witness adjustment according to the status of whole behavior of the group. If the safety noncompliance behavior in the group is relatively serious, employees therein will be inclined to the safety compliance strategy in the new round of the game, and vice versa as it is easy for an enterprise to relax the safety inspection when the safety compliance is better, and employees will be inclined to the safety

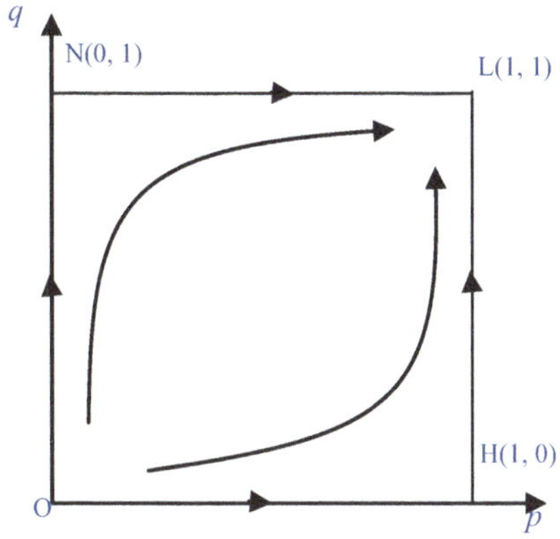

Fig. 1. Dynamic interactive process of case 1

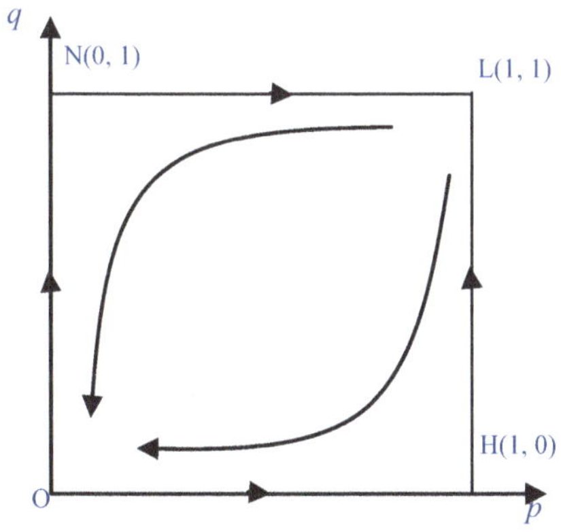

Fig. 2. Dynamic interactive process of case 2

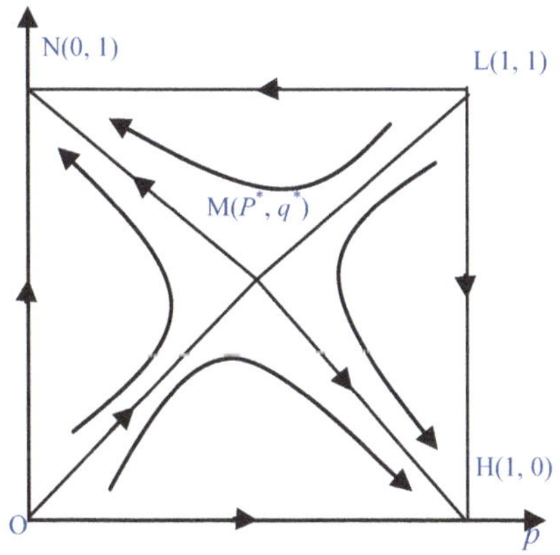

Fig. 3. Dynamic interactive process of case 3

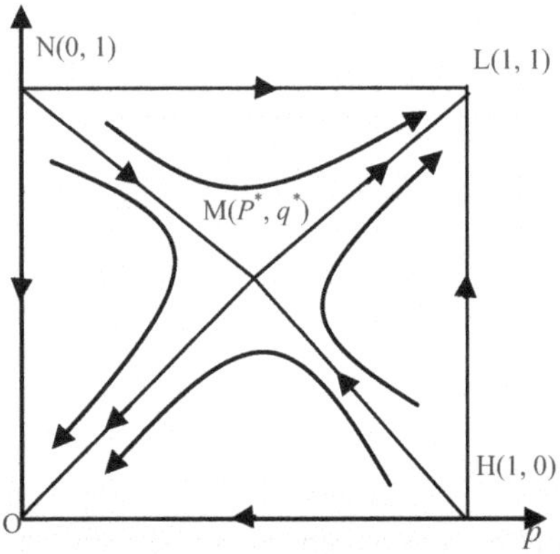

Fig. 4. Dynamic interactive process of case 4

noncompliance behavior to meet their own interests, at which time employees choosing safety compliance behavior will be inclined to choose the safety noncompliance strategy as they obtain smaller gains; however, with the seriousness of noncompliance, enterprise will strengthen the inspection, at which time employees will turn to choose the safety compliance behavior due to worrying about being punished and having their workmates get involved upon exposure of safety noncompliance, and those employees will not only get punished due to noncompliance with the standard for safety, but also the safety rating of their group will be affected and their workmates' bonus will be deducted, then employees choose safety compliance behavior will get gains higher than those of the employees choosing safety noncompliance behavior. To avoid the evolution of system into such a mode, the probability for the discovery of one employee's safety noncompliance behavior is required to be improved, which means enterprise shall immediately take measures to prevent and punish employees choosing safety noncompliance behavior instead of doing so until the safety noncompliance behavior in the group becomes serious.

Case 4: When $a - b > 0$ (or $e - g > 0$) and $c - d < 0$ (or $f - h < 0$), the system has five equilibriums, among which (0,1) and (1,0) are source point, (0,0) and (1,1) are stable point, while (p^*,q^*) is saddle point, and Fig. 4 shows the dynamic interactive process and Table 2 for analytic results of local stability.

$$p^* = q^* = \frac{(\alpha_2 - \alpha_1) + (\lambda - 1)S_Z + P_{C1}\varepsilon B - P_{C2}B + V + D}{P_{C1}B + P_{C1}E + P_{C1}\varepsilon B + \lambda S_Z + D - P_{C2}B} \tag{10}$$

In Fig. 4, the broken line between two unstable equilibriums (0,1) and (1,0) and saddle point (p^*,q^*) is the boundary where the system converges to the two different modes, (compliance, compliance) and (noncompliance, noncompliance), and the evolutionary process and stable state of group are affected by the proportion of employees choosing the different strategies in the group and relative location of saddle point M. The stable equilibrium result in the game between employees and their workmates may

Table 2. Analytic results of local stability

Equilibriums	Det J		Tr		Result
$p = 0, q = 0$	$-(c-d)(f-h)$	+	$(c-d)+(f-h)$	–	ESS
$p = 0, q = 1$	$-(c-d)(e-g)$	+	$-(c-d)+(e-g)$	+	Unstable
$p = 1, q = 0$	$-(a-b)(f-h)$	+	$(a-b)-(f-h)$	+	Unstable
$p = 1, q = 1$	$(a-b)(e-g)$	+	$-(a-b)-(e-g)$	–	ESS
$p = p^*, q = q^*$	$-p^*q^*(1-p^*)(1-q^*)$ $(a-b-c+d)(c-g-f+h)$	–	0		Saddle point

be compliance or noncompliance behavior, and their evolutionary path and stable state depend on the area S_{HMNO} and S_{HLNM} of the region HMNO and HLNM respectively. To conduct the parameter analysis and regulation through analyzing the factors affecting the area S of regions. When $S_{HLNM} > S_{HMNO}$, the system will evolve into the compliance behavior along the path ML in a higher probability; when $S_{HLNM} < S_{HMNO}$, the system will evolve into the noncompliance behavior along the path MO in a higher probability; according to Fig. 4,

$$S_{HLNM} = 1 - \frac{1}{2}(p^* + q^*) = 1 - \frac{(\alpha_2 - \alpha_1) + (\lambda - 1)S_Z + P_{C1}\varepsilon B - P_{C2}B + V + D}{P_{C1}B + P_{C1}E + P_{C1}\varepsilon B + \lambda S_Z + D - P_{C2}B} \tag{11}$$

In the compliance model of employees of middle and small-sized enterprises, there are 11 parameters involved therein in total, based on which the evolutionary direction of safety compliance behavior of employee group is analyzed below, with the result shown in Table 3.

Proposition 1: Proportion p and q of employees choosing compliance behavior in the group is in direct proportion to the extra labor L required to be contributed by employees choosing safety compliance behavior, puppyish gains V, variations of gains $(\alpha_2 - \alpha_1)$ and probability P_C for the discovery of noncompliance behavior, while in inverse proportion to negative effect D and psychological blow E.

As the L, V, $(\alpha_2 - \alpha_1)$ and E have a monotonic relationship with p^* and q^*, the impact of increase and decrease on p^* and q^* can be seen obviously. By seeking partial derivatives of P_C and D through p^* and q^* respectively, we can get.

$$\frac{\partial p^*}{\partial p_{C2}} = \frac{\partial q^*}{\partial p_{C2}} = \frac{B(b-a)}{(a-b-c+d)^2} < 0 \tag{12}$$

$$\frac{\partial p^*}{\partial D} = \frac{\partial q^*}{\partial D} = \frac{a-b}{(a-b-c+d)^2} > 0 \tag{13}$$

The proposition is proved.

Proposition 2: The impact of punishment B upon employees due to noncompliance on the employees' choice for behavior is uncertain.

By seeking partial derivatives of B through p^* and q^*, we can get:

$$\frac{\partial p^*}{\partial B} = \frac{\partial q^*}{\partial B} = \frac{(P_{C1}\varepsilon - p_{C2})(a-b) - P_{C1}(d-c)}{(a-b-c+d)^2} \tag{14}$$

When

$$\varepsilon < \frac{1}{P_{C1}}\left[\frac{P_{C1}(d-c)}{a-b} + P_{C2}\right] \tag{15}$$

And

$$\frac{\partial p^*}{\partial S_Z} = \frac{\partial q^*}{\partial S_Z} < 0 \tag{16}$$

B is in inverse proportion to the proportion p and q of employees choosing compliance behavior in the group; when

$$\varepsilon > \frac{1}{P_{C1}}\left[\frac{P_{C1}(d-c)}{a-b} + P_{C2}\right] \tag{17}$$

And

$$\frac{\partial p^*}{\partial S_Z} = \frac{\partial q^*}{\partial S_Z} < 0 \tag{18}$$

B is in direct proportion to the proportion p and q of employees choosing compliance behavior in the group; therefore, the proposition is proved.

Proposition 3: Impact of potential loss S_Z on the employees' choice for behavior is uncertain.

By seeking partial derivatives of S_Z through p^* and q^*, we can get:

$$\frac{\partial p^*}{\partial S_Z} = \frac{\partial q^*}{\partial S_Z} = \frac{(\lambda-1)(a-b)+c-d}{(a-b-c+d)^2} \tag{19}$$

When

$$0<\lambda<\frac{d-c}{a-b}+1 \tag{20}$$

And

$$\frac{\partial p^*}{\partial S_Z}=\frac{\partial q^*}{\partial S_Z}<0 \tag{21}$$

The potential loss S_Z is in inverse proportion to the proportion p and q of employees choosing compliance behavior in the group; when

$$\lambda>\frac{d-c}{a-b}+1 \tag{22}$$

And

$$\frac{\partial p^*}{\partial S_Z}=\frac{\partial q^*}{\partial S_Z}>0 \tag{23}$$

The potential loss S_Z is in direct proportion to the proportion p and q of employees choosing compliance behavior in the group; therefore, the proposition is proved.

Table 3. Impact of variations of parameters on employees' behavior strategies

Variations of parameters		Variations of saddle point	S_{HLNM}	Evolutionary direction
$L\downarrow$		$p^*\downarrow, q^*\downarrow$	$\uparrow$	(Compliance, compliance)
$V\downarrow$		$p^*\downarrow, q^*\downarrow$	$\uparrow$	(Compliance, compliance)
$(\alpha_2-\alpha_1)\downarrow$		$p^*\downarrow, q^*\downarrow$	$\uparrow$	(Compliance, compliance)
$E\uparrow$		$p^*\downarrow, q^*\downarrow$	$\uparrow$	(Compliance, compliance)
$P_{C2}\uparrow$		$p^*\downarrow, q^*\downarrow$	$\uparrow$	(Compliance, compliance)
$D\uparrow$		$p^*\uparrow, q^*\uparrow$	$\downarrow$	(Noncompliance, noncompliance)
$B\downarrow$	$\varepsilon<\frac{1}{P_{C1}}\left[\frac{P_{C1}(d-c)}{a-b}+P_{C2}\right]$	$p^*\uparrow, q^*\uparrow$	$\downarrow$	(Noncompliance, noncompliance)
	$\varepsilon>\frac{1}{P_{C1}}\left[\frac{P_{C1}(d-c)}{a-b}+P_{C2}\right]$	$p^*\downarrow, q^*\downarrow$	$\uparrow$	(Compliance, compliance)
$S_Z\uparrow$	$0<\lambda<\frac{d-c}{a-b}+1$	$p^*\downarrow, q^*\downarrow$	$\uparrow$	(Compliance, compliance)
	$\lambda>\frac{d-c}{a-b}+1$	$p^*\uparrow, q^*\uparrow$	$\downarrow$	(Noncompliance, noncompliance)

4 Results

In order to reflect the impact of variations of parameters on the evolutionary result more intuitively, and as some variables cannot be analyzed through derivation of saddle point, numerical experiment is hereby conducted with Matlab simulation software against the complex evolution of Case 4 to analyze the impact of initial proportion of the group choosing one tactic and variations of parameters affecting the gains of game players on the system evolutionary result. By considering the preconditions of Case 4 and constraint from facts, the value of parameters besides the analyzed object parameters is as follows: $\alpha_1 = 1.2$, $\alpha_2 = 2$, $L = 1.4$, $\lambda = 0.2$, $S_Z = 2$, $P_{C1} = 0.5$, $P_{C2} = 0.6$, $\varepsilon = 0.6$, $B = 0.3$, $V = 0.5$, $D = 0.6$ and $E = 0.06$.

(1) See Fig. 5 for the result of numerical experiment on the impact of changes of initial proportion of the group choosing one tactic on the evolutionary result, among which p_0 and q_0 respectively represent the initial proportion of employees choosing the safety compliance behavior in Group A and Group B.

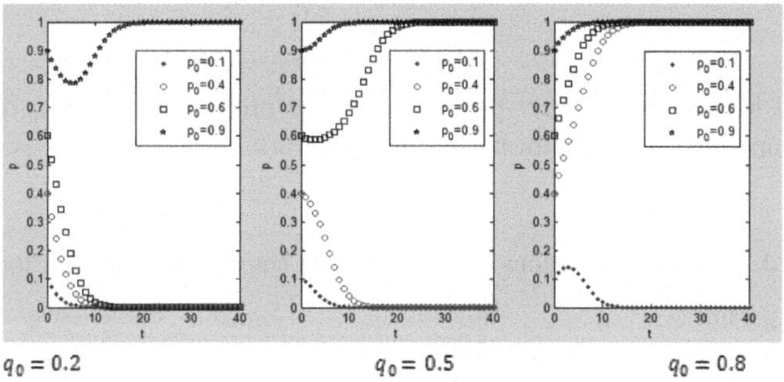

Fig. 5. Impact of changes of initial proportion's difference of the group choosing one tactic on the evolutionary result

The caption According to Fig. 5, the evolutionary process of strategy interaction behavior between employees and workmates has the path dependence. Path lines starting from different initial conditions will not be overlapped before converged to the equilibrium state. The time length for the convergence to the equilibrium state is related to the initial proportion of employees choosing different strategies in the group, when the proportion approaches to the equilibrium state, the rate of convergence becomes faster. For the circumstances where more employees choose safety noncompliance behavior, this essay will next focus on the analysis of the circumstance where the initial proportion of the group choosing safety compliance behavior which is lower (i.e. $q_0 = 0.2$) and analyze the impact of variations of different parameters on the evolutionary result of the group through numerical experiment to guide the system to evolve into the ideal stable state of (compliance, compliance).

(2) See Fig. 6 and Fig. 7 for the result of numerical experiment on the impact of extra labor L required for safety compliance behavior and changes of gains $(\alpha_2-\alpha_1)$ on the evolutionary result.

According to Fig. 6, when the extra labor L required for safety compliance behavior and gains difference between production and safety input $(\alpha_2-\alpha_1)$ rise to a certain extent, the employees' and their workmates' behavior will totally evolve into the mode of (noncompliance, noncompliance). Especially under the circumstance where the enterprise implements the piecework wage, grass-roots employees can clearly realize the gains from production, and work safety can be easily ignored for the saving of labor. Therefore, on one hand, enterprise should provide favorable conditions for work safety, improve production environment, have its safety facilities, equipment and products conform to the Standard for Ergonomics, and reduce additional operational steps and difficulty in operation to the greatest extent; on the other hand, enterprise should also provide the favorable cultural atmosphere for work safety instead of focusing on the improvement of production efficiency to make its employees fully realize the potential benefits of work safety for the enterprise, especially for themselves and their workmates.

(3) See Fig. 8 and Fig. 9 for the result of numerical experiment on the impact of puppyish gains V from safety noncompliance and psychological blow E after being discovered on the evolutionary result. V and E are the analysis of evolutionary result from the perspective of psychology, differently, the former refers to the psychological satisfaction of employees from following their workmates' noncompliance, while the later refers to the psychological damage caused to the employees due to their workmates' complaint and isolation as their noncompliance behavior causes injury and even death to their workmates. According to the figures, when the puppyish gains rise to a certain extent, the system will evolve into the mode of (noncompliance, noncompliance), which explains employees in the group can follow their workmates' safety noncompliance behavior and such behavior is infectious. Besides, psychological blow E has little influence on the evolutionary result, but the system will need a longer time to converge to the equilibrium state with the increase of E. Therefore, it is vital to improve the safety atmosphere of the group and to improve employees' sensitivity towards their workmates' noncompliance behavior to make the con-compliance behavior lack proper environment and thus try to reduce the psychological satisfaction to the employees from noncompliance behavior. Moreover, conduct training for the safety education to make employees fully realize the potential damage of noncompliance behavior to their workmates and improve their awareness to the cost of noncompliance behavior.

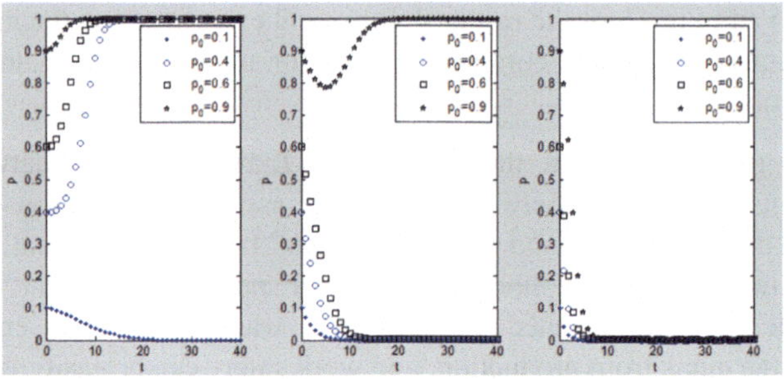

$L = 1$ $L = 1.4$ $L = 2$

Fig. 6. Impact of changes of extra lab of safety compliance behavior on the evolutionary result

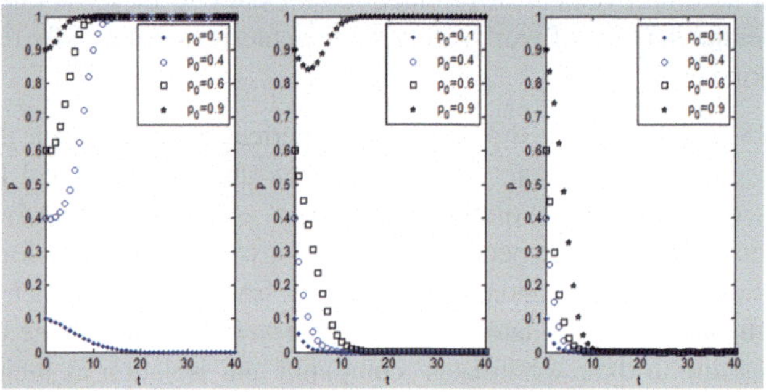

$\alpha_2 - \alpha_1 = 0.6$ $\alpha_2 - \alpha_1 = 0.8$ $\alpha_2 - \alpha_1 = 1.0$

Fig. 7. Impact of changes of the gains difference between production and safety input on the evolutionary result

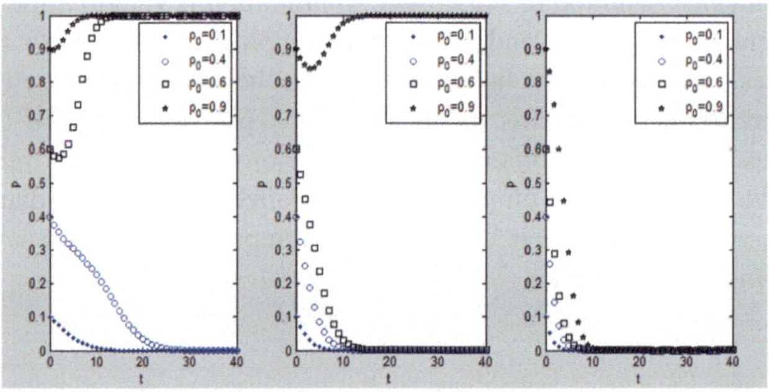

$V = 0.3$ $V = 0.5$ $V = 0.8$

Fig. 8. Impact of changes of puppyish gains on the evolutionary result

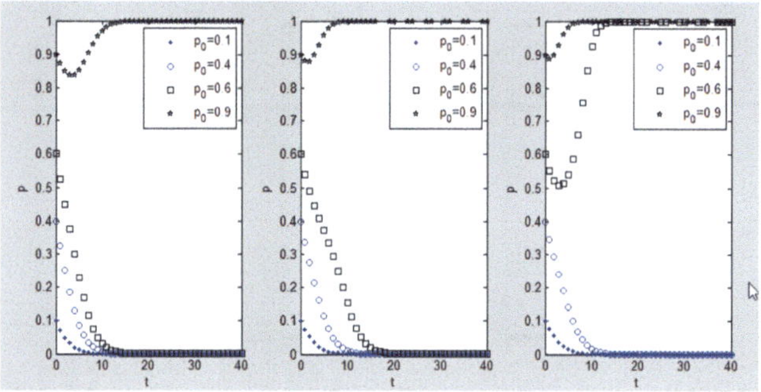

E = 0.3 E = 0.5 E = 0.8

Fig. 9. Impact of changes of psychological stress the evolutionary result

(4) See Fig. 10 to Fig. 12 for the result of numerical experiment on the impact of punishment B for and accident loss S_Z from safety noncompliance behavior and enterprise supervision on the evolutionary result. The trend of influence of B and S_Z, such two losses on the system evolutionary results are same. However, with respect to the punishment actually incurred, the expected accident loss is the potential loss, to which employees are insensitive. Changes of enterprise supervision exert a significant influence on the system evolutionary result, which means the system will evolve into the ideal mode of (compliance, compliance) when the supervision is greater. Moreover, it is important to note that when the probability for the discovery of safety noncompliance of employees of one player is lower, the system will eventually evolve into the adverse "locking" state of (noncompliance, noncompliance), even though the initial proportion of employees in the group choosing the safety compliance behavior is particularly high; when the probability for the discovery of safety noncompliance of employees of one player remains the same ($P_{C1} = 0.30$), the evolutionary direction of the system will witness changes when the probability for the discovery of safety noncompliance by the players is increased. Therefore, under the circumstance where the safety noncompliance is serious, enterprise shall strengthen the safety inspection to timely discover its employees' safety noncompliance behavior, and it is very necessary for the enterprise to impose strict punishment on employees having safety noncompliance behavior. When necessary, enterprise may introduce the system for the report of employees' safety noncompliance behavior, and drive its employees to participate in the management over work safety to make the safety noncompliance behavior discovered once it occurs and employees involved therein punished (Fig. 11).

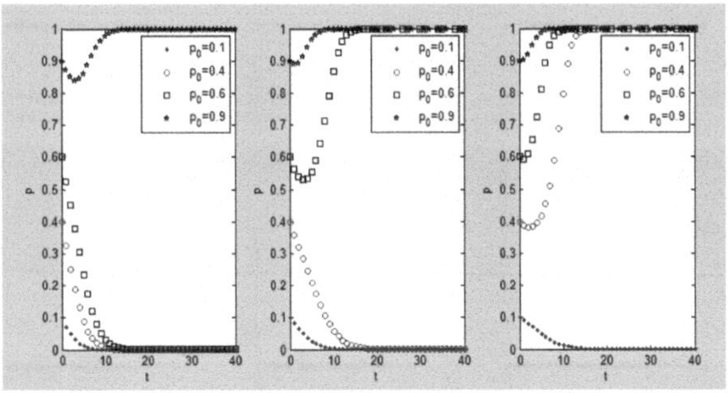

$B = 0.1$ $B = 0.3$ $B = 0.7$

Fig. 10. Impact of changes of fine on the evolutionary result

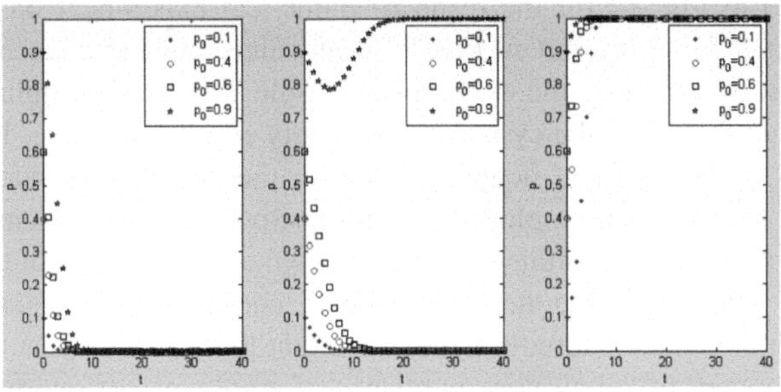

$S_z = 1.5$ $S_z = 2.0$ $S_z = 3.0$

Fig. 11. Impact of changes of potential loss on the evolutionary result

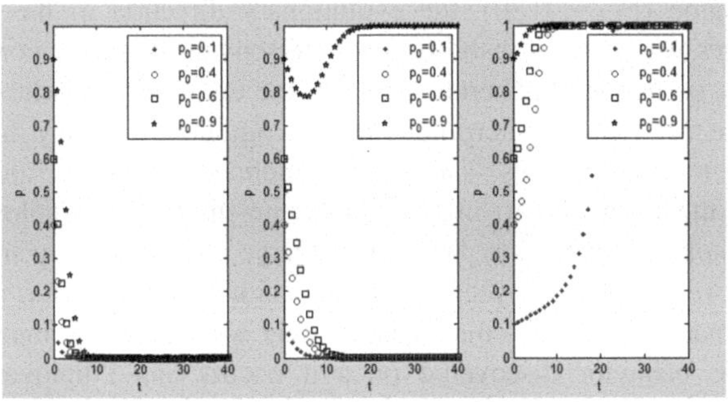

$P_{C1} = 0.15$ 且 $P_{C2} = 0.3$ $P_{C1} = 0.30$ 且 $P_{C2} = 0.30$ $P_{C1} = 0.30$ 且 $P_{C2} = 0.80$

Fig. 12. Impact of changes of safety check frequency on the evolutionary result

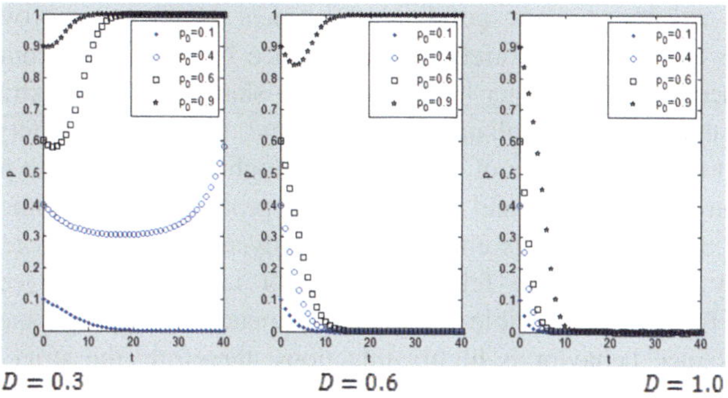

Fig. 13. Impact of changes of negative effect on the evolutionary result

(5) See Fig. 13 for the result of numerical experiment on the impact of negative effect *D* of safety noncompliance on the evolutionary result.

According to Fig. 13, the system will totally evolve into the mode of (noncompliance, noncompliance) when *D* reaches a certain degree. As employees can communicate with their workmates in the daily production and be influenced by their behavior, they will re-examine themselves when their workmates' noncompliance behavior does not cause any accident, nor discovered by enterprise; besides, more employees will follow their workmates as the noncompliance behavior is infectious; as a result, the poor safety atmosphere is formed and will bring a huge potential safety hazard and loss to the enterprise.

5 Conclusion

The grass-roots employees' safety behavior is the key to improve the level of enterprise in work safety, and the interaction effect of behavior among employees dynamically determines the safety state of group (workshop and team). In such group, the single employee will constantly adjust his or her strategy according to other employees' choice for behavior, and the system will eventually evolve into a certain stable state, which is the process for the employees to choose the group behavior after comprehensively measuring the factors, such as amount of labor necessary for safety compliance, safety gains, potential loss from safety noncompliance, reaction from workmates and enterprise supervision. During such process, enterprise can adjust and control relevant parameters to guide the system to evolve into an ideal state.

(1) Based on the facts that the safety awareness of most of grass-roots employees is low, and they will consider the gains from safety noncompliance behavior are greater than those form safety compliance behavior when they cannot realize the potential, indirect and non-economic benefits of safety compliance behavior, and then take the noncompliance behavior, which will exert an adverse demonstration effect in the group, on one hand, enterprise executives shall strengthen the safety inspection to timely discover, prevent and correct the safety noncompliance

behavior, and impose corresponding punishment thereupon, improve employees' awareness to the cost of safety noncompliance behavior and reduce their satisfaction thereto. On the other hand, enterprise shall reduce the extra labor necessary for the safety compliance behavior and the cost of safety compliance behavior to the knowledge of employees through normalizing the procedures for production and operation and optimizing the production environment.

(2) In the group of grass-roots employees, employee relationship is relatively close and many employees are fellow-villagers or relatives, employee's individual behavior is more susceptible to other workmates and group, and the group's noncompliance behavior is highly infectious, therefore, the strict prejob safety education and training is of great importance, through which employees can have a good command of Safety Operating Procedures and fully realize the importance of work safety, realize their own safety noncompliance behavior which may cause damage to themselves and their workmates, and other workmates' noncompliance may do harm to themselves, and can thus form the awareness of "no harm" [14], in this way, can employees refuse operation against rules and also actively stop their workmates from doing so.

(3) It is very important to completely eradicate the employees' safety noncompliance behavior and create a favorable safety atmosphere [15]. Employees break rules and regulations for saving labor and puppyism. Therefore, enterprise can build a favorable enterprise safety culture atmosphere to make employees break the rules unable to feel the psychological satisfaction and subject to the resistance from other workmates, to make employees clearly realize the safety noncompliance behavior does not conform to the enterprise requirements, but also the requirements of group norm so as to reduce the workmates' indifference or self-identity towards noncompliance behavior and lower the motivation of employees' safety noncompliance behavior and its adverse influence. In the meantime, enterprise shall reward and punish employees according to their performance in work safety, and impose strict punishment upon the employees having safety noncompliance behavior, provided that it shall cautiously use the employee's individual performance in work safety to measure other workmates' safety performance, otherwise, it may discourage the employees who take the safety compliance behavior strategy.

Acknowledgment. We are very grateful to the financial support provided by the National Natural Science Foundation of China under Grant [71403108]; the National Natural Science Foundation of China under Grant [71874012]; MOE Project of Humanities and Social Sciences in China [19YJAZH059] and Jiangsu Provincial Social Science Fund [16JD013].

References

1. Hackett, E.J., Parker, J.N.: Ecology reconfigured: organizational innovation, group dynamics and scientific change (Book style with paper title and editor). In: The Local Configuration of New Research Fields. Springer International Publishing, Heidelberg (2016)
2. Bandura, A.: Social learning theory of aggression (Periodical style). J. Commun. **28**(3), 12–29 (2010)
3. Rozin, P.: Social psychology and science: some lessons from solomonasch (Periodical style). Pers. Soc. Psychol. Rev. **5**(1), 2–14 (2001)
4. Degoey, P.: Contagious justice: exploring the social construction of justice in organizations (Periodical style). Res. Organ. Behav. **22**(00), 51–102 (2000)
5. Lingard, H., Cooke, T., Blismas, N.: Do perceptions of supervisors' safety responses mediate the relationship between perceptions of the organizational safety climate and incident rates in the construction supply chain? (Periodical style). J. Constr. Eng. Manag. **138**(2), 234–241 (2012)
6. Mohamed, S.: "Safety climate in construction site environments" (Periodical style). J. Constr. Eng. Manag. **128**(5), 375–384 (2002)
7. Chen, Y., Wong, L., Fang, D.: Safety climate in construction industry: a case study in Hong Kong (Periodical style). J. Constr. Eng. Manag. **132**(6), 573–584 (2006)
8. Zhou, Q., Fang, D., Mohamed, S.: Safety climate improvement: case study in a Chinese construction company (Periodical style). J. Constr. Eng. Manag. **137**(1), 86–95 (2011)
9. Zohar, D.: Thirty years of safety climate research: reflections and future directions (Periodical style). Acci. Anal. Prev. **42**(5), 1517–1522 (2010)
10. Fugas, C.S., Silva, S.A., Meliá, J.L.: Another look at safety climate and safety behavior: deepening the cognitive and social mediator mechanisms (Periodical style). Accid. Anal. Prev. **45**(1), 468–477 (2012)
11. Zhou, Q., Fang, D., Wang, X.: A method to identify strategies for the improvement of human safety behavior by considering safety climate and personal experience (Periodical style). Saf. Sci. **46**(10), 1406–1419 (2008)
12. Schütte, N., Blickle, G., Frieder, R.E., Wihler, A., Schnitzler, F., Heupel, J., Zettler, I.: The role of interpersonal influence in counterbalancing psychopathic personality trait facets at work (Periodical style). J. Manag. **27**(8), 1–34 (2015)
13. Goh, E., Ritchie, B., Wang, J.: Non-compliance in national parks: an extension of the theory of planned behaviour model with pro-environmental values (Periodical style). Tourism Manag. **59**(4), 123–127 (2017)
14. Bonafede, M., Corfiati, M., Gagliardi, D., Boccuni, F., Ronchetti, M., Valenti, A., Iavicoli, S.: OHS management and employers' perception: differences by firm size in a large Italian company survey (Periodical style). Saf. Sci. **89**, 11–18 (2016)
15. Chib, S., Kanetkar, M.: Safety culture: the buzzword to ensure occupational safety and health (Periodical style). Procedia Econ. Finan. **11**(14), 130–136 (2014)

Safety Risk Assessment and Management of Military Physical Fitness Training Based on Triangular Fuzzy Mathematics and Fault Tree Analysis

Qing-qing Cao(✉) and Wen-juan Huai

Department of Force Management, Officers College of Chinese People's Armed Police Force, Chengdu, China
caoqinqing08@qq.com

Abstract. In order to meet the needs of modern warfare, have higher requirements for the military physical fitness. Military physical fitness training subjects and difficulty increased, greater security risk. Therefore, based on the analysis of the relevant factors affecting the safety of military physical fitness training, combine the triangle fuzzy math and the accident tree, determining the order of fuzzy importance of military physical fitness training safety risk factors, finally, it proposes countermeasures to prevent the risk of military physical fitness training from the aspects of insufficient understanding of safety and single training methods.

Keywords: Accident tree analysis · Physical fitness training · Security risk assessment · Triangular fuzzy mathematics

1 Introduction

Accident tree analysis is one of the important analytical methods for safety system engineering, is a deductive security system analysis method. The accident tree analysis was first proposed by the American Bell Telephone Research Institute in 1961 for the study of the militia missile launch control system [1, 2]. The US Atomic Energy Commission used FTA to conduct a risk assessment of nuclear power plant accidents in 1974, published the famous Ramson Report, the report made a large-scale and effective application of accident tree analysis. Since then, it has caused great repercussions in all walks of life and has received extensive attention, which has been rapidly applied and promoted in many countries. The accident tree analysis starts with the specific accident or failure to be analyzed, analyzes the cause at various levels, until the basic cause of the accident is found [3]. Therefore, it can identify and evaluate the dangers of various systems, not only can directly analyze the direct cause of the accident, but also reveal the potential causes of the accident. However, in practical applications, due to the ambiguity of basic events and the inadequacy of data, the traditional accident tree analysis method is limited, it is difficult to accurately calculate the probability value of its occurrence, and fuzzy theory is an excellent tool to deal with this problem [4]. At present, military training is the central task of strategic position in army building and

C.-F. Chien et al. (Eds.): *IE&EM 2019*, pp. 94–100, 2020.
https://doi.org/10.1007/978-981-15-4530-6_10

the basic way to generate and improve combat effectiveness. However, as the training requirements increase, the number of training subjects' increases, and the safety risks appearing in physical fitness training become more apparent. Therefore, it is necessary to identify the main causes of risks and provide a reliable basis for determining safety measures, so as to prevent accidents.

2 Triangular Fuzzy Number Correlation Theory

2.1 Triangular Fuzzy Number

The fuzzy theory was proposed by Zadeh in 1965 to deal with the problem of inaccuracy and ambiguity. Set a fuzzy set u_λ on the domain U, for any $x \in U$, have a number $u_\lambda(x) \in [0, 1]$ corresponding to it, then the membership function of $u_\lambda(x)$ is u_λ. Write u_λ as $u_\lambda = \lim_{x \to u} \int (u_\lambda(x)/x)dx$, indicates the degree to which x belongs to u_λ, and the two are positively correlated. If the membership function of a fuzzy number consists of a linear function, expressed as

$$u_\lambda(x) = \begin{cases} \frac{x-m+l}{l} & 1 \leq x \leq m \\ \frac{m+u-x}{u} & m \leq x \leq u \\ 0 & other \end{cases} \tag{1}$$

Where m is the core of u_λ, $u+l$ is called the blindness of u_λ [5], its membership function is shown in Fig. 1.

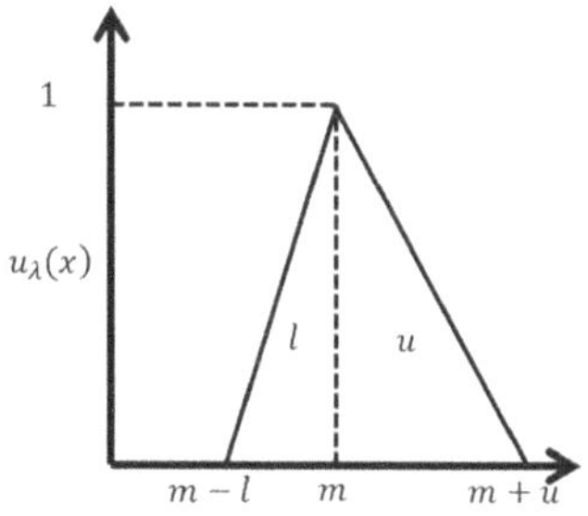

Fig. 1. Triangular fuzzy number

2.2 Triangular Fuzzy Number Operation

(1) 3σ method
The 3σ method is generally used to represent the fuzzy probability of a basic event. For the fuzzy probability of basic events without statistical data, the expert scoring method is used for evaluation. The expert group generally consists of three or more. Each expert gives the estimated value of each basic event, and the

mean value of each estimated probability is m, and the variance is σ^2, and the estimated probability value obeys the standard normal distribution, according to statistical knowledge, its probability of falling within the interval $[u-3\sigma, u+3\sigma]$ is 99.7%, so set $l = u = 3\sigma$, vaguely express each probability value as $\langle 3\sigma, m, 3\sigma\rangle$. There are n experts who score the score of an event. The mathematical expectation is the mean value $E(x)$. The variance is $D(x)$, then there are [6]:

$$m = E(x) = \left({}^{1}\!/_{n}\right) \times (x_1 + x_2 + \cdots + x_i) \tag{2}$$

Where $i = 1, 2, \cdots, n$.

$$\text{Variance } D(x) = \sigma^2 = \sum_{k=1}^{n} [x_k - E(x)]^2 P_k \tag{3}$$

$$\text{Standard deviation } \sigma = \sqrt{\sum_{k=1}^{n} [x_k - E(x)]^2 P_k} \tag{4}$$

Where x_k is the kth probability value.

(2) Basic event relationship processing in the accident tree
The AND gate fuzzy operator of the accident tree fuzzy processing [7] is:

$$\begin{aligned}\tilde{q}_{and} &= \langle l_{and}, m_{and}, u_{and}\rangle = \prod_{i=1}^{n} q_i = \tilde{q_1} \otimes \tilde{q_2} \otimes \tilde{q_3} \otimes \cdots \otimes \tilde{q_n} \\ &= \left\langle \prod_{i=1}^{n} l_i, \prod_{i=1}^{n} m_i, \prod_{i=1}^{n} u_i \right\rangle\end{aligned} \tag{5}$$

The OR gate fuzzy operator of the accident tree fuzzy processing is:

$$\begin{aligned}\tilde{q}_{or} &= \langle l_{or}, m_{or}, u_{or}\rangle = \langle 1 - \prod_{i=1}^{n}(1 - \tilde{q}_i)\rangle \\ &= \langle 1 - \prod_{i=1}^{n}(1 - \tilde{l}_i), 1 - \prod_{i=1}^{n}(1 - \tilde{m}_i), 1 - \prod_{i=1}^{n}(1 - \tilde{u}_i)\rangle\end{aligned} \tag{6}$$

(3) Accidental tree basic event fuzzy importance
For Fig. 1:

Make $A_1 = \int_{m-l}^{m} u_{\tilde{A}}(x)dx$, $A_2 = \int_{m}^{m+u} u_{\tilde{A}}(x)dx$, $A = A_1 + A_2$, there is a point Z through which the boundary is the boundary, and the left and right portions under the fuzzy number curve are equal in area, and Z is said to be the median of the fuzzy number. Let the structure function of the accident tree be $\varphi(x_1, x_2, \ldots, x_n)$, x_i are the basic events of the accident tree, the probability distribution fuzzy number is recorded as $\tilde{x}_i$, top event probability distribution $\tilde{T} = \varphi(\tilde{x}_1, \tilde{x}_2, \ldots, \tilde{x}_n) = (3\sigma, m, 3\sigma)$, the median is recorded as T_Z. $\tilde{T}_i = \varphi(\tilde{x}_1, \tilde{x}_2, \ldots, \tilde{x}_{i-1}, \tilde{x}_{i+1}, \ldots, \tilde{x}_n) = (3\sigma, m, 3\sigma)$, the median is recorded as T_Z^i. Fuzzy importance of basic events x_i [8, 9]:

$$S_i = T_Z - T_Z^i \tag{7}$$

If $S_i > S_j$, it is considered that x_i is important to x_j, that is, x_i to x_j have a great influence on the system.

3 Accident Tree Establishment of Military Physical Fitness Training Safety Risk Analysis

3.1 Building an Accident Tree

In terms of military physical fitness, countries have put forward higher requirements for the military physical fitness in order to meet the needs of modern warfare. For example, how to play the role of military physical fitness training under modern high-tech military conditions, how to implement scientific physical fitness training to enhance the military physical fitness and so on. However, behind the higher requirements, more training intensity and more training subjects are needed. Although the units attach great importance to the safety work in training, they still expose problems such as insufficient implementation of the system and insufficient scientific analysis of risks. Military physical fitness training safety risk is the result of many factors, but the general summary comes from two aspects, one is training environmental problems, such as bad weather, poor medical security, equipment and equipment damage, etc. Trainer issues, such as inadequate education and guidance, lack of awareness of safety, single training methods, and inadequate physical activity [10, 11]. The specific accident tree is shown below (Fig. 2).

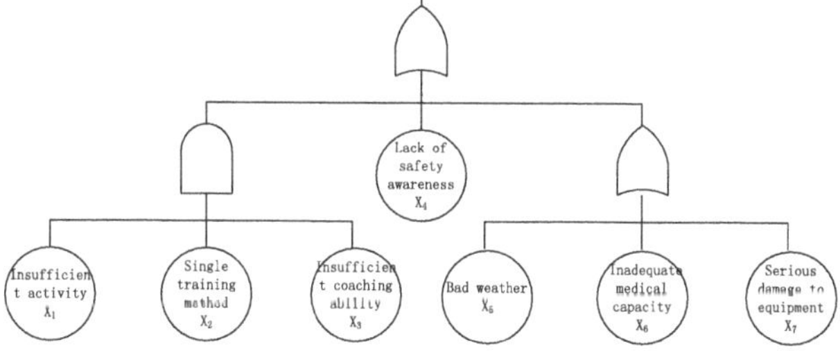

Fig. 2. Physical fitness training safety risk accident tree

3.2 Quantitative Analysis Based on Fuzzy Accident Tree

Since the basic event statistics work is very difficult, the expert group of the three groups is used for scoring. The members of the expert group are the teachers of the relevant subjects, the grassroots commanders and the squad leader of the grassroots units, and score the experts to find the fuzzy probability of the basic events. The results of the blurring according to Eq. 4 are shown in the following Table 1.

The probability of occurrence of $x_1 - x_7$ and the top event T using the above formula is shown in the following table.

Table 1. Data got by method of marks of experts and treated by fuzzy method

Basic event	Expert 1	Expert 2	Expert 3
X_1	0.015	0.011	0.014
X_2	0.022	0.020	0.015
X_3	0.002	0.006	0.008
X_4	0.026	0.023	0.016
X_5	0.004	0.008	0.011
X_6	0.015	0.017	0.023
X_7	0.014	0.012	0.011

Table 2. Statistical data treated by fuzzy method

Basic event	P_{x_j}	Importance S
X_1	<0.005, 0.013, 0.005>	0.013
X_2	<0.008, 0.019, 0.008>	0.019
X_3	<0.007, 0.005, 0.007>	0.005
X_4	<0.012, 0.021, 0.012>	0.021
X_5	<0.008, 0.007, 0.008>	0.007
X_6	<0.009, 0.018, 0.009>	0.018
X_7	<0.003, 0.012, 0.003>	0.012

4 Discussion

Through the 3σ method, the triangular fuzzy number graph is symmetric in the area of the mean value, so the median of the top event and the median of the basic event are equal to the respective mean values, namely $T_Z = m$, $T_Z^i = m_i$. Since the median T_Z of each basic event and top event is constant, the size of S can be compared by directly comparing m_i. The importance of the basic event S is as shown in the above table. Through Table 2, the final importance order of the basic events can be obtained as: $X_4 > X_2 > X_6 > X_1 > X_7 > X_5 > X_3$. Judging from the above sorting results, in the military physical fitness training, the lack of understanding of safety, the single training method is the main factor that causes the military physical fitness to train safety risks, and secondly, the medical ability is insufficient, the living body is not in place, and the equipment is seriously damaged, bad weather, lack of educational guidance.

5 Conclusion

From the order of importance, it can be seen that among the factors leading to the occurrence of top-level events, the factors of safety awareness have the greatest impact. Safety education is one of the important contents of the security work of the army. Thought is the forerunner of behavior. As long as it is recognized by thought, it can establish a strong sense of security. Therefore, it is necessary for the troops to actively publicize the relevant protective measures for prevention of training injuries, as well as to analyze and summarize past cases, sort out common training safety risks, and put the work to the ground, to be practical, and to integrate the contents of safety education into the usual training. Secondly, the training method is single. The single training method not only makes the military feel dull and boring, but also increases the training injury. In the usual training, different training methods should be used flexibly according to different actual situations, so that the training can be divided into shallow and deep. Finally, the purpose of scientific training is achieved. A strong health care team can best minimize military training injuries. Therefore, it is necessary to strengthen medical supervision and medical support, establish personal health records for military personnel, and achieve early prevention, early detection, and early treatment. In addition, an important factor that is easy to occur in training injuries is that the warm-up activities are not enough. The human body needs an adaptive process from static to dynamic. Failure to warm up can result in physical discomfort and discomfort. Physical training is an important part of the military training of the military. The assessment and management of the safety training of the human body can effectively reduce and prevent physical training injuries, and help improve the safety of the troops.

References

1. Chen, Y., Tian, S., Sun, B.: Defining of safety assessment index based on fault tree analysis method. Shaanxi Coal (03) (2003)
2. Huang, Y.: On the application of tree diagram analysis to fire safety management in production. J. Chin. People's Armed Police Force Acad. **24**(8) (2008)
3. Bu, Q., Wang, Y., Wang, D.: Application of fault tree analysis. J. Southwest Pet. Univ. (4) (2007)
4. Hou, F., Xiao, G., Yang, S.: Study on fuzzy fault tree analysis and its application. J. Hebei Norm. Univ. (Nat. Sci. Ed.) (12) (2001)
5. Jia, Z., Jing, G., Zhang, Q.: Fault tree analysis of mining fire occurrence based on triangle fuzzy numerals. J. Saf. Environ. **4**(6) (2004)
6. Li, D.: Dissertation research on risk analysis of shooting training in a regiment as an example based on the accident tree method. Graduate School of National University of Defense Technology, Hunan (2016)

7. Wuzhuang, S., He, X.: A fault tree analysis method based on fuzzy set theory and its application. Syst. Eng. Electron. Technol. **22**(9) (2000)
8. Du, J., Wang, R.: Analysis of air traffic fire accident tree of lithium battery based on triangular fuzzy number. Traffic Inf. Secur. **32**(3) (2014)
9. Wang, K.: Research on risk assessment method of pressure pipeline based on fault tree analysis. Wuhan University of Technology, Wuhan (2009)
10. Wei, G.: Safety Principle and Accident Prediction. Metallurgical Industry Press, Beijing (1995)
11. Chen, Y., Wang, H., Wang, X.: Analysis of risk management in military training. Natl. Def. (4) (2014)

A Hybrid Similarity-Aware Clustering Approach in Cloud Manufacturing Systems

Jian Liu(✉) and Youling Chen

College of Mechanical Engineering, Chongqing University, Chongqing, China
liujiancqu@126.com

Abstract. With the rapid development of cloud manufacturing (CMfg), a lot of cloud services are emerging on the Internet, which leads to cloud service clustering a critical topic. However, most existing approaches suffer from the low clustering quality due to the data sparsity condition, and are thus prone to the unreal result. To handle this problem, we put out a hybrid approach called HCA for cloud service clustering. At the first, we utilize Pearson Correlation Coefficient (PCC) and Proximity-Significance-Singularity (PSS) to compute the user similarity. Then, a similar group of users can be obtained using K-medoids algorithm, in which an ensemble model is established by incorporating those two user similarities. Based on two real-world data sets, the results show that the effectiveness of HCA.

Keywords: Cloud manufacturing · Clustering · Manufacturing systems · User similarity

1 Introduction

Nowadays, cloud manufacturing (CMfg) [1, 2] is becoming an emerging platform, has been put forward utilizing the modern information technologies, involving Big Data [3], Cloud Computing [4], Internet of Things [5]. The main purpose of CMfg are to put the distributed resources and idle capabilities into cloud pool, and provide cloud service to meet user's requirements (e.g., faster execution time, higher reliability, lower cost and so on.) [6].

However, the growing number of similar cloud services available in the CMfg systems raises a new research problem: it is costly and time consuming to make a cloud service selection of the large range of options. Under this circumstance, cloud service clustering is becoming a critical topic recently.

Clustering is an unsupervised classification algorithm, which aims to reduce the search space and find similar grouping of pattern [7]. In general, there are three classic clustering algorithms in the CMfg systems: K-means [7], K-medoids [8] and Hyper-graph partition algorithm [9].

For example, Ghazanfar et al. [7] found a similar group of users by integrating the user similarity into clustering algorithm, namely K-means. However, the results of this clustering algorithm mentioned above may be unreliable by averaging all the attribute values. To cope with the problem, Guo et al. [8] utilized a novelclustering algorithm called K-medoids to develop a multi-view recommendation system.

C.-F. Chien et al. (Eds.): *IE&EM 2019*, pp. 101–108, 2020.
https://doi.org/10.1007/978-981-15-4530-6_11

But it also shows a relatively poor clustering quality. Then, Yang et al. [9] designed a vertex hypergraph partitioning algorithm, EQHyperpart, to generate better partitioning results. The hypergraph partitioning algorithm has a faster execution time, but limited to low coverage and partitioning quality.

Above all, we put out a novel hybrid clustering approach, HCA, whose main purpose is to improve the clustering quality. In the HCA, firstly, we integrate the PCC and PSS into our improved K-medoids algorithm to achieve a better result. And the experiment part demonstrates the superiority of HCA.

2 Methodology

Considering the fact that PCC is more understandable and easy to implement, PCC becomes a common method to compute the user similarity, and has been widely used in the CMfg systems recently [10, 11]. And PCC is computed as follows:

$$PCC(u, v) = \frac{\sum_{s \in S} (r_{u,s} - \bar{r}_u)(r_{v,s} - \bar{r}_v)}{\sqrt{\sum_{s \in S} (r_{u,s} - \bar{r}_u)^2} \sqrt{\sum_{s \in S} (r_{v,s} - \bar{r}_v)^2}} \quad (1)$$

where $S = S_u \cap S_v$ is the set of co-rated services by user u and v, $r_{u,s}$ represents the value of service s rated by u, and $\bar{r}_u$ denotes the average value of all service s rated by u.

Despite high accuracy of PCC, there are still some problems remained that will lead to unsatisfactory results, since the user similarity is computed only based on the co-invoked users or services.

To overcome the problem and better compute the non-linear relationship between users, we also employ the Proximity-Significance-Singularity (PSS) model [12] to compute the user similarity which not only relies on the co-invoked services.

And the PSS model in this paper is defined as below:

$$\begin{aligned} PSS(u, v) &= Proximity(r_{ui}, r_{vj}) \times \\ Significance(r_{ui}, r_{vj}) &\times Singularity(r_{ui}, r_{vj}) \end{aligned} \quad (2)$$

where $PSS(u, v)$ denotes the similarity between u and v, r_{ui} and r_{vj} denote the rating value of i and j, respectively.

Here, *Proximity* is the absolute difference between r_{ui} and r_{vj}. *Significance* aims to measure the impact of the rating pair to the similarity values. *Singularity* denotes the difference between one rating pair to others. And the formulas of three functions are presented as below:

$$Proximity(r_{ui}, r_{vj}) = 1 - \frac{1}{1 + \exp(-|r_{ui} - r_{vj}|)} \quad (3)$$

$$Significance(r_{ui}, r_{vj}) = \frac{1}{1 + \exp(-|r_{ui} - \bar{r}_u| \cdot |r_{vj} - \bar{r}_v|)} \quad (4)$$

$$Singularity\,(r_{ui},\, r_{vj}) = 1 - \frac{1}{1 + \exp\left(-\left|\frac{r_{ui} + r_{vj}}{2} - \frac{\bar{r}_i + \bar{r}_j}{2}\right|\right)} \tag{5}$$

Based on the PCC and PSS, we can establish a novel ensemble model to compute user similarity, and the formula is presented as below:

$$Sim\,(u, v) = \lambda \times PCC(u, v) + (1 - \lambda) \times PSS(u, v) \tag{6}$$

where $Sim\,(u, v)$ is the hybrid user similarity between u and v. $\lambda\,(0 < \lambda < 1)$ denotes a specific parameter to affect how much the hybrid user similarity focus on the $PCC\,(u,\, v)$ and $PSS\,(u,\, v)$, and the impact of λ will be further studied in the following section.

Then, we use the improved clustering algorithm to find the similar subgroups of users by incorporating hybrid user similarity into K-medoids (see in Fig. 1). Different from the existing approaches, this improved clustering algorithm considers a true user as the center and then keep up the group characteristics [8].

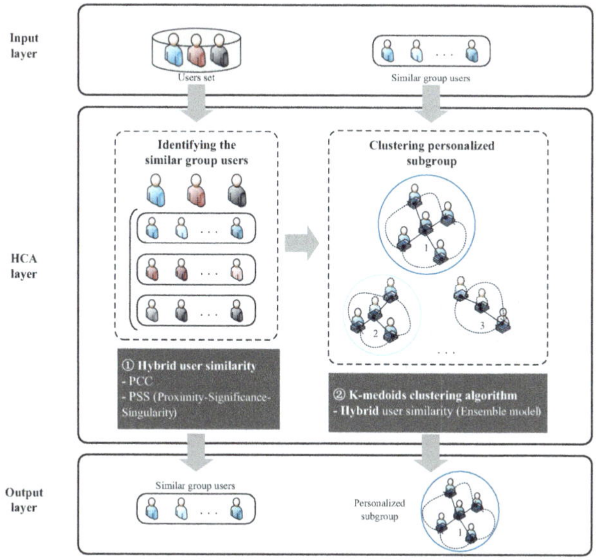

Fig. 1. Overview of the main steps of HCA.

The objective function is defined as below:

$$J = \min \sum_{c \in C} \sum_{u,v \in c} d(u, v) \tag{7}$$

Algorithm 1. K-medoids clustering algorithm

Input: distance matrix D_t; number of clusters K
Output: user subgroups C
$p \leftarrow 0$;
randomly select K medoids m_t from users, $\theta_t^0 \leftarrow m_t$;
$C_t^0 \leftarrow u$, given $\min(d(u,m_t))$; /* Calculate the distance using Eq. (8) */
while medoids changed **and** $< maxiterations$ **do**
    $p \leftarrow p+1$;
    $\theta_t^p \leftarrow \theta_t^{p-1}$;
    $swap(m_t,v)$, $v \in C_t^{p-1}$;
    calculate $sum_t(v)=\sum_u d(u,v)$, $u \in C_t^{p-1}$;
    if $sum_t(v) < sum_t(m_t)$ **then**
        $m_t \leftarrow v$;
        $\theta_t^p \leftarrow m_t$;
    $C_t^p \leftarrow u$, for $\forall u$, find m_t $s.t.$ $min(d(u,m_t))$;
return $C \leftarrow C_t^p$;

where C denotes the user subgroups in which u and v belong to the subgroup $c \in C$, and $d(u, v)$ denotes the key distance between u and v.

In order to find the user subgroups, hybrid user similarity is replaced as the distance metric to cluster the similar users. More specifically, the higher user similarity means the users are closer than others [6].

Therefore, the distance can be calculated as follows:

$$d(u, v) = 1 - Sim(u, v) \tag{8}$$

Based on the above analysis, the pseudocode of our clustering algorithm is clearly presented in Algorithm 1.

3 Results

In this part, like our previous paper [6], we carry out a series of experiments to validate the effectiveness of our novel approach HCA.

To verify the performance of HCA, we collect two data sets in real-world. One is about web service called WSDream, which is gathered by Zheng et al. [10]. The WSDream-dataset includes 339 × 5825 rating records of two attributes (response time and throughput). Another is about CMfg service collected by Xiang et al. [13], including 463 users and 7548 services.

All the experiments were processed with Intel Core i5-4210 M 2.60 GHz processors and 4 GB RAM on a Lenovo E540 computer. The experimental parameters are presented in Table 1.

Table 1. Experimental parameters

Parameters	Web	CMfg
m: the number of users	100	100
n: the number of services	500	500
λ: how much HCA relies on PCC and PSS	0.45	0.45
K: maximum cluster number	6	8

In this section, like our previous paper [6] and [11], we employ the Normalized Mean Absolute Error, namely NMAE, as a metric to evaluate the clustering quality. Considering the fact that the huge ranges of response time in two different data sets, NMAE is computed as below:

$$\begin{cases} MAE = \dfrac{\sum_{u,i} (r_{ui} - \hat{r}_{ui})}{W} \\ NMAE = \dfrac{MAE}{\sum_{u,i} {r_{ui}}/{W}} \end{cases} \tag{9}$$

where MAE represents the average absolute deviation between the predicted value and real value according to the clustering users, r_{ui} denotes the real value of service i rated by user u, $\hat{r}_{ui}$ is the predicted rating value. W denotes the number of whole predicted values.

In order to achieve a performance comparison, we compare our hybrid clustering approach (HCA) with other popular approaches:

K-Mean: K-Mean clustering algorithm [7] is designed to search the similar group to improve the recommendation quality.

K-medoids: K-medoids [8] is another clustering algorithm to find the similar group using the PCC similarity model.

EQHyperpart: EQHyperpart [9] is a hypergraph partitioning method to obtain high quality clustering results based on information entropy modularity.

We also explore the impact of different parameters on the performance of HCA. Generally, we change a specific parameter under different condition (i.e., λ and K) while holding the others consistent. And the whole analysis process is concluded as below.

4 Discussion

The experimental results in Fig. 2 give that the value of NMAE in above approaches are decreasing as the matrix density increase, which means that the clustering quality can be significantly improved by employing more data in the rating matrix.

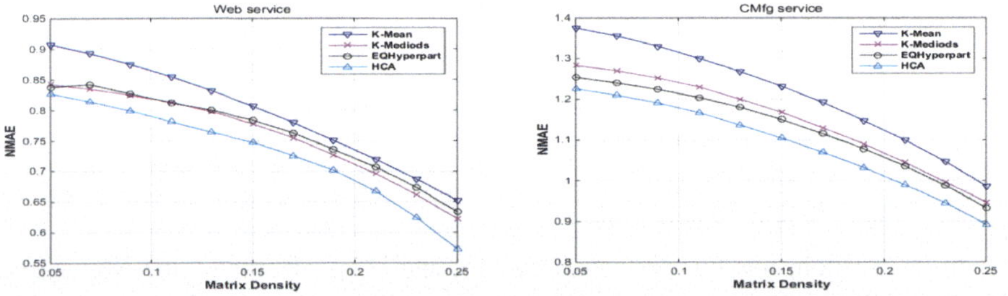

Fig. 2. Performance comparison of HCA.

Obviously, the NMAE value of HCA is relatively lower compared with other algorithms K-Mean, K-medoids and EQHyperpart. This is because HCA not only focus on the co-invoked rating records, but also the other rating information. By combining the both PCC and PSS to compute a hybrid user similarity seems to be helpful when making cloud service clustering. In addition, K-medoids show a better performance than other approaches when integrating a hybrid user similarity into it.

In Fig. 3, we can find that HCA obtains the best clustering quality in two data sets, namely web service and CMfg service, indicating that the optimal value of λ are affected by the matrix density to some extent. Since when $\lambda < 0.5$, the impact of PCC are less to be considered, while when $\lambda > 0.5$, the impact of PSS is also decreased. Thus, we set $\lambda = 0.45$ as the unchanged value in this part.

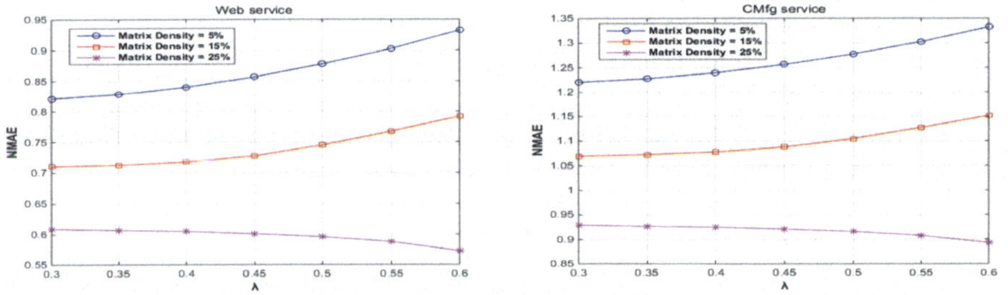

Fig. 3. Impact of λ.

As shown in Fig. 4, the experimental results of K does influence the clustering quality of our approach, an appropriate K value will give out better solution in two data sets. Besides, we set $K = 6$ in the web service data set, and $K = 8$ in the CMfg service data set.

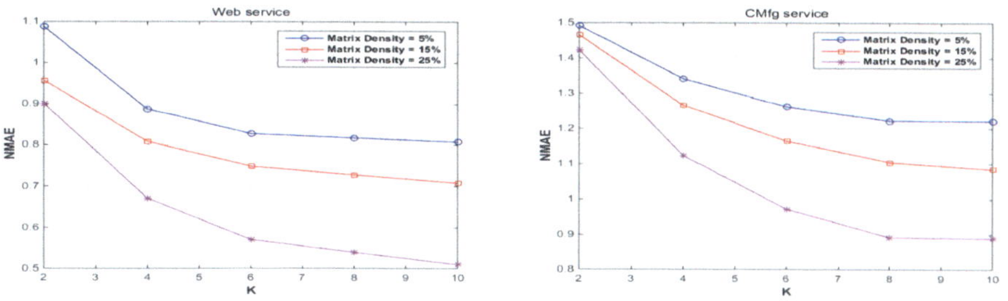

Fig. 4. Impact of *K*.

5 Conclusion

In cloud manufacturing systems, cloud service clustering is important for successful implementation of CMfg application. For a better clustering quality, we put out a hybrid clustering approach, called HCA. The main contribution of this manuscript lie in three-folds: (1) we establish an ensemble model to take advantage of both PCC and PSS. (2) we insert the hybrid user similarity into K-medoids clustering algorithm to acquire a high clustering quality. (3) we carry out a series of experiments on two data sets to validate the performance of HCA.

In the future, we plan to establish a comprehensive model to compute the user similarity by considering more factors in the CMfg systems.

Acknowledgment. The author would like to appreciate the editors and experts for their greatful and helpful comments which encouraged to improve the quality of the paper. And, this paper was supported in part by National Key Research and Development Program of China under grant No. 2018YFB1703002, and in part by the Fundamental Research Funds for the Central Universities under grant No. 2019CDCGJX222.

References

1. Li, B.H., Zhang, L., Wang, S.L., Tao, F., Cao, J.W., Jiang, X.D., Song, X., Chai, X.D.: Cloud manufacturing: a new service-oriented networked manufacturing model. Comput. Integr. Manuf. Syst. **16**(1), 1–7 + 16 (2010)
2. Zhang, L., Luo, Y.L., Tao, F., Li, B.H., Ren, L., Zhang, X.S., Guo, H., Cheng, Y., Hu, A.R., Liu, Y.K.: Cloud manufacturing: a new manufacturing paradigm. Enterp. Inf. Syst. **8**(2), 167–187 (2014)
3. Li, J.R., Tao, F., Cheng, Y., Zhao, L.J.: Big Data in product lifecycle management. Int. J. Adv. Manuf. Technol. **81**(1–4), 667–684 (2015)
4. Xu, X.: From cloud computing to cloud manufacturing. Robot. Comput. Integr. Manuf. **28** (1), 75–86 (2012)
5. Lu, Y.J., Cecil, J.: An Internet of Things (IoT)-based collaborative framework for advanced manufacturing. Int. J. Adv. Manuf. Technol. **84**(5–8), 1141–1152 (2016)
6. Liu, J., Chen, Y.: A personalized clustering-based and reliable trust-aware QoS prediction approach for cloud service recommendation in cloud manufacturing. Knowl.-Based Syst. **174**, 43–56 (2019)

7. Ghazanfar, M.A., Prügel-Bennett, A.: Leveraging clustering approaches to solve the gray-sheep users problem in recommender systems. Exp. Syst. Appl. **41**(7), 3261–3275 (2014)
8. Guo, G., Zhang, J., Yorke-Smith, N.: Leveraging multiviews of trust and similarity to enhance clustering-based recommender systems. Knowl.-Based Syst. **74**, 14–27 (2015)
9. Yang, W., Wang, G., Bhuiyan, M.Z.A., Choo, K.K.R.: Hypergraph partitioning for social networks based on information entropy modularity. J. Netw. Comput. Appl. **86**, 59–71 (2017)
10. Zheng, Z., Ma, H., Lyu, M.R., King, I.: QoS-aware web service recommendation by collaborative filtering. IEEE Trans. Serv. Comput. **4**(2), 140–152 (2011)
11. Liu, J., Chen, Y.: HAP: a hybrid QoS prediction approach in cloud manufacturing combining local collaborative filtering and global case-based reasoning. IEEE Trans. Serv. Comput. (2019)
12. Wang, Y., Deng, J., Gao, J., Zhang, P.: A hybrid user similarity model for collaborative filtering. Inf. Sci. **418–419**, 102–118 (2017)
13. Xiang, F., Jiang, G.Z., Xu, L.L., Wang, N.X.: The case-library method for service composition and optimal selection of big manufacturing data in cloud manufacturing system. Int. J. Adv. Manuf. Technol. **84**, 59–70 (2016)

Improvement of M Assembly Line Based on Assembly Line Balancing and Overall Equipment Effectiveness

Man Zhao, Yan-hua Ma, Xin-chen Wang(✉), Yi-xin Zhang, and Ruo-lan Xu

Department of Industrial Engineering, Jilin University, Changchun, China
wangxc18@mails.jlu.edu.cn

Abstract. The improvement of assembly line is a crucial issue for the mass production in factory, which can effectively increase the balancing rate of production line and overall equipment effectiveness (OEE) for product. Therefore, the improvement of assembly line plays an important role in the management and cost control of factories. In this paper, the process analysis method is adopted to analysis the assembly line to find the bottleneck process firstly. Furthermore, the content of the bottleneck process and the stations' equipment are optimized to satisfy the requirements of Takt. Meanwhile, the occasional factors which cause to reduce the OEE are founded and corresponding work standards are established to eliminate these influences. The F factory example can verify that improvement of assembly line can effectively increase the OEE and reduce the operate time of bottleneck process. Finally, the improvement achieves the line balance in F factory.

Keywords: Assembly line balancing · Overall equipment effectiveness · Process analysis

1 Introduction

Production line balancing is an indispensable part for the improvement of the on-site in assembly line factory [1]. The aim of line balance is to average all the time of production processes and adjust the operating load for the whole production line [2]. Finally, it can eliminate the efficiency loss and overproduction which caused by unbalance between work stations and increase the overall equipment effectiveness (OEE) for factory. Therefore, production line balance is widely studied in practice and theory.

The assembly line balancing problem is a non-deterministic polynomial type-planning problem for mass production. The heuristic based on critical path method has been proposed for the assignment of tasks to the workstations on U-shaped assembly line layout [3]. Meanwhile, A hybrid ant colony algorithm for U-line balancing is proposed to satisfy the just-in-time production environment [4]. A genetic algorithm proposed to balance the mixed model assembly line [5, 6]. Balancing the production line can also improve the OEE for the whole factory [7, 8]. OEE is widely accepted by

C.-F. Chien et al. (Eds.): *IE&EM 2019*, pp. 109–116, 2020.
https://doi.org/10.1007/978-981-15-4530-6_12

whole word factories to evaluate the effectiveness of product [9]. Therefore, Assembly Sequence Planning (ASP) and Assembly Line Balancing (ALB) problems play an important role in the optimization of product [10]. Currently, the cost and productivity is used to evaluate the effect of line balance [11] and the simulation model is adopted to analyze the bottleneck process of the production line to achieve continue improvement [12]. For more detail, the tasks in the workstations are redistributed to achieve the line balance [13] and unnecessary stations are eliminated in the whole assembly line.

The line balance accounts high weight for the future development of factory. In this paper, the work processes are reanalyzed to find the bottleneck for the whole production line. Furthermore, the content of the workstation and the employees' actions are studied to modify based on the human factors engineering standard. The F company example can verify that the improvement of production line can finally achieve the synchronization of processes, improvement of OEE and balance of assembly line.

2 Current Situation of Production Line

Company F is a company that produces auto parts. Its main product is Electronic Stability Program (ESP). ESP products are mainly composed of electronic control unit (ECU), pump shell, and motor. Pump shell is the main part of ESP, and the assembly of other parts is based on pump shell. Currently, seven production lines in the company's workshop are used to produce ESP with overall equipment efficiency of approximately 90% or the M assembly line has 23 workstations and 8 workers, one of whom is the team leader, responsible for the normal operation of the line. The assembly line layout is shown in Fig. 1. The green arrow represents the position and movement route of the worker, and a Loop represents the operation cycle of a worker. All the 23 workstations and 7 loops are regarded as processes, and the takt time is 21 seconds/piece. The whole M assembly line is U-shaped production line, which starts from the upper right pump shell input port.

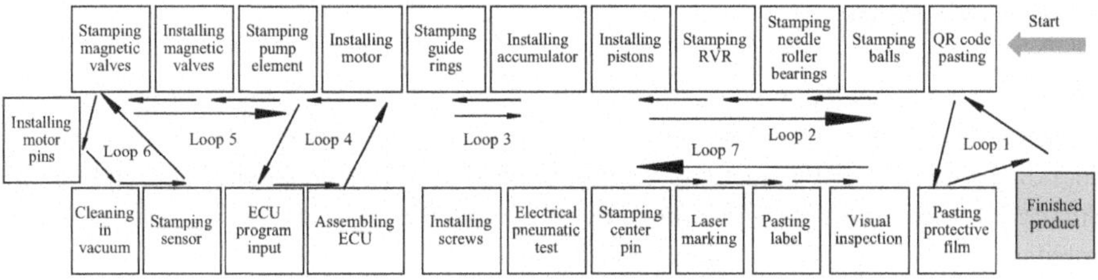

Fig. 1. Layout and process of M assembly line

Although the standard operating time of each process of the M assembly line is lower than the cycle time of production line, the investigation found that the actual production time of Loop 1 and Loop 4 exceeded the cycle time of production line, which was the bottleneck of the assembly line. Meanwhile, each process has different degrees of time loss, which leads to low OEE index. It is necessary to find out the reasons for these time losses and improve them.

3 Improvement of M Assembly Line

3.1 Improvement of Bottleneck Process Loop 4

The cycle time of Loop 4 is 19.6 s. For device-centric workstations, it is safe as long as cycle time is below cycle time of production line of 21 s. Loop 4 is the work of workers, with an allowance rate of 13%. Therefore, the actual cycle time of Loop 4 is 22.1 s, which exceeds the cycle time of production line. After decommissioning the process of Loop 4, it was found that the workers spent 1.6 s in the last work of ram pump stamping workstation, which could be transferred to the neighboring Loop 5. After the transfer, the cycle time of Loop 5 is 19.6 s, which becomes the new bottleneck process.

The job contents for Loop 5 are shown in Table 1. The most time-consuming operation is the installation of magnetic valve, which needs 12.9 s. In this operation, workers assemble 12 magnetic valves on the pump shell, with a total of 6 models, so there are 6 material containers for storing magnetic valves, as shown in Fig. 2.

Table 1. Standard work instruction of Loop 5

	Loop 5		
	Process	Time (s)	Total Time (s)
1	Take the finished pump shell	2.1	19.6
2	Go to workstation of installing magnetic valves	1.1	
3	Install magnetic valves	12.9	
4	Go to workstation of stamping magnetic valves	1.1	
5	Assemble the pump shell	0.9	
6	Walk back to the workstation of stamping pump element	1.6	

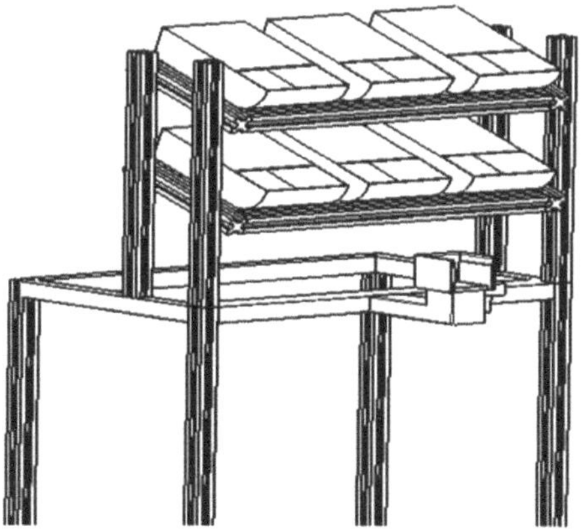

Fig. 2. Workstation of storing magnetic valves

The time of assembling 12 magnetic valves was estimated by MOD method, and the measured time was 14.9 s. At the same time, the result is correct when measured by stopwatch. It was found that the actual time for workers to preinstall magnetic valves was more than 12.9 s of operation standard time. Improvements will be made through the following measures.

(1) Design to the Material Container: The container of magnetic valves was modified as shown in Fig. 3. Two slides were added and the end of each slide was provided with a bayonet. Since the material container is inclined at a certain angle, the materials in the container can slide out by gravity. Compared with taking magnetic valves from the container, the design made it easier for workers to access magnetic valves.

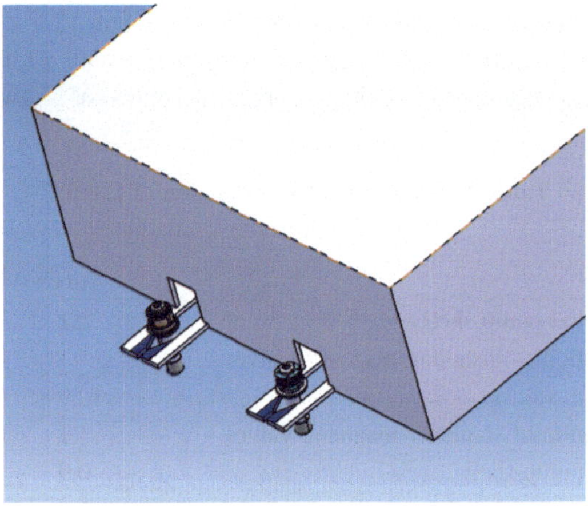

Fig. 3. The new container of magnetic valves

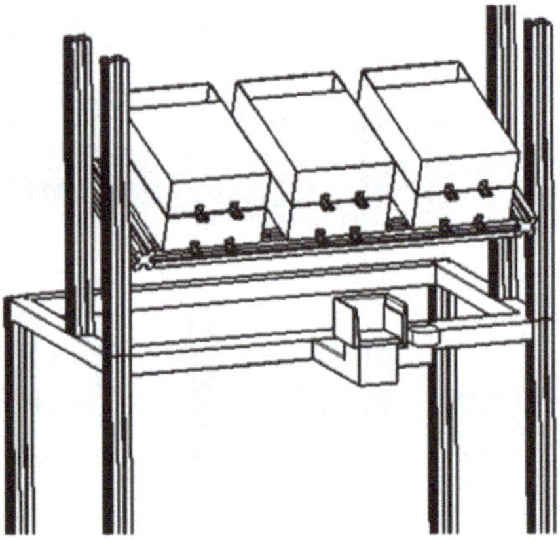

Fig. 4. Placement position of containers

(2) Placement of Material Containers: It can be seen from the Fig. 2 that some containers were too high for workers to take magnetic valves from it. After redesigning these containers which is shown as Fig. 4, the placement of it can also be modified more concentrated. Workers would use shorter movement distance and less time to take magnetic valves.

(3) Design to Error Prevention Device: Figure 5 presents there are twelve holes on the pump shell, corresponding to six types of magnetic valves. Workers need to select the corresponding hole when assembling the magnetic valve. To save time and prevent the magnetic valve from being installed in the wrong hole, the error prevention device was designed as shown in Fig. 6.

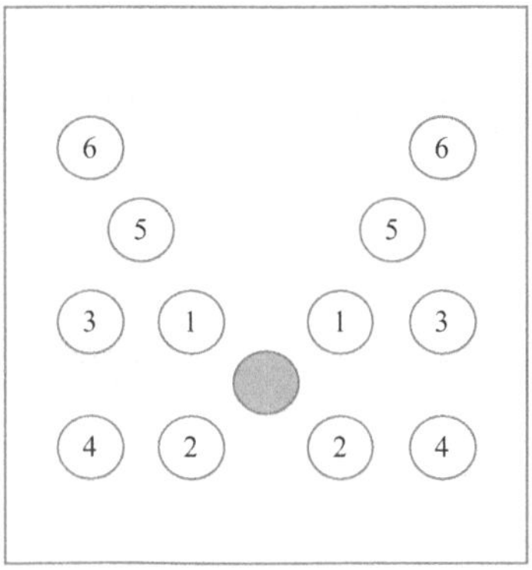

Fig. 5. Holes of magnetic valves on the pump shell

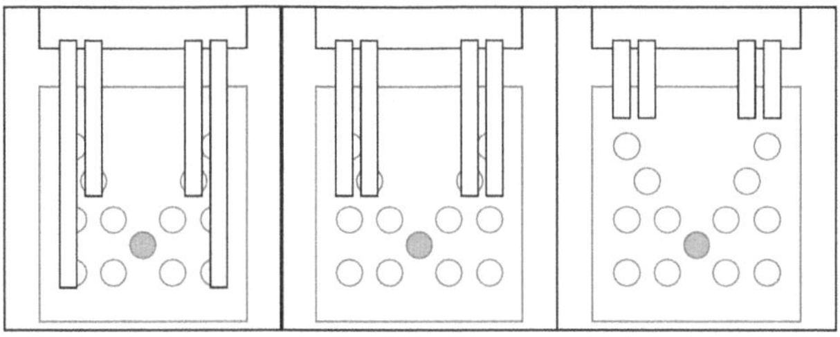

Fig. 6. The error prevention device

Six types of magnetic valves were divide into three groups. When installing one of the groups, other holes of magnetic valves would be covered by four sticks. Figure 6 presents three working modes of the error prevention device. Work modes can be switched automatically when the sensor detected that the group of magnetic valves were installed. In this way, Workers have increased the speed and accuracy of installing magnetic valves.

Cycle time of the Loop 5 have been reduced from 19.6 s to 16.8 s which time met the cycle time of production line.

3.2 Reducing the Lost Time of Equipment

Statistics indicates that major equipment time loss occurred at the workstation with screws installed on the ECU. In this workstation, screws were hooked up by the device and transferred to the screw gun through a pipe. Screws were placed in the machine as shown in Fig. 7. When there were too many screws the entrance of the pipe would be blocked. While screws cannot be hooked up when it's amount was too few. Therefore, a proper standard of screws quantity should be set to make sure the device is working well.

Fig. 7. Machine of supplying screws

Firstly, the lowest level of the screws quantity should be determined. That means the device can at least pick up one screw at a time. Set six lines on the side of the container for different quantity of screws. Observe how many screws the device picked at one time at different levels of screws quantity. Calculate the average of the data after multiple trials and the results are shown in the Table 2. It indicates that the lowest level of screws quantity corresponding to line 5.

Secondly, considering that workers add screws every 20 min. During this time, 57 products were produced and 228 screws were consumed. So on the lowest level of screws quantity, 228 more screws should be added into the container. In this way, there would be no shortage of screws and the pipe would not be blocked. Therefore, the highest level of screws quantity was determined which corresponding to line 3.

Table 2. Average of screws number

Lines	1	2	3	4	5	6
Screws	2.4	2.1	2	1.7	1.1	0.8

4 Conclusion

This paper presented a study which applied assembly line balancing to increase the balancing rate of production line and overall equipment effectiveness. With the improvement of tools and method of operation, the cycle time of bottleneck process reduced from 19.6 s to 16.8 s that satisfied the requirements of cycle time of produce line. Operating standards were developed through experimentation which avoided time loss of equipment. The overall equipment effectiveness increased from 83% to 89.6%. Those achievements prove that assembly line balancing and the OEE play an important role in the improvement of production line.

References

1. Battaïa, O., Dolgui, A.: A taxonomy of line balancing problems and their solution approaches (Periodical style). Int. J. Prod. Econ. **142**(2), 259–277 (2013)
2. Kumar, N., Mahto, D.: Assembly line balancing: a review of developments and trends in approach to industrial application (Periodical style). Global J. Res. Eng. (2013)
3. Avikal, S., Jain, R., Mishra, P., Yadav, H.: A heuristic approach for U-shaped assembly line balancing to improve labor productivity (Periodical style). Comput. Ind. Eng. **64**(4), 895–901 (2013)
4. Zha, J., Yu, J.-J.: A hybrid ant colony algorithm for U-line balancing and rebalancing in just-in-time production environment (Periodical style). J. Manuf. Syst. **33**(1), 93–102 (2014)
5. Akpınar, S., Bayhan, G.M.: A hybrid genetic algorithm for mixed model assembly line balancing problem with parallel workstations and zoning constraints (Periodical style). Eng. Appl. Artif. Intell. **24**(3), 449–457 (2011)
6. Chutima, P., Chimklai, P.: Multi-objective two-sided mixed-model assembly line balancing using particle swarm optimisation with negative knowledge (Periodical style). Comput. Ind. Eng. **62**(1), 39–55 (2012)
7. Rajput, H.S., Jayaswal, P.: A total productive maintenance (TPM) approach to improve overall equipment efficiency (Periodical style). Int. J. Modern Eng. Res. **2**(6), 4383–4386 (2012)
8. Zuashkiani, A., Rahmandad, H., Jardine, A.K.: Mapping the dynamics of overall equipment effectiveness to enhance asset management practices (Periodical style). J. Qual. Maintenance Eng. **17**(1), 74–92 (2011)
9. Puvanasvaran, P., Teoh, Y., Tay, C.: Consideration of demand rate in Overall Equipment Effectiveness (OEE) on equipment with constant process time (Periodical style). J. Ind. Eng. Manage. **6**(2), 507–524 (2013)
10. Rashid, M.F.F., Hutabarat, W., Tiwari, A.: A review on assembly sequence planning and assembly line balancing optimisation using soft computing approaches (Periodical style). Int. J. Adv. Manuf. Technol. **59**(1–4), 335–349 (2012)

11. Zupan, H., Herakovic, N.: Production line balancing with discrete event simulation: a case study (Periodical style). IFAC-PapersOnLine **48**(3), 2305–2311 (2015)
12. Jevgeni, S., Eduard, S., Roman, Z.: Framework for continuous improvement of production processes and product throughput (Periodical style). Procedia Eng. **100**, 511–519 (2015)
13. Tuncel, G., Aydin, D.: Two-sided assembly line balancing using teaching–learning based optimization algorithm (Periodical style). Comput. Ind. Eng. **74**, 291–299 (2014)

Storage Optimization of Low-Level Manual Picking Warehouse Considering Vertical Travel

Ai-ping Wu[1,2], Hua Li[1,2(✉)], Hai-mei Li[1,2], and Rui-peng Liu[1,2]

[1] School of Economics and Management, Xidian University, Xi'an 710126, China
lihua@xidian.edu.cn
[2] Shaanxi Soft Science Institute of Informatization and Digital Economy, Xi'an 710126, China

Abstract. Order picking is considered the most critical operation in warehousing. In the low-level manual picking warehouse, in order to increase the warehouse capacity and make full use of the rack space, the overall height of the rack is increased by increasing the number of layer and reducing the height of each layer of the rack, or directly increasing the number of layers. These all lead to take more time to pick up the goods at the bottom or top of the rack. If the picking time of the vertical travel is not taken into account and the best-selling goods are placed on the bottom or top layer, the order picking efficiency will be greatly affected. Therefore, based on class-based storage strategy, we propose a picking time estimation model that considers the vertical travel. A heuristic algorithm is constructed to optimize the warehousing location and determine the optimal warehousing boundary of goods of different types on each shelf. The results show that the proposed algorithm can better optimize the storage location and reduce the picking time.

Keywords: Order picking · Low-level manual warehouse · Vertical travel · Storage optimization

1 Introduction

With the continuous development of economic globalization and e-commerce, consumption demand of consumers is becoming more and more personalized and diverse. As a result, enterprises receive more and more orders, and the types of goods are also more and more abundant. In this context, enterprise warehouses need more space to store goods, but as the soaring value of land, expanding the warehouse area requires significant costs. Therefore, for low-level manual picking warehouses, especially for e-commerce enterprise warehouses with a large variety of products, it is possible to expand the warehouse area and enhance the warehouse space utilization by increasing the number of shelves plies and reducing the height of each shelf layer, or directly increasing the number of shelves plies.

Order picking can be described as the process of picking up multiple items from a warehouse storage to complete a customer order. In low-level manual picking

C.-F. Chien et al. (Eds.): *IE&EM 2019*, pp. 117–133, 2020.
https://doi.org/10.1007/978-981-15-4530-6_13

warehouses, order sorting operations are cumbersome and time-consuming. According to Topkkins et al. [1], the cost of sorting orders may account for 55% of all operating costs of the warehouse; and the labor of sorting orders accounts for about 60% of the total workload [2]. There are many factors to affect the efficiency of order picking, such as facility layout, order picking path, and storage allocation and so on. Among them, the optimization of storage places has been concerned by many scholars. Some scholars carried out a series of studies in the context of low-level manual picking warehouses without considering vertical travel. Le-Duc et al. [3] constructed an approximate model to estimate the picking distance in a dual-area warehouse, and proposed a heuristic algorithm to determine the optimal storage boundary problem with the goal of minimizing the distance of picking. Wu et al. [4] took a storage system as the research object, carried out a reasonable division of the storage area, and established an integer linear programming model with minimum picking time, then obtained the method of optimizing the location allocation. Pan et al. [5] considered the situation where multiple pickers work in the same area, and proposed a throughput model to optimize the order picking efficiency by considering a trade-off between the picking distance and the blocking-caused delay for the storage assignment. Bottani et al. [6] used genetic algorithm (GA) to optimize the storage location in the warehouse, which ultimately reduced the picking time and simplified the order picking operation. Glock et al. [7] created a U-shaped order picking system and proposed an alternative strategy for distributing products to shelves in it. Grosse et al. [8] considered the influence of human factors on the efficiency of the order picking system. Their results showed that learning and forgetting should be considered in order to properly plan the storage allocation strategy. Calzavara et al. [9] combined cost and ergonomics to conduct comprehensive planning of shelves with different layouts. Their results showed that the traditional pallet storage system could be replaced with half pallet on lower shelves to improve ergonomics and reduce costs. Li et al. [10] proposed an approximation algorithm to optimize the allocation of storage considering the turnover rate of goods and the correlation of demand. Zhang et al. [11] introduced the concept of demand-related models to describe the correlation between goods, and proposed a determining DCP method from historical data. On this basis, a new model was constructed to solve the problem of storage allocation. To solve this model, they developed heuristic and simulated annealing methods.

These scholars studied storage optimization from the perspective of models and algorithms. Some scholars carried out related research from the perspective of the combination of routing strategy and storage strategy, such as Dekker et al. [12] used computer simulation to combine storage and routing strategies, and then discussed the ways to improve the efficiency of warehouse operations. Dijkstra et al. [13] constructed the precise average path length mode of multi-lane and multi-cargo picking under the return, S-shaped, maximum gap and midpoint routing methods, and proposed a dynamic programming algorithm to allocate storage locations in the warehouse.

In addition, many scholars have done research on high-level shelf warehouses and AS/RS system warehouses with considering vertical travel. For example, Parikh et al. [14] pointed out that when vertical travel can be considered, models can be used to determine the optimal storage system configuration. Chan et al. [15] conducted research on storage allocation strategies and optimization orders, such as warehouse division, goods classification and division, to make the warehouse layout more reasonable and the sorting operation more efficient; Chen et al. [16] studied the optimization of shelf allocation and access path in automated warehouse based on mixed integer programming model, and proposed a two-stage decision-making algorithm based on graphics. Pan et al. [17] constructed a multi-layer "artificial" picking time estimation model based on the classification storage strategy and various routing strategies, with considering the shelf height, the number of layers and the vertical motion of the picking. Their experimental results showed that there is a sufficiently accurate practical use in the model. Kou et al. [18] used a random storage strategy in a parallel storage system based on belt conveyor to construct a travel time model for automatic retrieval and automatic storage systems, and used Matlab to verify the model and optimize the automatic storage system.

In summary, previous studies focused on low-level manual picking warehouses without considering vertical travel or high-rise warehouses and AS/RS system warehouses that consider vertical travel. However, in some low-level manual picking warehouses, in order to increase warehouse capacity, people tend to make full use of the shelf space by increasing the number of shelves and reducing the height of each layer of the shelf, or directly increasing the number of shelves to increase the overall height. This leads to the need to squat and other additional actions to pick up goods, and people need to step on the lower shelves to pick up the goods when picking up the top shelf goods. With the rapid development of e-commerce, the warehouse receives more and more orders every day. Under such circumstances, the order picking efficiency will be greatly affected if the goods with high order frequency are placed at the bottom or top layer, which takes more time for picking, without considering the vertical picking time. Therefore, this paper constructs an average picking time estimation model considering vertical travel, and proposes a heuristic algorithm to optimize the shelf storage.

2 Average Picking Time Estimation Model

2.1 Problem Description

The warehouse has a horizontal roadway (in the middle of the warehouse) and multiple working roadways. The layout of the warehouse is symmetric with respect to the lateral roadway, that is, i.e. the storage layout of the symmetric working roadway is completely consistent. The return path strategy is adopted, which is derived from Caron et al. [19]. The storage allocation uses a classification storage strategy. Figure 1 shows the warehouse layout of a horizontal roadway and six working roadways; each shelf consists of several columns and several layers, and the storage capacity on each shelf is equal to the number of columns. Figure 2 shows the layout of five shelves.

The order picking process of such warehouses can be described as follows: the pickers start picking the goods from the warehouse entrance/exit. First, they enter and the nearest working roadway where goods needed to be picked up. Then they travel to the position of the required goods, and pick up the goods from the shelf level until the farthest goods are picked up. Then they return to the middle of the horizontal roadway.

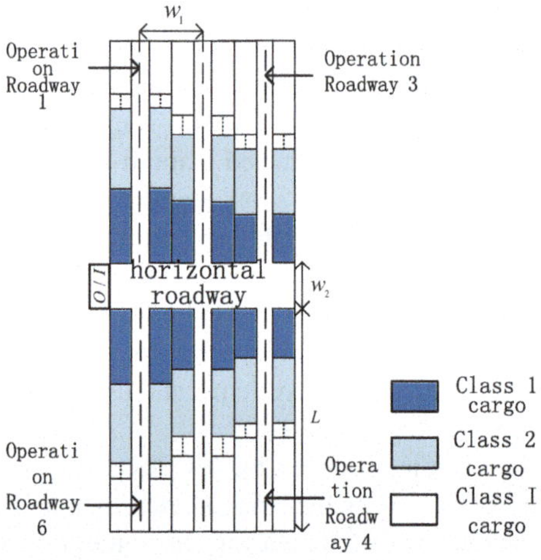

Fig. 1. Warehouse layout

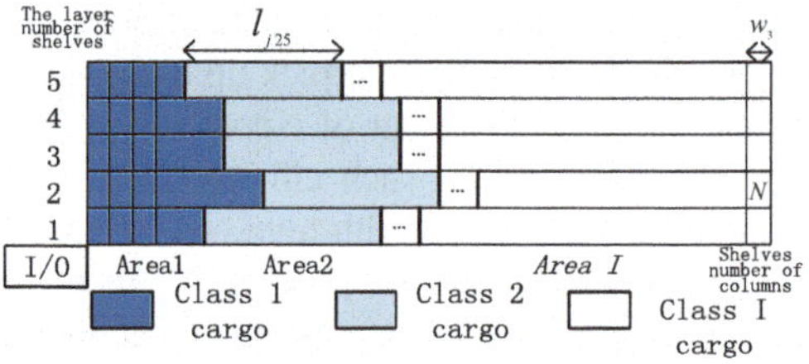

Fig. 2. Shelf layout of operation roadway

Next, the pickers enter the nearest working roadway where the goods needed to be picked up. Repeat the above process. After completing the farthest goods picking in the operation roadway, the pickers return to the warehouse entrance/exit. The picker first picks up the goods on one side of the warehouse and then picks the other side of the goods have the same total travel time with picking up the goods on both sides exactly.

2.2 Modeling Assumption, Parameter and Decision Variable

This paper proposes the following hypotheses:

(1) The storage shelves in the warehouse are identical, each storage space is of the same size, and each shelf can store multiple types of goods.
(2) The length, width and height of all operation roadway are equal.
(3) The working roadway is narrow enough that the picker can pick the goods on both shelves at the same time without changing the position.
(4) The picker can pick up the goods on the shelves regardless of the height of the shelves.

(5) Order frequency of the same category goods is equal. The order frequency of each category is defined as the number of goods needed for that category in a certain period of time.

The parameters in model as follows:

j represents the sequence number of operation roadway $(j = 1, 2, \ldots, J)$;
i represents the category of goods $(i = 1, 2, \ldots, I)$;
k represents the layer number of shelves $(k = 1, 2, \ldots, K)$;
n represents the layer number of columns $(n = 1, 2, \cdots, N)$;
(j, n, k) represents the space coordinates of storing shelves;
w_1 represents the distance between two adjacent operation roadway center lines;
w_2 represents the width of horizontal roadway;
w_3 represents the length of each storage;
v represents the speed of the picker;
t_k represents the time of picking k layer numbers of the goods;
O_i represents the order frequency of i goods.
S_i represents the proportion of storage space of i goods, that is, the proportion of the total storage space;
q_{ji} represents the numbers of i goods picked in j operation roadway;
p_{ji} represents the probability of i goods picked in j operation roadway;
P_{ji} represents the probability of picking the farthest goods in the category i goods area in the j operation roadway;
l_{jik} represents the storage length of i goods on the k shelves in the j operation roadway;
HT_{ji} represents the horizontal travel picking time in category i cargo area of j operation roadway;
VT_j represents the vertical travel picking time in j operation roadway;
TT_j represents the total picking time in j operation roadway

Decision variable:

n_{jik} represents the numbers of storage space of i goods stored in k shelves of j operation roadway.

2.3 Single Working Roadway Model

First, we consider the situation of single working roadway (all goods are in j th operation roadway), and the shelf layout of operation roadway is shown in Fig. 2, region 1, region 2,…, region I store separately 1 goods, 2 goods… I kinds of goods, the quantity of storage space for i goods in each layer is n_{jik}, within each class of cargo area, the probability of picking each stored cargo is equal.

Expected picking time TT_j in operation roadway includes the vertical travel expected picking time VT_j and the horizontal travel expected picking time. We can determine expected picking time by adjusting the farthest reach area, that is, the probability of farthest pick position in region i multiply the corresponding desired horizontal expected picking time, multiply 2 is the time of entry and exit in the operation roadway. Expected picking time in operation roadway can be obtained as follow:

$$TT_j = 2\sum\nolimits_{i=1}^{I} P_{ji} \times HT_{ji} + VT_j \tag{2 – 1}$$

(1) Vertical travel expected picking time

When picking one good, we can calculate the probability that the goods on each shelf in the category are picked up by $l_{jik}/\sum_{k=1}^{K} l_{jik}$, and then multiply the picking time required to pick each shelf t_k to get the horizontal expected picking time of the goods on each shelf. Then add the time spent on all shelves and multiply the probability of each category of goods will be picked so as to get the vertical travel desired picking time. When the expected quantity of picking goods is q_j, the vertical travel expected picking time is:

$$VT_j = q_j \sum_{i=1}^{I} p_{ji} \sum_{k=1}^{K} \frac{l_{jik}}{\sum_{k=1}^{K} l_{jik}} t_k \tag{2 – 2}$$
$$l_{jik} = n_{jik} \times w_3$$

(2) Horizontal travel expected picking time

We consider the horizontal travel expected picking time in the following situation:

(1) If the farthest picking goods are in region 1, this means that all goods to be picked are in region 1, i.e. $\sum_{i=1}^{I} q_{ji} = q_{j1} = q_j$, its probability is:

$$P_{j1} = p_{j1}^{q_j} \tag{2 – 3}$$

According to the ratio of the storage length l_{jik} of each shelf to the total storage length of the cargo in the operation lane, to obtain the probability of picking each shelf cargo, and then multiply the horizontal expected picking time of each shelf, adding them together to get the total horizontal expected picking time HT_{ji}

$$HT_{j1} = \sum_{k=1}^{K} \left[\left(\frac{l_{j1k}}{\sum_{k=1}^{K} l_{jik}} \right) \left(\frac{l_{j1k}}{v} \times \frac{q_{j1}}{q_{j1}+1} \right) \right] \tag{2 – 4}$$

Remark: q_{j1}random variables x are uniformly distributed in the interval $[0, 1]$. Its maximum expected value is $q_{j1}/(q_{j1}+1)$ [3]. The maximum expected value (the time to get to the furthest position) is $(l_{j1k}/v) \times (q_{ji}/(q_{ji}+1))$ when pick q_j goods in region 1. Therefore, Eq. (2-4) is valid when the situation is generalized to each shelf.

(2) If the farthest picking goods are in region 2, this means that all goods to be picked are in region 1 or region 2, but not only in region 1, i.e. $\sum_{i=1}^{I} q_{ji} = q_{j1} + q_{j2} = q_j$, its probability is:

$$P_{j2} = (p_{j1} + p_{j2})^{q_j} - (p_{j1})^{q_j} \tag{2-5}$$

Calculating the horizontal stroke picking time HT_{ji} is basically the same as the calculation method of HT_{j2}:

$$HT_{j2} = \sum_{k=1}^{K} \left[\left(\frac{l_{j2k}}{\sum_{k=1}^{K} l_{j2k}} \right) \left(\frac{l_{j1k}}{v} + \frac{l_{j2k}}{v} \times \frac{q_{j2}}{q_{j2}+1} \right) \right] \tag{2-6}$$

q_{j2} is the number of goods to be picked in the area 2. It is difficult to calculate HT_{J2} by formula (2-6) directly, so we can calculate HT_{J2} by estimating. First, we calculate the expectation of picking quantity $E(q_{j2})$ as formula (2-7):

$$\begin{aligned} E(q_{j2}) &= \sum_{q_{j2}=1}^{q_j} q_{j2} P(B/A) \\ &= \sum_{q_{j2}=1}^{q_j} q_{j2} \frac{P(AB)}{P(A)} \\ &= \sum_{q_{j2}=1}^{q_j} q_{j2} \frac{P(AB)}{P_{j2}} \\ &= \frac{\sum_{q_{j2}=1}^{q_j} q_{j2} \binom{q_{j2}}{q_j} \left(((p_{j2})/(p_{j1}+p_{j2})) \right)^{q_{j2}} \left(((p_{j1})/(p_{j1}+p_{j2})) \right)^{q_j - q_{j2}}}{\left((p_{j1}+p_{j2})^{q_j} - p_{j1}^{q_j} \right)} \\ &= \frac{(q_j p_{j2})/(p_{j1}+p_{j2})}{(p_{j1}+p_{j2})^{q_j} - p_{j1}^{q_j}} \end{aligned} \tag{2-7}$$

As shown in the Eq. (2-7), event B refers to the quantity of goods picked in region 2, event A refers to all of the goods are in region 1 or region 2, but not only in region 1; so event AB refers to the quantity of goods picked in region 2 is q_{j2}, and the quantity of goods picked in region 1 is $q_j - q_{j2}$. The final step of the formula is derived from the expectation of the binomial distribution.

We can get the HT_{j2} by replacing q_{j2} with $E(q_{j2})$ [3]:

$$\begin{aligned} HT_{j2} &\approx \sum_{k=1}^{K}\left[\left(\frac{l_{j2k}}{\sum_{k=1}^{K} l_{j2k}}\right)\left(\frac{l_{j1k}}{v}+\frac{l_{j2k}}{v}\times\frac{E(q_{j2})}{E(q_{j2})+1}\right)\right] \\ &= \sum_{k=1}^{K}\left[\left(\frac{l_{j2k}}{\sum_{k=1}^{K} l_{j2k}}\right)\left(\frac{l_{j1k}}{v}+\frac{l_{j2k}}{v}\times\frac{(q_j p_{j2}/(p_{j1}+p_{j2}))/\left((p_{j1}+p_{j2})^{q_j}-p_{j1}^{q_j}\right)}{(q_j p_{j2}/(p_{j1}+p_{j2}))/\left((p_{j1}+p_{j2})^{q_j}-p_{j1}^{q_j}\right)+1}\right)\right] \\ &= \sum_{k=1}^{K}\left[\left(\frac{l_{j2k}}{\sum_{k=1}^{K} l_{j2k}}\right)\left(\frac{l_{j1k}}{v}+\frac{l_{j2k}}{v}\times\frac{q_j p_{j2}/(p_{j1}+p_{j2})}{(q_j p_{j2}/(p_{j1}+p_{j2}))+(p_{j1}+p_{j2})^{q_j}-p_{j1}^{q_j}}\right)\right] \\ &= \sum_{k=1}^{K}\left[\left(\frac{l_{j2k}}{\sum_{k=1}^{K} l_{j2k}}\right)\left(\frac{l_{j1k}}{v}+\frac{l_{j2k}}{v}\times\frac{q_j p_{j2}}{q_j p_{j2}+(p_{j1}+p_{j2})\left((p_{j1}+p_{j2})^{q_j}-p_{j2}^{q_j}\right)}\right)\right] \end{aligned} \quad (2-8)$$

(3) If the farthest picking goods are in region i ($i = 3, 4, \ldots I$)

If the farthest picking goods are in region i, this means that all goods to be picked are in region 1 to region i, but not only in region 1 to region i-1, i.e. $\sum_{a=1}^{I} q_{ja} = \sum_{a=1}^{i} q_{ja} = q_j$, its probability is:

$$\begin{aligned} P_{\substack{ji \\ i\geq 3}} &= (p_{j1}+p_{j2}+\cdots+p_{ji})^{q_j} \\ &\quad-(p_{j1}+p_{j2}+\cdots+p_{j,i-1})^{q_j} \\ &= \left(\sum\nolimits_{a=1}^{i} p_{ja}\right)^{q_j}-\left(\sum\nolimits_{a=1}^{i-1} p_{ja}\right)^{q_j} \end{aligned} \quad (2-9)$$

Let $H(p_{ja}, q_j) = \left(\sum_{a=1}^{i} p_{ja}\right)^{q_j} - \left(\sum_{a=1}^{i-1} p_{ja}\right)^{q_j}$. The expectation picking time of horizontal stroke $HT_{ji,i\geq 3}$ is calculated in a similar way HT_{J2} is calculated:

$$HT_{\substack{ji \\ i\geq 3}} \approx \sum_{k=1}^{K}\left[\left(\frac{l_{jik}}{\sum_{k=1}^{K} l_{jik}}\right)\left(\sum_{a=1}^{i-1}\frac{l_{jak}}{v}+\frac{l_{jik}}{v}\times\frac{q_j p_{ji}}{q_j p_{ji}+\left(\sum_{a=1}^{i} p_{ja}\right)H(p_{ja}, q_j)}\right)\right] \quad (2-10)$$

Finally, we can get the total expected picking time in the operation roadway TT_J:

$$\begin{aligned} TT_j &= 2\sum_{i=1}^{I} P_{ji}\times HT_{ji}+VT_j \\ &= 2\times p_{j1}^{q_j}\times HT_{j1}+2\sum\nolimits_{i=2}^{I}\left[H\left(p_{ja}, q_j\right)\times HT_{ji}\right]+VT_j \end{aligned} \quad (2-11)$$

2.4 Multiple Operation Roadway Model

The average picking time estimation model for multi-job roadway includes two parts: the expected picking time of all operation roadway and the expected picking time of the horizontal roadway.

First, the probability of i goods being picked in j th operation roadway p_{ji} can be obtained by the ratio of the storage length of i goods stored in j operation roadways to the storage length of i goods stored in all operation roadways, multiplied by the probability of i goods be picked O_i:

$$p_{ji} = O_i \frac{\sum_{k=1}^{K} l_{jik}}{\sum_{j=1}^{J} \sum_{k=1}^{K} l_{jik}}, \forall j \in J, i \in I$$

Then, we calculate the probability of entering each operation roadway. If a cargo is picked, the probability of entering the working roadway j is the sum of the probability of picking all the goods picked in the roadway $\sum_{i=1}^{I} p_{ji}$. If q items cargo is picked, the probability of entering the working roadway can j be obtained by the probability that all the selected items are not in the working roadway j is:

$$b_j = 1 - \left(1 - \sum_{i=1}^{I} p_{ji}\right)^q$$

Next, calculate the expected quantity q_j of goods picked up in the working roadway j based on entering the working roadway j:

$$q_j = \frac{q \sum_{i=1}^{I} p_{ji}}{b_j}$$

TT_j is the time taking to pick up q_j goods from the center-line of the horizontal roadway to the farthest picking location in the operation roadway j. In formula (2-12), $p'_{ji} = p_{ji} / \sum_{i=1}^{I} p_{ji} (\forall j \in J)$; $w_2/2v$ is the time taking from the center line of the horizontal roadway to the entrance of the working roadway.

$$\begin{aligned} TT_j &= 2 \times \frac{w_2}{2v} + 2 \times \left(p'_{j1}\right)^{q_j} \times HT_{j1} + 2 \sum_{i=2}^{I} \left[H\left(p'_{ja}, q_j\right) \times HT_{ji}\right] + VT_j \\ &= \frac{w_2}{v} + 2 \times \left(p'_{j1}\right)^{q_j} \left(\sum_{k=1}^{K} \left(\frac{l_{j1k}}{\sum_{k=1}^{K} l_{j1k}} \times \frac{l_{j1k}}{v} \times \frac{q_j}{q_j + 1}\right)\right) + 2 \times \\ & \sum_{i=2}^{I} \left\{ \begin{array}{l} H\left(p'_{ja}, q_j\right) \times \\ \left[\sum_{k=1}^{K} \left(\frac{l_{jik}}{\sum_{k=1}^{K} l_{jik}} \left(\sum_{a=1}^{i-1} \frac{l_{jak}}{v} + \frac{l_{jik}}{v} \times \frac{q_j p'_{ji}}{q_j p'_{ji} + H\left(p'_{ja}, q_j\right) \sum_{a=1}^{i} p'_{ja}}\right)\right)\right] \end{array} \right\} \\ & + q_j \sum_{i=1}^{I} p'_{ji} \sum_{k=1}^{K} \frac{l_{jIk} t_k}{\sum_{k=1}^{K} l_{jIk}} \end{aligned} \quad (2-12)$$

We can get the total expected picking time in each working roadway by adding the expected picking time in each working roadway:

$$\begin{aligned} TT^a &= \sum_{j=1}^{J} TT_j b_j \\ &= \sum_{j=1}^{J} \left[TT_j \left(1 - \left(1 - \sum\nolimits_{i=1}^{I} p_{ji}\right)^q\right)\right] \\ &= \sum_{j=1}^{J} \left\{ \left[\frac{w_2}{v} + 2 \times \left(p'_{j1}\right)^{q_j} \times HT_{j1} + 2 \times \sum_{i=2}^{I} \left[H\left(p'_{ja}, q_j\right) \times HT_{ji} \right] + VT_j \right] \times b_j \right\} \end{aligned} \quad (2-13)$$

Next, we calculate the expected picking time in the horizontal roadway. Due to the symmetry of the warehouse layout, we assume that the probability of at least one of the working roadways j or the working roadways $(J-j+1)$ being accessed is n_j, which can be obtained by calculating the probability that neither of the working lanes are accessed (m_j and m_{J-j+1}):

$$n_j = 1 - (1 - m_j)(1 - m_{J-j+1}) = 1 - (1 - m_j)^2 = 1 - (1 - \sum_{i=1}^{I} p_{ji})^{2q}.$$

The expected picking time TT^c in the horizontal roadway is obtained by multiplying the probability of the furthest picking goods in the working roadway j with the time spent on arriving working roadway j on the horizontal roadway:

$$\begin{aligned} TT^c &= 2\sum_{j=1}^{J/2} \left[(2j-1)(w_1/2v) \left[\left(\sum_{e=1}^{j} n'_e \right)^q - \left(\sum_{e}^{j-1} n'_e \right)^q \right] \right] \\ &= \sum_{j=1}^{J/2} \left[\frac{(2j-1)w_1}{v} \left[\left(\sum_{e=1}^{j} n'_e \right)^q - \left(\sum_{e=1}^{j-1} n'_e \right)^q \right] \right] \end{aligned} \quad (2-14)$$

And $n'_j = n_j \Big/ \sum_{j=1}^{J/2} n_j (\forall j \in J/2)$. Finally, we can estimate the average picking time TT according to TT^a and TT^c.

3 Storage Location Optimization

The average picking time can be estimated using a 2-section model for a given layout $(L, J, I, K, N, w_1, w_2, w_3)$, order frequency O_i, storage allocation scheme n_{jik}, pick quantity q, and picking time in every layer t_k. In this section, if a storage space ratio S_i is given, the formulas are used to optimize the storage location.

3.1 Model Building

In many cases, the problem of determining the optimal storage space for each category of goods on each shelf floor in each working roadway often occurs, for example, when the warehouse first starts to be used, and the goods classification or order mode change. This section defines the problem as follows: based on various given parameters of the warehouse, we determine the optimal storage boundary of the types of goods on each shelf in each working roadway, so as to get minimize the total average picking time, based on the classification storage strategy (probability of picking each category of goods and the proportion of total storage required). The objective function is:

$$\min TT = TT^a + TT^c \tag{3-1}$$

The constraint conditions are:

$$\sum\nolimits_{i=1}^{I} n_{jik} = N, \forall j \in J, k \in K \tag{3-2}$$

$$\sum\nolimits_{J=1}^{J} \sum\nolimits_{k=1}^{K} n_{jik} = s_i KNJ, \forall i \in I \tag{3-3}$$

$$p_{ji} = O_i \sum\nolimits_{k=1}^{K} n_{jik} \Big/ s_i KNJ, \forall j \in J, i \in I \tag{3-4}$$

$$n_{jik} = n_{J-j+1,ik}, \forall j \in J, i \in I, k \in K \tag{3-5}$$

$$n_{jik} \geq 0, \forall j \in J, i \in I, k \in K \tag{3-6}$$

The objective function is to pick up q goods with the minimum picking time. The picking time consists of two parts: the operation roadway and the horizontal roadway picking time. Use the formula (2-13) to calculate the picking time of the operation roadway, and use formula (2-14) to calculate the picking time of the horizontal roadway.

In the constraint conditions, formula (3-2) indicates the number of storage spaces each shelf is N; formula (3-3) indicates the storage space required of each category goods; formula (3-4) indicates relationship between p_{ji} and n_{jik}, formula (3-5) represents the symmetry of the warehouse layout; the non-negative property of n_{jik} is represented by Eq. (3-6).

With the increase of the quantity of picking goods, working roadways, goods categories and shelf layers, the objective function will become very complex. Therefore, this paper tends to propose the following heuristic algorithm to solve this problem.

3.2 Heuristic Algorithm

First, define the following terms: the same roadway layout means that in all the operation roadway, the quantities n_{jik} of goods of the same category on all shelves are the same. We defined that a and i are adjacent categories if $|a - i| = 1$.

The heuristic algorithm proposed in this paper is a 2-opt switching technology, which belongs to the neighbor search heuristic algorithm. In the algorithm, the picking time of each shelf is sorted in descending order. If it takes the same time, they will be sorted in ascending order according to the number of shelves. The highest ranking of shelves in the longest picking time is rank as 1, the lowest ranking of shelves in the shortest picking time is rank as K, and the corresponding numbers are $1, 2 \ldots, K$. The heuristic algorithm in this paper is divided into two parts. The first part is horizontal optimization of storage location. The initial layout is the same roadway layout. Considering the picking time of horizontal travel, the adjacent goods are exchanged between the operation roadways step by step. The goods of high order frequency far from the starting point of the warehouse and the goods of low order frequency near the starting point of the warehouse are exchanged. This part refers to the heuristic algorithm of Le-Duc et al. [3]. The second part is storage vertical optimization, as shown in Fig. 3. On the basis of the first part optimization, considering the time spent on vertical travel, goods with long picking time and high shelf order frequency are exchanged with goods with short picking time and low shelf order frequency between adjacent categories in each operation roadway, so as to determine the optimal storage boundary of each category of goods on each shelf in each operation roadway and minimize picking time. Since the warehouse layout is symmetrical, only half of the warehouse needs to be considered.

To illustrate this method, assume there is a semi-layout of three-layer shelves in three operation roadways (numbered as operation lanes 1 to 3 and shelf layers 1 to 3, in which shelf layer 3 takes the most picking time, shelf layer 2 takes the least picking time, and shelf layer 1 takes the middle picking time) and three categories of goods (category A, B, and C). First of all, only consider the picking time in the horizontal direction, let start from the farthest working lane 3, the class A goods in the working lane 3 and the class B goods in the working lane 1 are exchanged, then the class B goods in the working lane 3 and the class C goods in the working lane 1 are exchanged, and then the exchange is carried out between the working lane 3 and the working lane 2. Finally, consider the exchange between operation lane 2 and operation lane 1. After completing the previous steps, start to consider the vertical travel picking time. Between adjacent categories in each operation roadway, first exchange the storage locations of class A goods on third shelf and class B goods on second shelf one by one, then exchange the storage locations of class A goods on third shelf and class B goods on first shelf one by one, finally exchange the storage locations of class A goods on first shelf and class B goods on second shelf, and so on until the exchange between all adjacent categories and all operation lanes is completed.

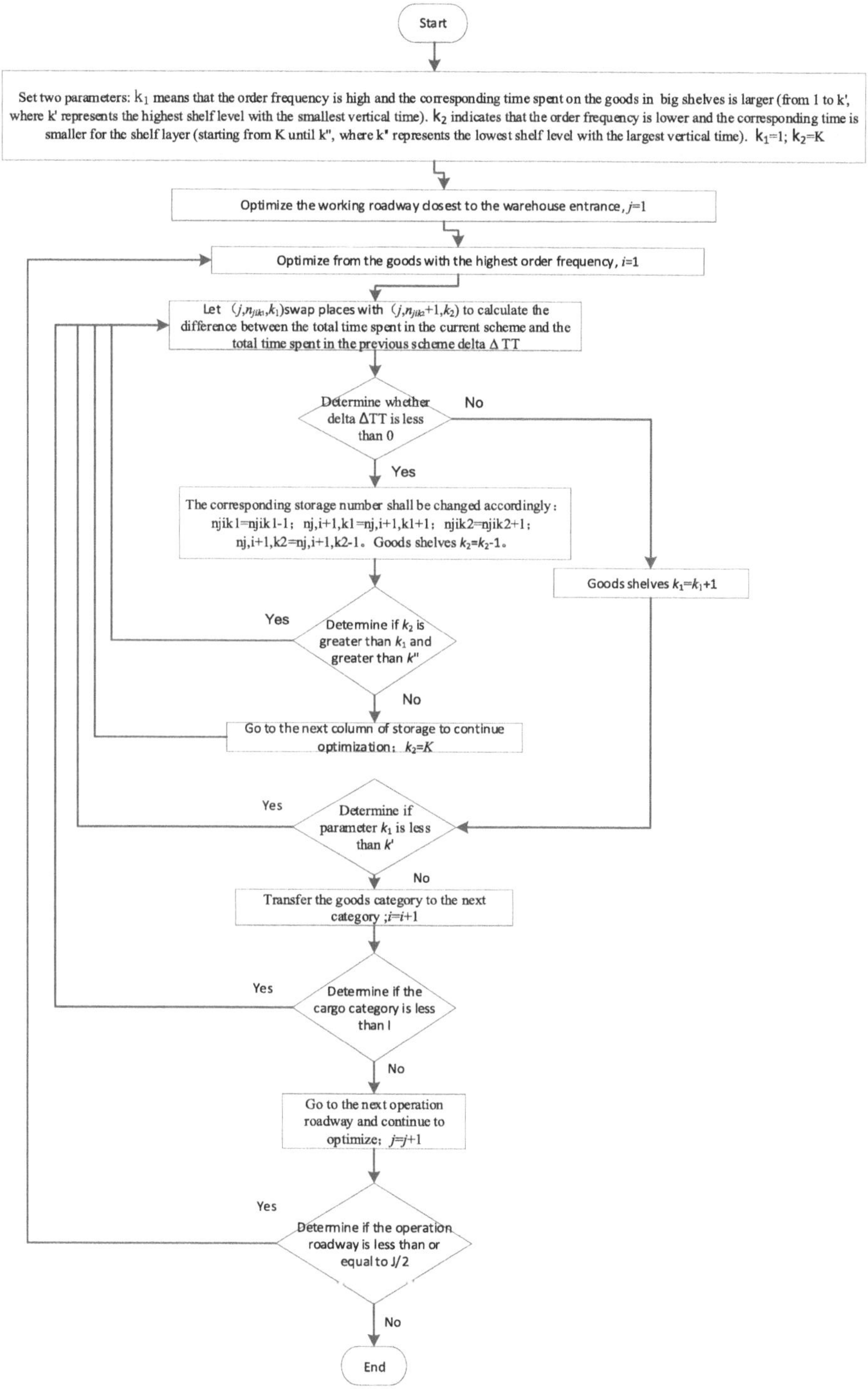

Fig. 3. Flow chart of storage vertical optimization

3.3 An Illustrative Example

In order to evaluate the performance of heuristic algorithm and determine the best storage area in the warehouse, assume that an average of 100,000 pieces of goods are picked up in the warehouse A every day, and the picking time required for each shelf is estimated (in reality, according to the actual situation the time required for each shelf can be measured in the warehouse). Refer to the parameters in literature [3], various numerical experiments are carried out to compare with the heuristic algorithm in literature [3], which did not optimize the storage location in the vertical direction. In these experiments, this paper considers the classified storage strategy of three layouts (with 6, 12 and 24 work lanes) and different picking quantities (6, 12, 24 and 48 goods). The warehouse layout parameters are as follows:

$$N = 90, K = 6, w_1 = 15, w_2 = 10,$$
$$w_3 = 1, v = 2, t_k = (2, 1, 1, 1, 1, 4)$$

Goods are divided into three categories according the order frequency and ratio of storage space, namely ABC storage strategy, as shown in Table 1.

Table 1. Classified storage strategy (order frequency/storage space ratio)

Classified storage strategy	Class A(%)	Class B(%)	Class C(%)
	80/20	15/30	5/50

According to the above parameters, the comparison results of the average picking time are shown in Table 2:

Table 2. The comparison results of the average picking time

Numbers of operation roadway	Quantity to be picked each time	Total quantity to be picked	Average picking time		The time saved by the algorithm in this paper
			Le-Duc' heuristic algorithm	This paper heuristic algorithm	
6	4	100000	2797531.11	2794564.96	2966.15
	12	100000	1782121.96	1776317.91	5804.05
	24	100000	1254183.73	1237251.41	16932.32
	48	100000	864585.33	818266.00	46319.33
12	4	100000	3730103.36	3727897.67	2205.69
	12	100000	2449103.45	2445371.48	3731.97
	24	100000	1803769.53	1797844.51	5925.02
	48	100000	1268285.28	1251245.43	17039.85
24	4	100000	5168603.00	5167128.97	1474.03
	12	100000	3316805.86	3314125.63	2680.23
	24	100000	2475759.60	2471967.24	3792.36
	48	100000	1814884.50	1808899.98	5984.52

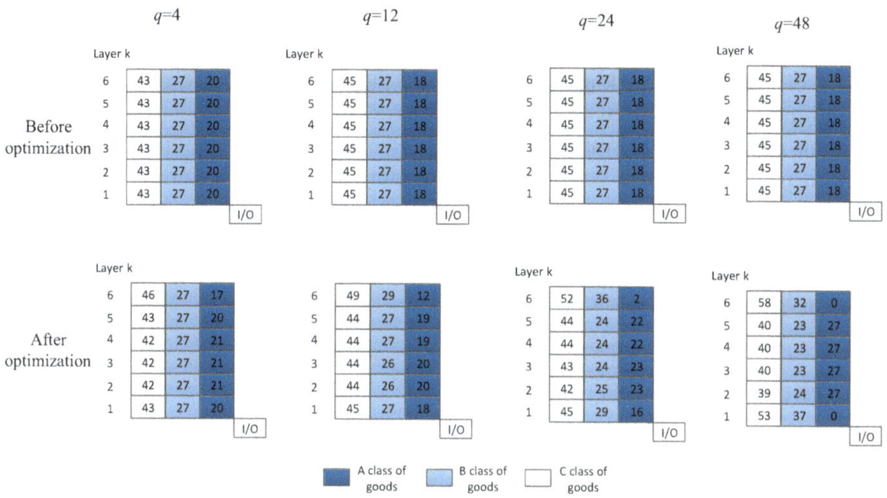

Fig. 4. Operation roadway layout before and after optimization

As we can see from Table 2, compared with the heuristic algorithm from literature [3], the heuristic algorithm proposed by us can reduce the picking time by optimizing the storage location in the vertical direction, and can save more time as the picking quantity increases. Therefore, considering the picking time on the vertical travel in the low-level manual picking warehouse can better optimize the storage, reduce the picking time and improve the picking efficiency.

Figure 4 shows the comparison of 4, 12, 24, 48 goods in the 6 operation roadways before and after vertical optimization of storage location of operation roadway 1. Between adjacent categories, the optimized operation roadway layout will exchange the higher order frequency goods with longer picking time in a shelf with lower order frequency goods with shorter picking time in a shelf, we place more A goods in 2–5th layer with the least time spending, and place fewer A goods in 1st and 6th layers. We should place more B type goods on the 6th and 1st floors than the middle layer, this is because Class A goods occupy more storage locations of 1st and 6th layers. More Class C cargo is placed on the 6th and 1st floors, while on the 2–5th floor there are fewer Class C cargoes. With the number of picks increasing, this phenomenon becomes more apparent as the number of picks increases, the picker needs to take a complete operation roadway to pick up all the goods. At this time, reducing the vertical pick picking can reduce the total picking time, so more and more goods with high order frequency are placed on the shelf layer with less picking time.

4 Conclusion

Based on the existing research, this paper optimizes the storage position in the vertical direction for the low-level manual picking warehouse that takes a long time to pick up goods at the bottom or top of the shelf. The goods with high order frequency and the longer picking time on the shelf are exchanged with those with low order frequency and the shorter picking time. The results show that the optimization of the storage position can reduce the picking time. And, with the number of picking increasing, the picking time can be reduced more significantly.

In this paper, only one warehouse layout is considered, and the influence of vertical travel on picking time can be further studied later in other low-level manual picking warehouse layouts.

Acknowledgement. The authors acknowledge the financial support from Fundamental Research Funds for the Central Universities (No. RW180215).

References

1. Tompkins, J.A., White, J.A., Bozer, Y.A., et al.: Facilities Planning. Wiley, Hoboken (2010)
2. Petersen, C.G.: An evaluation of order picking routeing policies. Int. J. Oper. Prod. Manag. **17**(11), 1098–1111 (1997)
3. Le-Duc, T., De Koster, R.M.B.M.: Travel distance estimation and storage zone optimization in a 2-block class-based storage strategy warehouse. Int. J. Prod. Res. **43**(17), 3561–3581 (2005)
4. Wu, L., Zhou, Z., Zheng, J.: The analysis of storage zone and storage location of a specific corporation. Ind. Eng. Manag. **4**, 24 (2006). (in Chinese)
5. Pan, J.C.H., Shih, P.H.: Evaluation of the throughput of a multiple-picker order picking system with congestion consideration. Comput. Ind. Eng. **55**(2), 379–389 (2008)
6. Bottani, E., Cecconi, M., Vignali, G., et al.: Optimisation of storage allocation in order picking operations through a genetic algorithm. Int. J. Logist. Res. Appl. **15**(2), 127–146 (2012)
7. Glock, C.H., Grosse, E.H.: Storage policies and order picking strategies in U-shaped order-picking systems with a movable base. Int. J. Prod. Res. **50**(16), 4344–4357 (2012)
8. Grosse, E.H., Glock, C.H., Jaber, M.Y.: The effect of worker learning and forgetting on storage reassignment decisions in order picking systems. Comput. Ind. Eng. **66**(4), 653–662 (2013)
9. Calzavara, M., Glock, C.H., Grosse, E.H., et al.: Analysis of economic and ergonomic performance measures of different rack layouts in an order picking warehouse. Comput. Ind. Eng. **111**, 527–536 (2017)
10. Li, M., Jiang, X.: A method for storage location assignment considering COI and demand correlations. Oper. Manag. **27**(09), 22–23 (2018). (in Chinese)
11. Zhang, R.Q., Wang, M., Pan, X.: New model of the storage location assignment problem considering demand correlation pattern. Comput. Ind. Eng. **129**, 210–219 (2019)
12. Dekker, R., De Koster, M.B.M., Roodbergen, K.J., et al.: Improving order-picking response time at Ankor's warehouse. Interfaces **34**(4), 303–313 (2004)
13. Dijkstra, A.S., Roodbergen, K.J.: Exact route-length formulas and a storage location assignment heuristic for picker-to-parts warehouses. Transp. Res. Part E: Logist. Transp. Rev. **102**, 38–59 (2017)
14. Parikh, P.J., Meller, R.D.: A travel-time model for a person-onboard order picking system. Eur. J. Oper. Res. **200**(2), 385–394 (2010)

15. Chan, F.T.S., Chan, H.K.: Improving the productivity of order picking of a manual-pick and multi-level rack distribution warehouse through the implementation of class-based storage. Expert Syst. Appl. **38**(3), 2686–2700 (2011)
16. Chen, C., Lu, Z.: Optimization for storage location assignments and interleaving problems in an automated storage/retrieval system. J. Manag. Eng. **26**(01), 42–47 (2012). (in Chinese)
17. Pan, J.C.H., Wu, M.H., Chang, W.L.: A travel time estimation model for a high-level picker-to-part system with class-based storage policies. Eur. J. Oper. Res. **237**(3), 1054–1066 (2014)
18. Kou, X., Xu, G., Yi, C.: Belt-conveyor based efficient parallel storage system design and travel time model analysis. Int. J. Prod. Res. **56**(23), 7142–7159 (2018)
19. Caron, F., Marchet, G., Perego, A.: Routing policies and COI-based storage policies in picker-to-part systems. Int. J. Prod. Res. **36**(3), 713–732 (1998)

Research on Furniture Workshop Layout Optimization Based on SHA and SLP

Qing Zeng(✉), Xue-wen Dong, and Wan-lin Cao

Institute of Intelligent Manufacturing, Department of Industrial Engineering, Panzhihua College, Panzhihua, China
852404237@qq.com

Abstract. This paper took plate workshop of A furniture production enterprise as research object, and analyzed logistics situation of workshop according to data materials such as material flow of workshop and distance between production processes of workshop, and found out problems existing in layout of workshop facilities of the company. Then, targeting at these problems, system layout planning method was used to analyze logistics relationship, non-logistic relationship, and comprehensive relationship, diagram of work area was drew and an improvement plan for the company's workshop layout was proposed. Finally, simulations of the front and post improvement scenarios were simulated by flexsim simulation software and effects were evaluated.

Keywords: Facility layout · Flexsim simulation · System layout planning method (SLP)

1 Introduction

Workshop layout, as a key factor affecting logistics cost, directly affects enterprise logistics efficiency, the rate of workshop equipment utilization, production efficiency, production balancing and safety. Therefore, it is of great significance in practice to study the layout of workshop to improve the economic benefit of enterprises. this paper discussed the optimization of the board workshop in B area of A furniture company based on SHA and SLP methods, and improvement schemes are proposed. Flexsim software is used to simulate the schemes before and after improvement and evaluate the effects.

2 Methodology

2.1 Analysis of the Layout of the Plate Workshop

2.1.1 Analysis of the Production Process

The plate workshop of company A mainly produces medium fiber plate and triamine plate. The processing process of medium fiber plate is as follows: cutting, bonding, fine cutting, edge sealing, modeling, special-shaped edge sealing, modeling, drilling, sanding, finishing, sand dyeing and final inspection. The technological process of

C.-F. Chien et al. (Eds.): *IE&EM 2019*, pp. 134–144, 2020.
https://doi.org/10.1007/978-981-15-4530-6_14

triamine plate is as follows: blanking, edge sealing, drilling, sanding and final inspection. According to the production process flow and process characteristics of the plate workshop, the plate workshop can be divided into 13 work units, as shown in Table 1.

Table 1. Work unit division

Serial number	Job unit name	Content	Area (m^2)
1	Raw material area	Stacking of plates, hardware and packaging materials	3700
2	Cutting area	Xianlin electronic saw, vertical and horizontal saw, push table saw, electronic cutting saw	1200
3	Edge banding area 1	Triamine plate parts: four-end edged drilling line, double-end edge line, narrow part edge banding, single edge banding line, shaped edge banding area	1550
4	Edge banding area 2	Thickened parts: edge sealing line, straight edge banding machine	1000
5	Drilling area	Ten row drill, six row drill, hinge drill, double end drill	980
6	Thickened area	Binding, bonding, cold pressing, hot pressing	1250
7	Modeling area	Machining center, engraving machine, vertical axis milling, router	1080
8	Sanding area 1	Triamine plate parts: coarse sand, fine sand, knocked embedded, hand drill	400
9	Sanding area 2	Thickened parts: coarse sand, fine sand, knocked embedded, hand drill	360
10	Finishing area	Paint process	3100
11	Detection area	Hand sand, inspection, labeling	850
12	packaging area	Trial assembly, packing	450
13	Finished product library	Packed parts classified storage	2800
14	Workspace	Production plan release, logistics management	340

2.1.2 Analysis of Material Handling Route

Workshop efficiency and production cost control are closely related to material handling. It is difficult to control the logistics and information flow in the production process because of the many influencing factors in the layout of furniture processing workshop facilities. Therefore, the study and analysis of material handling route is an important step in the layout optimization of workshop facilities. The layout of the plate workshop is shown in Fig. 1. It can be seen from the figure that there are two major problems: in the area of drilling area and sanding area 1, there are round-trip and cross-transport; repeated handling and excessively long handling distance exist in the thickening zone and blanking interval.

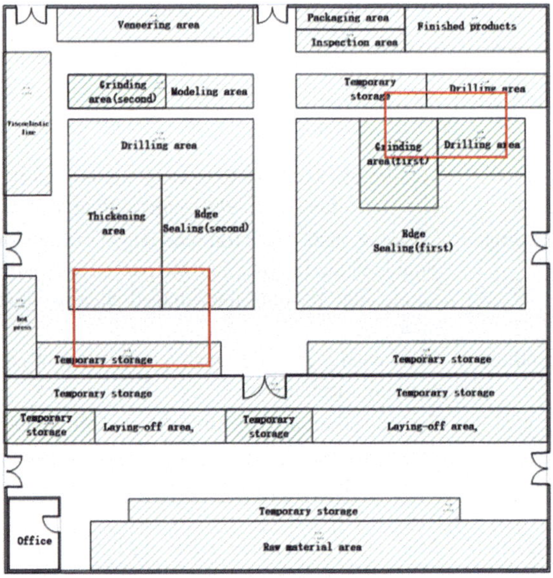

Fig. 1. Layout of plate workshop

2.1.3 F—D Analysis of Workshop Logistics

According to the relationship between the flow rate and the distance between each operation unit, the F—D diagram is shown in Fig. 2.

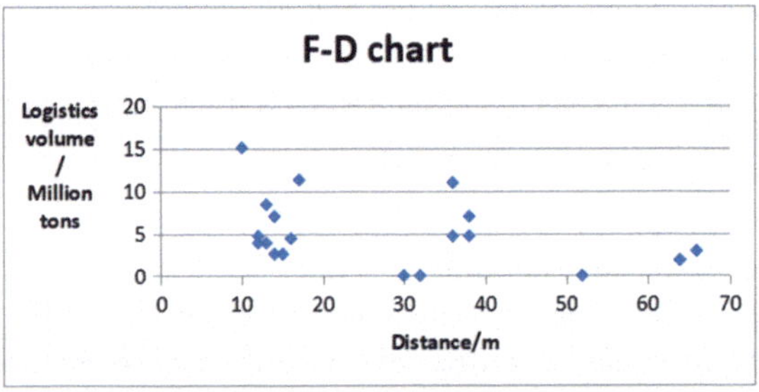

Fig. 2. F-D diagram. Note: Longitudinal axis represents logistics volume (F), horizontal axis represents distance (D).

From the F-D diagram, it can be seen that there are some problems in the middle of the diagram, such as several points with large flow rate and far distance, that is, in the feeding area to the thickening area, from the feeding area to the sealing edge area 1, and from the thickening area to the feeding area.

2.1.4 Problems Existing in the Current Layout of Corporate Facilities

Through the analysis of the process flow, material handling route and material flow rate of the plate workshop, as well as the field investigation, the main problems in the layout of the workshop facilities are as follows: (1) there are the following problems in the layout of the workshop facilities: (1) there are long handling routes, cross round trips and too many handling times. (2) the layout of the workshop facilities is not in accordance with the requirements of logistics distance. (3) there are more open spaces in the workshop and the space utilization rate is low. (4) the large equipment in the workshop has not been put into production for a long time, and the equipment utilization rate is low.

2.2 Optimization of Plant Facilities Layout Based on SLP Method

2.2.1 Analysis of Logistics Relationship of Operating Units

For the subjects studied, the logistics intensity was transformed into five grades, which were expressed by the letters A, E, I, O, U, that is, A was the strongest, and the order from high to low was E, I, O, U.

According to the characteristics of logistics intensity, the proportion of material flow is calculated, and the standard table of logistics intensity grade is used to determine the intensity level of the operation unit. Then the analysis table of the logistics intensity grade of the operation unit is obtained, so as to draw the logistics analysis of the operation unit as shown in Fig. 3:

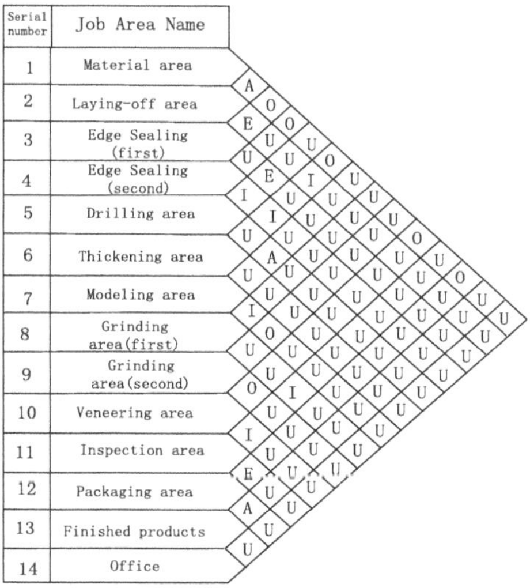

Fig. 3. Logistics intensity level analysis table

2.2.2 Analysis of Non-logistics Relations in Operating Units

Through the established benchmark relationship, the relationship table between the operating unit pairs is obtained, and the non-logistics correlation diagram of the operating unit is established, Fig. 4.

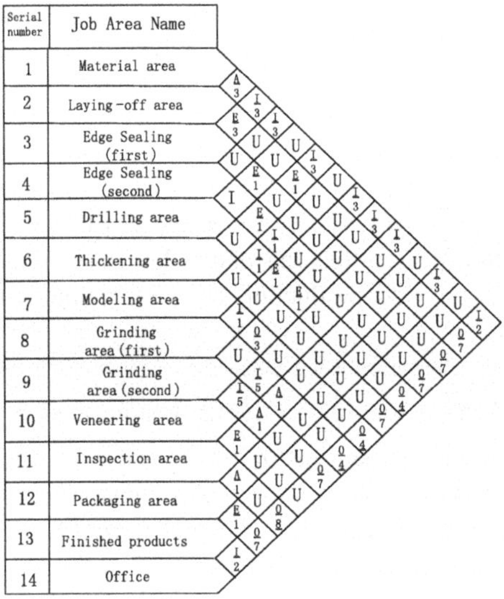

Fig. 4. Non-logistic relationship correlation graph

2.2.3 Analysis of the Comprehensive Relationship of Operating Units

The correlation diagram of the comprehensive relationship of the operation unit is established by Figs. 3 and 4. As shown in Fig. 5.

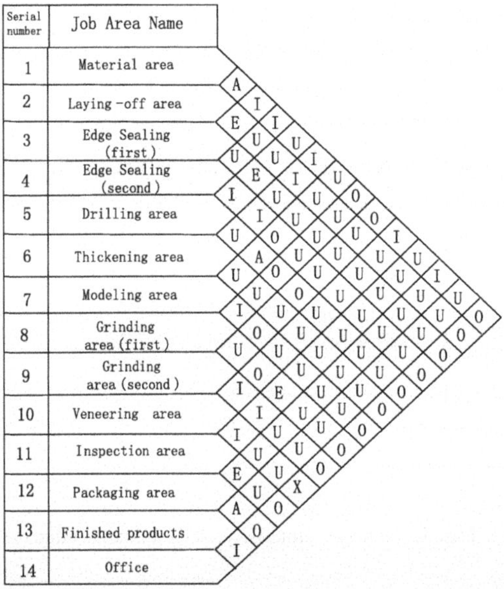

Fig. 5. Comprehensive relationship correlation graph

2.2.4 Correlation Diagram of Operation Unit Location

For a more reasonable layout of facilities, the operation area with high comprehensive proximity should be in the central position, which can greatly shorten the movement times and distance of materials. Through the analysis of the intuitive data of comprehensive proximity, the values of comprehensive proximity are arranged in descending order. The ranking order of each pair of operation units is: 1, 14, 5, 11, 2, 3, 12, 7, 8, 4, 6, 9, 10, 13. The correlation diagram of the position of operation units is drawn as shown in Fig. 6.

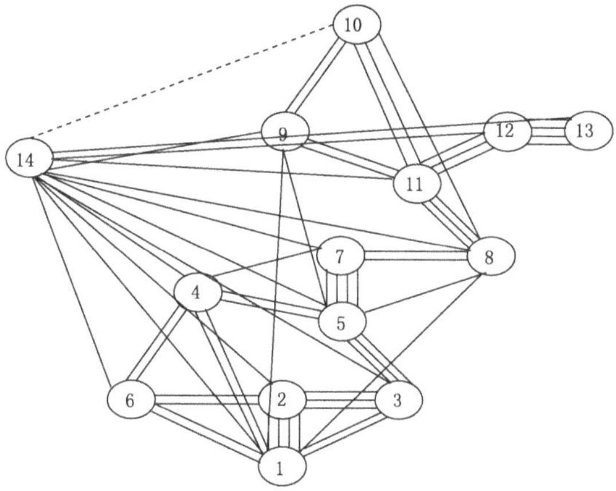

Fig. 6. Job unit location correlation graph

2.2.5 Correlation Diagram of Working Unit Area

Consider the following two aspects of the job unit position correlation diagram: the area of each operation unit and the shape of the equipment involved in each operation unit, and draw the area correlation diagram of the job unit, as shown in Fig. 7 below:

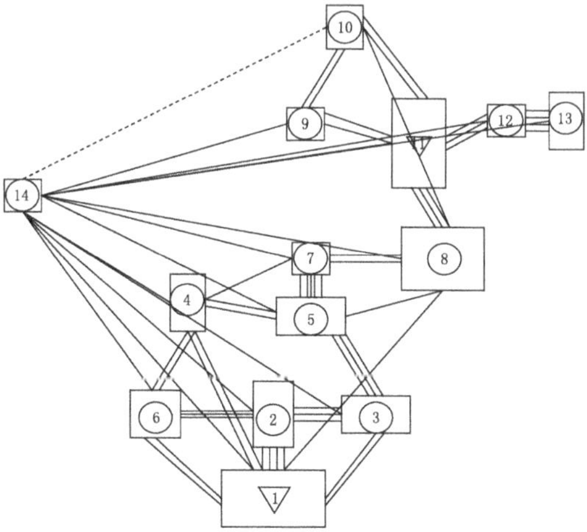

Fig. 7. Job unit area correlation graph

2.2.6 The Optimization of the Layout of the Workshop Facilities

According to the above analysis, the relative position between the operating units is rearranged, and the layout is further optimized in accordance with the actual situation of the plate workshop of Company A, and the improvement is shown in Fig. 8. There are five points to be improved as follows: (1) through field investigation and interview method, it can be seen that the adhesive line on the left side of the workshop is no longer suitable for the existing production demand, so it is changed to rest culture hall and office area 2; (2) in the blanking area and edge sealing area 1, a push table saw is added to meet the finishing process of unilateral edge sealing parts; (3) in order to reduce the problem of reciprocating transportation between the thickening area and the feeding area in the finishing process, two additional sets are added in the thickening area. The fine cutting equipment is electronic saw and push table saw, which are simple in operation and low in cost, which can meet the existing production needs of the company from the point of view of cost and technology. (4) in order to reduce the problem of round trip and cross handling between grinding area 1 and drilling area, the position layout of drilling area and grinding area 1 is adjusted, and the handling of small equipment is considered from the point of view of cost and production practice. (5) in order to shorten the handling distance from the blanking area to the thickened area and the closed edge area 1, two doors and two slide tracks were added to the corresponding position of the temporary storage area to improve the handling efficiency, but at the same time, the cost of improving the company was also increased.

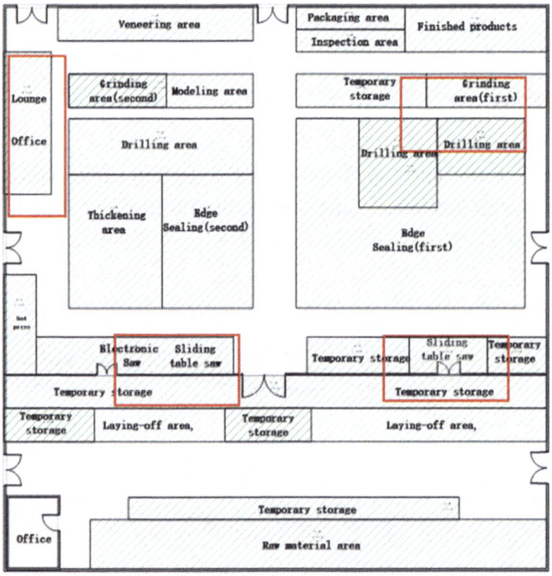

Fig. 8. Improvement solution

3 Discussion

3.1 Modeling of the Workshop Layout and Evaluation on the Solution Based on Flexsim

3.1.1 Modeling and Simulation of Workshop Layout Based on Flexsim

In the process of modeling and verification, the on-line production components of the two lines (bedside table and panel, tea cabinet side board) are modeled and simulated respectively.

According to the different processes of the two production lines, the operating time of each process is derived from the SAP system, as shown in Table 2:

Table 2. Work time of part process

Product parts	Process name	Single processing quantity	Single processing time
Bedside table	Cutting	5	30
	Fixed thickness	1	33
	Frame	1	78
	Hot pressing	1	24
	Fine cut	1	33
	Drilling	1	30.96
	Edge banding	1	3.54
	modeling	1	93.06
	Sand dyeing	1	103.44

The models of both the old and new layouts are shown as follows: Fig. 9 is about the old layout and Fig. 10 the new layout.

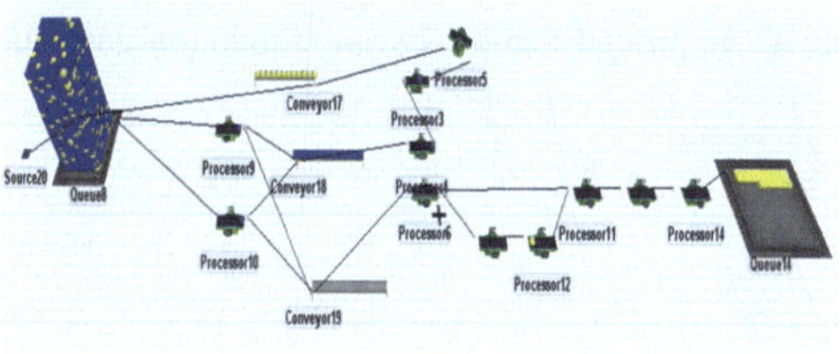

Fig. 9. The model of the old layout

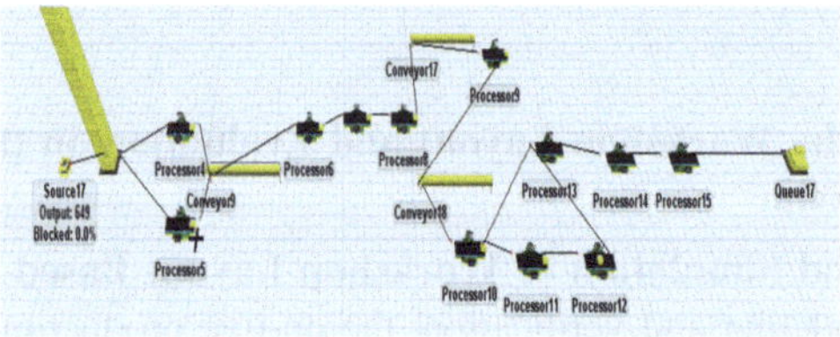

Fig. 10. The model of the new layout

3.1.2 Analysis of Simulation Results

The simulation outputs of facility layout before and after the improvement are shown in Tables 3 and 4.

Table 3. Pre-improvement facility layout simulation data output

Flexsim Summary Report							
Time:	1526582.909						
Object	Class	stats_input	stats_output	idle	conveying	processing	stats_staytimeavg
Processor3	Processor	5200	5200	885.84893	0	405541.624	179.851692
Processor4	Processor	5200	5200	905.413097	0	171556.1261	179.833056
Processor5	Processor	5200	5200	9399.952655	0	124901.9511	178.228966
Processor6	Processor	5200	5200	1361487.422	0	18478.96002	31.638122
Processor7	Processor	1358	1358	1167203.54	0	69803.2177	264.257764
Queue8	Queue	10400	10400	0		0	580365.205
Processor9	Processor	6505	6505	12.877266	0	194980.264	234.494586
Processor10	Processor	3895	3895	26.187525	0	117064.1627	391.508086
Processor11	Processor	5200	5200	1120997.138	0	66925.01604	77.959531
Processor12	Processor	1358	1358	849310.7905	0	420414.6092	498.57378
Processor13	Processor	5200	5200	1001716.438	0	483959.8395	100.915957
Processor14	Processor	5200	5200	988657.0979	0	537925.8114	103.447271
Queue14	Queue	5200	0	0		0	0
Conveyor17	Conveyor	5200	5200	0	936310.2058	0	900
Source20	Source	0	5200	0		0	0
Conveyor18	Conveyor	5200	5200	0	349608.9932	0	899.446268
Conveyor19	Conveyor	5200	5200	0	754156.2063	0	600.009951

Table 4. Improved facility layout simulation data output

Flexsim Summary Report							
Time:	1299220.248						
Object	Class	stats_output	stats_input	idle	processing	conveying	stats_staytimeavg
Queue2	Queue	5200	5200	0	0		618431.3932
Processor4	Processor	4	4	12.877266	108.881367	0	323378.7262
Processor5	Processor	5196	5196	26.187525	155770.0188	0	248.882244
Processor6	Processor	5200	5200	363.078921	171698.5365	0	248.923433
Processor7	Processor	5200	5200	396.488624	405641.2419	0	248.976486
Processor8	Processor	5200	5200	46795.63135	124711.3249	0	240.113499
Processor9	Processor	5200	5200	957.520216	171603.7128	0	249.166558
Conveyor9	Conveyor	5200	5200	0	0	63795.3042	1244.2713
Processor10	Processor	5200	5200	695.659291	18479.95922	0	249.57388
Processor11	Processor	4191	4191	773.707677	215373.2936	0	309.714708
Processor12	Processor	4191	4191	603.292227	1297587.321	0	309.829043
Processor13	Processor	5200	5200	1148955.622	66869.64535	0	28.876004
Processor14	Processor	5200	5200	815247.7878	483956.8651	0	93.068628
Processor15	Processor	5200	5200	1214902.803	84317.44461	0	16.214893
Queue17	Queue	0	5200	0	0		0
Source17	Source	5200	0	0	0		0
Conveyor17	Conveyor	5200	5200	0	0	3284.68	1244.645248
Conveyor18	Conveyor	5200	5200	0	0	3209.56	1243.659264

3.1.3 Effectiveness Evaluation

According to the output results of simulation data, the processing time before and after the improvement of facility layout can be known. Before the improvement, it takes the workshop about 1526582.909 s to reach the production volume of 5200 pieces, while adopting the improving solution, the whole processing time reduces to 1299220.248 s, and the time utilization rate is obviously increased by 17%. At the same time, we can indirectly get the utilization rates of the workshop equipment of the old layout and the new one from the above simulation results. The utilization rates of equipment before and after improvement are shown in Fig. 11. The utilization rate of improved equipment is much higher than that before improvement, among which, the utilization rate of machining center equipment increases by 12.01% and the utilization rate of special-shaped edge sealing equipment increases by 72.33%. In general, the improved scheme has significantly improved the efficiency of time utilization and equipment utilization, thus reducing the logistics cost of the enterprise and improving the production efficiency.

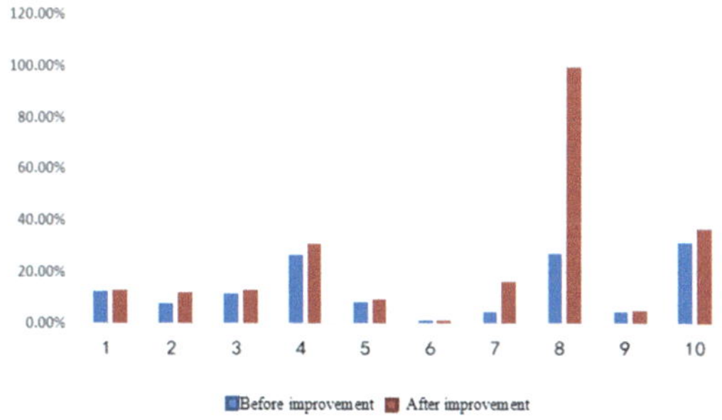

Fig. 11. Comparing table of equipment utilization rate before and after improvement. Note: 1; Blanking/2; Blanking/3; Sanding/4; Nail frame/5: Hot pressing/6: Edge closure/7: Machining center/8: Special-shaped edge sealing/9: Drilling holes/10: Modeling.

4 Conclusion

For the present furniture manufacturing enterprises, whether the workshop layout and space utilization are reasonable or not will directly affect the production efficiency and economic benefits of the whole enterprise. This paper takes the board workshop of A furniture enterprise as the research object, applies the material handling and system layout planning method to optimize the workshop logistics and layout, and puts forward the improvement plan. In addition, Flexsim software was used to simulate the schemes before and after improvement and evaluate the effects. After layout optimization, the time utilization efficiency of plate workshop was increased by 17%, and the utilization rate of equipment was also greatly improved. The utilization rate of machining center equipment was increased by 12.01%, and the utilization rate of special-shaped edge sealing equipment was increased by 72.33%. The results show that the study on the optimization of workshop layout can greatly improve the logistics efficiency, workshop equipment utilization rate and production efficiency of furniture enterprises, thus reducing the logistics cost and enhancing the competitive advantage of enterprises in the market.

References

1. Zhuo, Y.: Optimal design of production line based on Flexsim. J. Beijing Univ. Inf. Technol. (Nat. Sci. Ed.) (03) (2011)
2. Bao, S., Chen, H., Jiang, G., Zhou, X.: Optimization of assembly line production logistics based on Flexsim. Ind. Eng. Manag. (04) (2008)
3. Li, Q., Li, Z., Wen, Z.: Optimal design of facility layout in a workshop based on SLP. Coal Mine Mach. (05) (2011)
4. Dong, X.: The application of SLP in the optimization of the vertical rectification of C company. Tianjin University (2012)
5. Ma, C., Yan, Z.: Analysis and optimization of assembly workshop production system based on SLP. Logist. Eng. Manag. (07) (2009)
6. Wilkinson, J.P.: Nonlinear resonant circuit devices (Patent style). U.S. Patent 3 624 12, 16 July 1990
7. Monzon, J.E.: The cultural approach to telemedicine in Latin American homes (Published Conference Proceedings style). In: Proceedings of 3rd Conference Information Technology Applications in Biomedicine, ITAB 2000, Arlington, VA, pp. 50–53 (2000)
8. Saunders, F.A.: Electrotactile sensory aids for the handicapped (Presented Conference Paper style). Presented at the 4th Annual Meeting Biomedical Engineering Society, Los Angeles, CA (1973)
9. Moodie, C.L., Young, H.H.: A heuristic method of assembly line balancing for assumptions constant or variable work element times. J. Ind. Eng. **16**(1), 23–29 (1965)

Option Game Analysis in IT Investment Strategy of B2B E-Intermediary

Hua Zhang, Li Li(✉), Xingzhen Zhu, and Xiang He

School of Economics and Management, Nanjing University of Science and Technology, Nanjing 210094, China
13155533006@163.com

Abstract. Based on the option game theory, we build up the asymmetric duopoly B2B platforms' information technology (IT) investment decision-making model under the uncertain market demand environment. By adopting the method of game analysis, we obtain B2B electronic intermediary companies' optimal investment strategy under simultaneous equilibrium and sequential equilibrium respectively. Equilibrium analysis shows that B2B platform's user capacity is inversely proportional to its own marginal operation cost and directly proportional to its rival's marginal cost. Numerical analysis shows that competition between B2B platforms encourages preemptive investment while the appropriate delayed investment strategy can bring higher profits to the intermediary enterprises. this research result can provide scientific theoretical guidance for the IT investment decision-making of B2B electronic intermediary enterprises.

Keywords: B2B platform · IT investment strategy · Option game

1 Introduction

B2B electronic intermediary companies as the Internet provides third-party B2B platform services, e-commerce has become a dynamic, fast-growing market main body. However, with the rapid development, the competition of B2B e-intermediary industry is also intensifying. With the constant emergence and rapid diffusion of new technologies, B2B electronic intermediary enterprises face huge investment pressure and serious homogenization competition. The industry presents the situation that the operating income is soaring, but the net profit loss continues to expand. Therefore, B2B e-intermediary enterprises face a problem in operation and management practice, which is how to choose the most favorable IT investment strategy according to market demand, environment and other factors to achieve the long-term profit maximization.

When choosing IT investment strategy, B2B e-intermediary enterprises will consider the level of platform service quality, IT cost reduction and asymmetry with competitors. In general, IT investment includes human capital, hardware and software equipment investment in information technology research and development, etc. The empirical research shows that the B2B platform services are promoted quality mainly depends on the B2B electronic intermediary enterprise IT investments (Devaraj and Kohli 2000; Thatcher and Oliver 2001; Thatcher and Pingry 2004). The size of the IT

C.-F. Chien et al. (Eds.): *IE&EM 2019*, pp. 145–152, 2020.
https://doi.org/10.1007/978-981-15-4530-6_15

investment costs often affects the platform to the level of service quality of the existing research is to assume that the quality and cost of investment in a quadratic function relation (Demirhan et al. 2007). Due to the rapid development of information technology, the cost of IT investment over time deferred and a trend of rapid decline is a significant feature of IT investment, important influence on enterprise investment strategy. Demirhan et al. (2006) studied IT costs decline timing penetration into the market two oligopoly enterprise IT investment strategy, found that after the entrants can through low cost IT investment improve the quality of services beyond the first-mover. Xie and Li (2004) studied the investment strategy of e-commerce websites from the perspective of IT investment cost reduction, and the research results pointed out that when the later-in-market makes low-cost IT investment, the quality of the website can be improved to attract more users, considering the time sequence and technology of enterprise scale entering the market. Level of management level differences, asymmetry between B2B e-intermediaries are the important influencing factors of enterprise IT investment strategy, in the asymmetric duopoly model, the current scholars focus on the cost of asymmetric impact on enterprise technology innovation investment strategy. Pawlina and Kort (2006) analyzed the asymmetry of two enterprises investment cost, the choice of investment strategy in the results show that when the asymmetric cost level within a certain range, with a cost disadvantage enterprise by increasing investment to improve their income and reduce a competitor's earnings. Yang and Da (2005) researched that the asymmetric cost impact on enterprise technology innovation investment timing by the option game method, they found that the equilibrium results depends on the cost of the asymmetric degree of the existing IT investments decisions. Most studies only consider investing and not investing in both cases, ignore the possibility of deferred investment. However, in the real investment environment of electronic commerce, an information technology investment will continue for a period of time. Accordingly, B2B e-intermediary enterprises can choose the best investment time according to the changes of market demand, technology level and investment cost during this period of time. Therefore, it is of great significance to research on B2B platform technology investment decisions in multiple periods.

Three basic characteristics of technological innovation investment: irreversibility of investment cost, uncertain future returns and flexibility in investment decisions. The concept of real option was first proposed by Mayers (1977), and further developed by Brennan and Schwartz (1985) and McDonald and Siegel (1986). Real option has gradually become an important method of the research of technology innovation investment, and it admits that the value of management flexibility and information, compared with the traditional investment evaluation methods more in line with the actual situation of technology innovation investment. B2B e-intermediary industry is highly competitive, and the investment decisions of enterprises are not only affected by the uncertainty of the project itself and the rationality of the decision, but also affected by the investment behaviors of other enterprises. Considering the market competitive behavior, scholars gradually game analysis method into the real option theory (Pawlina and Kort 2006; Michele 2008), competition for enterprise investment decision analysis to provide more scientific theoretical guidance.

Considering two oligopoly B2B platform, we combined real option theory with game theory, discussed the information technology investment timing and the selection of investment strategy to reduce the B2B electronic intermediary companies in the uncertainty of technology investment, make up the lack of experience decision-making, improve the investment decision-making scientific nature and effectiveness of the enterprise operator.

2 Model

Assume that there are two B2B e-intermediaries in the market that are risk-neutral and pursue profit maximization, and they have the same IT investment opportunity to improve the service quality of their B2B platform. Assume that this new technology is inseparable and only requires one-time investment, but the investment opportunity can last n periods.

IT investment cost of B2B platform includes fixed cost of new technology investment and operation cost of platform maintenance due to quality improvement. As IT costs decrease over time, fixed costs are affected by the service quality level and investment time to be achieved by the new platform. The operation cost is related to the enterprise's own size and other factors.

Assumption 1: For technology investment projects in the same period, the higher the service quality of the new platform, the higher the investment cost. If the platform decides to invest in the technology to achieve a service quality of λ, the corresponding fixed investment cost can be described as $I = \lambda^2/2$.

Assumption 2: For investment projects of the same technical level, let I_0 be cost of investment in the t_0 period. The cost of investment in the t period can be described as $I_t = I_0(vt+1)^{-1}$, where v is the rate at which IT costs fall over time.

Assumption 3: The operating cost of the platform increased due to the investment in new technologies can be expressed as $C_i(q_i) = c_i q_i$ (I = A, B), where c_i as the marginal cost, represents the increase in the operating cost of B2B platform for each additional unit of users, and q_i is the number (capacity) of users after the upgrading of B2B platform. Due to differences in technology level, management level and enterprise strength, the operating costs of platform A and platform B are asymmetric, i.e. $c_A \neq c_B$.

Based on the above assumptions, it can be further concluded that the expected revenue of the intermediary in each period:

$$\pi_i(q_A, q_B) = P(\Theta_t, (q_A + q_B))q_i - C_i, \quad i = A, B$$

Where $P(\Theta_t, Q) = \Theta_t - b(q_A + q_B)$ is inverse demand function, Θ_t is a random demand drift parameter, which represents the uncertainty of market demand and follows a continuous log-normal distribution process. Parameter b measures elasticity of demand, which inversely relates to the number of users.

If the new intermediary platform continues to operate for $n - t + 1$ years, then the expected total return of B2B e-intermediary on this technology investment is $\Pi_i = \sum_t^n \frac{\pi_i}{(1+k)^t} - I_i$, $i = A, B$, where k is discount rate.

3 Single Period Investment Strategy Analysis

To maximize their expected returns, intermediaries A and B need to decide whether to invest or delay in each period, and if an electronic intermediary decides to invest in the technology during A certain period, it also needs to decide the user capacity of the new product. Therefore, there is room for strategy for each e-intermediary $S_{i,t} = (I, D\,; q_i|I)$, $i = A, B$. Until the investment or project is completed. The following paper presents the equilibrium results of simultaneous investment and sequential investment of platform A and B in A single period decision, and analyzes the impact of the marginal operating cost of the platform on the optimal number of users and the equilibrium revenue.

Proposition 1: Under the balance of investment strategy at the same time, the optimal number of users of two B2B platforms is

$$\begin{cases} q_A^{*1} = \dfrac{1}{3b}(\Theta_t - 2c_A + c_B) \\ q_B^{*1} = \dfrac{1}{3b}(\Theta_t + c_A - 2c_B) \end{cases} \tag{1}$$

The equilibrium revenue of each B2B e-intermediary is

$$\begin{cases} \pi_A^{*1} = \dfrac{1}{9b}(\Theta_t - 2c_A + c_B)^2 \\ \pi_B^{*1} = \dfrac{1}{9b}(\Theta_t - 2c_B + c_A)^2 \end{cases} \tag{2}$$

According to Eqs. (1) and (2), when simultaneously investing, the optimal number of users and platform revenue of either platform are not only affected by market demand and its own marginal cost, but also by the marginal cost of competitors.

Corollary 1: The optimal user capacity of B2B platforms decreases with the increase of their own marginal operating costs and increases with the increase of the marginal operating costs of rival platforms.

Corollary 1 shows B2B electronic intermediary companies in the investment decision-making when considering their rival between enterprises of the differences in terms of size, strength, and to infer the opponent platform operating costs, on the basis of this, choose this enterprise B2B platform can achieve the user capacity and platform of IT investment, if the relative to other competitors have high marginal costs, enterprises will choose smaller user capacity, on the contrary, will choose the larger user capacity of IT investments on B2B platform. In the B2B electronic intermediary management practice, we observe the actual situation of consistent with Corollary 1.

Proposition 2: The optimal user capacity of two B2B platforms is respectively

$$\begin{cases} q_A^{*2} = \frac{1}{2b}(\Theta_{t_2} - 2c_A + c_B) \\ q_B^{*2} = \frac{1}{4b}(\Theta_{t_2} + 2c_A - 3c_B) \end{cases} \tag{3}$$

The equilibrium profits of the two intermediary enterprises are respectively

$$\begin{cases} \pi_A^{*2} = \frac{1}{8b}(\Theta_{t_2} - 2c_A + c_B)^2 \\ \pi_B^{*2} = \frac{1}{16b}(\Theta_{t_2} + 2c_A - 3c_B)^2 \end{cases} \quad (4)$$

It can be seen from Eqs. (3) and (4) that, similar to the situation of simultaneous investment, the optimal number of users and platform revenue of either platform in sequential investment are not only affected by market demand and its own marginal cost, but also by the marginal cost of competitors.

Corollary 2: The optimal number of users selected under the sequential investment of two platforms is a decreasing function of their own marginal operating costs and an increasing function of competitors' marginal operating costs.

Corollary 3: The optimal number of users and equilibrium income of the invested B2B platform are larger than the optimal number of users and equilibrium income of the invested B2B platform, and B2B e-intermediary enterprises have the incentive to invest in advance.

Corollary 3 shows that compared with the condition of simultaneous investment, leading B2B e-intermediary enterprises can choose larger user capacity and obtain higher investment returns. Therefore, enterprises have the incentive to invest in new technologies to improve platform quality and expand user scale. Seize the chance to beat to invest in new technology to gain high monopoly profits, can make the enterprise get a competitive advantage and quickly from many market competitors. Because there is a strong incentive for B2B e-intermediaries to invest first, each B2B e-intermediary makes IT investment at the cost of loss. As a result, a new technology soon becomes the standard match of each B2B platform, which is also the reason for serious homogeneous competition in B2B market at the present stage.

4 Multi-period Investment Analysis

If the final intermediary B decides to invest in the technology that can last for n years with intermediary A, the expected income of intermediary A and B is respectively

$$\begin{cases} \Pi_A^1 = \sum_{t=t_1}^{n} \frac{\pi_A^{*1}}{(1+k)^t} - I_A^1 = \sum_{t=t_1}^{n} \frac{(\Theta_t - 2c_A + c_B)^2}{9b \cdot (1+k)^t} - \frac{\lambda^2}{2(\upsilon t_1 + 1)} \\ \Pi_B^1 = \sum_{t=t_1}^{n} \frac{\pi_B^{*1}}{(1+k)^t} - I_B^1 = \sum_{t=t_1}^{n} \frac{(\Theta_t - 2c_B + c_A)^2}{9b \cdot (1+k)^t} - \frac{\lambda^2}{2(\upsilon t_1 + 1)} \end{cases} \quad (5)$$

If intermediary B decides to extend the investment in the technology operated for n years and exercise the option in t_2 period, then the expected income of intermediary A and B is respectively

$$\begin{cases} \Pi_A^2 = \sum_{t=t_1}^{n} \frac{\pi_A^{*2}}{(1+k)^t} - I_A^2 = \sum_{t=t_1}^{n} \frac{(\Theta_{t_2} - 2c_A + c_B)^2}{8b \cdot (1+k)^t} - \frac{\lambda_A^2}{2(\upsilon t_1 + 1)} \\ \Pi_B^2 = \sum_{t=t_2}^{n} \frac{\pi_B^{*2}}{(1+k)^t} - I_B^2 = \sum_{t=t_2}^{n} \frac{(\Theta_{t_2} + 2c_A - 3c_B)^2}{16b \cdot (1+k)^t} - \frac{\lambda_B^2}{2(\upsilon t_2 + 1)} \end{cases} \quad (6)$$

From Eqs. (5) and (6), IT investment projects of B2B expected revenue is related to marginal operating cost, platform service level, rate of decline of information technology cost with time, investment opportunity and project operation years. Due to the complexity of function form, the following numerical examples will be further used to analyze the influence of each parameter on investment timing, investment strategy selection and total return.

5 Result

For B2B e-intermediaries investing in technology, market competition brings advantages to the pioneers of investment, so each intermediary has the incentive to invest first.

When $t = 2$, that is, in the second phase of the investment, intermediary A predicts that the expected return in this phase is higher than that in each subsequent phase, whether it invests the project simultaneously with intermediary B or in sequence, so intermediary A decides to exercise the option in this phase. Intermediary B decides to exercise the option in the second period when it expects that the expected income obtained from the simultaneous investment with intermediary A is higher than that of each subsequent period. In the opinion of intermediary A, if the option is exercised in the second period, the expected return of the pre-investment is 49.98 million yuan higher than the expected return of the simultaneous investment is 44.42 million yuan, so the incentive of pre-investment is generated. From the perspective of intermediary B, if the option is exercised in the second period, the pioneer advantage and follower disadvantage of the market are very obvious, so there is an incentive to invest in advance. Therefore, the final investment decision is that B2B electronic intermediary enterprises A and B invest in this information technology at the same time in the second phase, so as to achieve an equilibrium.

For B2B e-intermediaries investing in technology, the choice of investment timing affects their investment returns, and delaying investment may yield more returns.

If he does not consider delay investment option value, that is, intermediary A and B only at the beginning of the project the first phase of investment opportunities, so, because of the competition in the market, agents A and B are realized the first mover advantage and follower disadvantage, also has A preemptive investment motives, the final investment decision and investment at the same time, the expected return for 18.03 million, the earnings significantly less than consider the balance of the delay option income 44.42 million, so this research adopts multiple period investment decision model based on the theory of option games to A certain extent, improved the traditional investment decision model is a single period, This point has important guiding significance to the investment decision in the real market.

6 Conclusion

Based on the option game theory, this paper constructs the information technology (IT) investment decision model of asymmetric B2B platform under uncertain market demand, and obtains the optimal investment strategy of B2B e-intermediary enterprises under the condition of simultaneous investment equilibrium and sequential investment equilibrium by using the game analysis method. Different from the previous research on technology innovation investment decision-making of e-commerce, this paper considers investment as an executable or deferred option on the basis of considering the competitive nature of investment, and analyzes the decision-making process of technology investment projects of B2B e-intermediary enterprises in various periods. The result shows that the market competition will lead B2B e-intermediary enterprises to invest in advance, so as to gain more market share and expected investment income. In a single period, the user capacity selected at the beginning of B2B platform investment is inversely proportional to its own marginal operating cost, and directly proportional to the marginal operating cost of competitors' platforms. The competition between platforms makes both parties have the incentive to invest in advance. Due to uncertain market demand, appropriate delay investment strategy can bring higher returns to intermediary enterprises. In the further research, we could consider the B2B platform IT investment strategy under incomplete information closer to practice, and the research on investment decisions in multiple periods can adopt dynamic simulation technology to simulate the decision-making process of competitors in a more intuitive and reliable way.

Acknowledgment. This research is supported by the National Natural Science Foundation of China (No. 71771122).

References

Devaraj, S., Kohli, R.: Information technology payoff in the health-care industry: a longitudinal study. J. Manag. Inf. Syst. **16**(4), 41–68 (2000)

Thatcher, M., Oliver, J.: The impact of technology investments on a firm's production efficiency, product quality, and productivity. J. Manag. Inf. Sys. **18**(2), 17–46 (2001)

Thatcher, M., Pingry, D.: Understanding the business value of information technology investments: theoretical evidence from alternative market and cost structures. J. Manag. Inf. Syst. **21**(2), 61–85 (2004)

Demirhan, D., Jacob, V., Raghunathan, S.: Strategic IT investments: the impact of switching cost and declining IT cost. Manag. Sci. **53**(2), 208–226 (2007)

Demirhan, D., Jacob, V., Raghunathan, S.: Information technology investment strategies under declining technology cost. J. Manag. Inf. Syst. **22**(3), 321–350 (2006)

Xie, Z.X., Li, L.: IT investment strategy of a duopoly with sequential entry for B2B e-commerce website. Chin. J. Manag. Sci. **22**(2), 56–64 (2004)

Pawlina, G., Kort, P.: Real options in an asymmetric duopoly: who benefits from your competitive disadvantage? J. Econ. Manag. Strat. **15**(1), 1–35 (2006)

Yang, Y., Da, Q.L.: Study on investment in technology innovation of asymmetric duopoly. Chin. J. Manag. Sci. **13**(4), 95–99 (2005)

Myers, S.: Determinants of corporate borrowing. J. Financ. Econ. **5**(2), 147–175 (1977)

Brennan, M., Schwartz, E.: Evaluating natural resource investments. J. Bus. **58**(2), 135–157 (1985)

McDonald, R., Siegel, D.: The value of waiting to invest. Q. J. Econ. **101**(4), 707–728 (1986)

Michele, M.: Competition and irreversible investment under uncertainty. Inf. Econ. Policy **20**(1), 75–88 (2008)

Research on Path Planning of Mobile Robot Based on Improved A* Algorithm

Xiu-hong Wang, Xue-hao Liu(✉), Yong-cheng Wang, and Shuai-peng Liang

School of Management Engineering, Zhengzhou University of Aeronautics School, Zhengzhou, China
liu781899823@qq.com

Abstract. By analyzing the shortcomings of Manhattan distance and Euclidean distance algorithm commonly applied in A* algorithm, this paper is to theoretically solve the traditional A* algorithm with problems, imprecise search results and long response time when being applied in path planning process, as well as improving and obtaining the complex diagonal distance with the heuristic function distance to simulate the path planning. Meanwhile, it proposes the eight-neighbor search optimization with the path planning simulation of the improved A* algorithm. The simulated results indicate that the improved A* algorithm can achieve the ideal effect of reducing and improving the search rate to finally locate the optimal path with precise search results and prompt response.

Keywords: A* algorithm · Mobile robot · Path planning

1 Introduction

In recent years, with the continuous development of science and technology and computer level, the wave of scientific and technological revolution with intelligence as the core has swept the world. As an intelligent product, mobile robots have been rapidly developed under technological innovation and market demand, and are widely used in manufacturing, medical services, military and other fields, effectively improving people's production and lifestyle. Since path planning is the key technology to realize intelligent control of mobile robots, mobile robot path planning has become a hot research topic for scholars.

Path planning refers to finding an optimal or sub-optimal collision-free path from the starting node to the target node according to an evaluation index (such as time, distance, search efficiency, etc.) in the working environment with obstacles. Mobile robots can safely pass all obstacles in a complex work environment [1, 2]. Common path planning algorithms are Dijkstra algorithm [3], A* algorithm [4], Artificial potential field method [5], Fuzzy logic method [6], Genetic algorithm [7], Ant colony algorithm [8], etc. Among them, A* algorithm has the advantages of high search efficiency and short planning path in global path planning, and is widely used in mobile robot path planning.

C.-F. Chien et al. (Eds.): *IE&EM 2019*, pp. 153–161, 2020.
https://doi.org/10.1007/978-981-15-4530-6_16

At present, most scholars study mobile robot path planning mainly focusing on algorithm improvement, and make different improvements for different application requirements. In [9], Wang solved the problem of redundant points in the traditional A* algorithm path search and found the minimum corner at the path turning point. In [10], Chen improved the obstacle search method and heuristic function in the traditional A* algorithm, improved the path-finding speed of the path search, and narrowed the search range. In [11], Lu reduced the number of path search nodes by adding a threshold to the open list, and optimized the planned path to obtain a smooth path.

The rest of this article is organized as follows. In Sect. 2, introduce the basic principles and algorithm flow of the traditional A* algorithm. In Sect. 3, according to the shortcomings of traditional A* algorithm path planning, the optimization of heuristic function distance algorithm and path planning search neighborhood is improved. In Sect. 4, simulation experiment analysis for improved A* algorithm. In Sect. 5, introduce the conclusion of the paper.

2 Traditional A*Algorithm

The A* algorithm is a typical heuristic search algorithm, first proposed by Hart [12] in 1968. The algorithm combines the advantages of both Dijkstra algorithm and best-first search algorithm. By adding heuristic function in Dijkstra algorithm to determine the search direction, it solves the problems of large path search range and long response time, and has high search efficiency [13].

Cost function expression of A* algorithm:

$$F(\mathrm{n}) = G(\mathrm{n}) + H(\mathrm{n}) \tag{1}$$

In the formula (1), $F(\mathrm{n})$ is the value of the value from the starting node to the target node through the current node. $G(\mathrm{n})$ is the actual value of the starting node to the current node. $H(\mathrm{n})$ is the estimated cost of the current node to the target node.

When using the A* algorithm for path search. Firstly, the extended node that can be reached except the obstacle point around the starting node is added to the open list, and the value of all the nodes in the open list is calculated. The node with the smallest F value is added to the close list, which is recorded as the parent node, and as the starting node for the next search. Secondly, continue to calculate until the search target node stops, thus getting all path nodes. Finally, the parent node returns the starting node from the target node, and sequentially connects all the passed nodes to obtain the planned path. The open list is used to store extended nodes in the search process, and the close list is used to store the barrier points and the nodes with the lowest value F in the search process.

The flow chart of the A* algorithm is shown in Fig. 1.

Using A* algorithm for path planning, the path search process is shown in Fig. 2. The side of the square is 10, The *F*, *G*, *H* values are calculated using the Manhattan distance formula. The red star indicates the starting point position, the green star indicates the target point position, the black square indicates the obstacle point position, and the red solid circle indicates the planned path node position. The values in the red

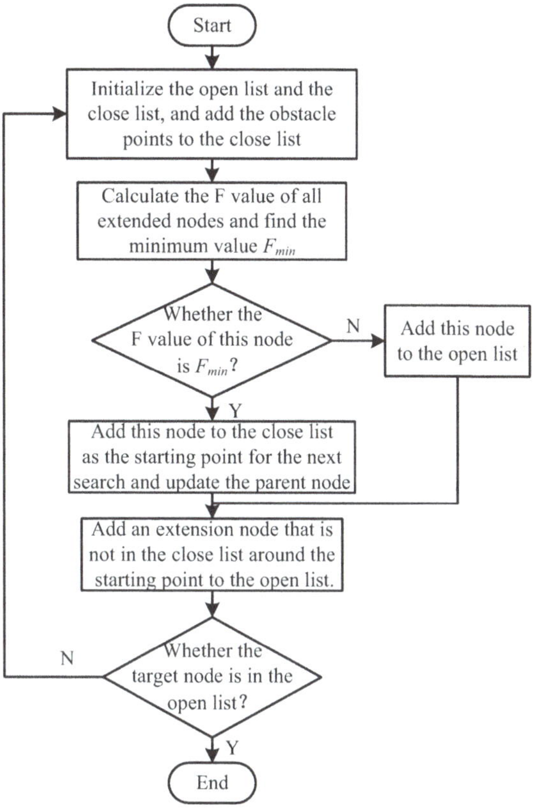

Fig. 1. A* algorithm flow chart

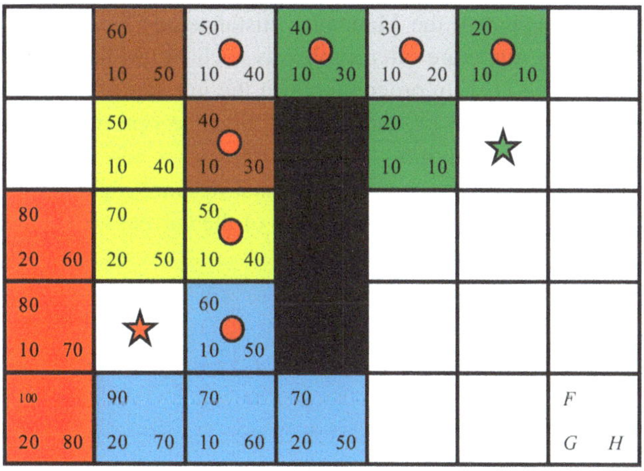

Fig. 2. A* algorithm path search

square are the *F*, *G*, *H* values obtained in the first search, and the values in the blue square are the *F*, *G*, *H* values obtained in the second search, and the latter search will overwrite the value obtained in the previous search. The values in the yellow, brown, gray, and green squares represent the values obtained by the third, fourth, fifth, and sixth searches.

3 Improved A* Algorithm

The traditional A* algorithm planning path has many problems such as many turning points, large search range, and long response time [14]. In order to optimize the planning path of the A* algorithm and optimize the planning path, the traditional A* algorithm is improved by optimizing the heuristic function distance algorithm and the path planning search neighborhood.

3.1 Heuristic Function Distance Algorithm

In the traditional A* algorithm cost function, the commonly used distance algorithm of the heuristic function *H*(n) is the Manhattan distance and Euclidean distance algorithm. Suppose the starting node coordinate is $S(S_x, S_y)$, the current node coordinate is $N(N_x, N_y)$, and the target node coordinate is $T(T_x, T_y)$.

The Manhattan distance formula:

$$H(\mathrm{n}) = |N_x - T_x| + |N_y - T_y| \tag{2}$$

The Euclidean distance formula:

$$H(\mathrm{n}) = \sqrt{(N_x - T_x)^2 + (N_y - T_y)^2} \tag{3}$$

Because mobile robots use the Manhattan distance and Euclidean distance algorithm for path planning, there are often problems such as too many search nodes and too long calculation time [15]. In order to reduce the number of search nodes for path planning, narrow the search scope, and quickly find the optimal planning path. The heuristic function distance algorithm is improved, and a new distance algorithm is proposed, which is recorded as a complex diagonal distance algorithm. The algorithm considers the moving distance and moving steps of the current node to the target node, and performs weighting calculation.

The complex diagonal distance formula:

$$H(\mathrm{n}) = \left(\sqrt{2} - 2\right) \times \min\{|N_x - T_x|, |N_y - T_y|\} + (|N_x - T_x| + |N_y - T_y|)$$

3.2 Path Planning Search Neighborhood

In the raster map, if the edges are calculated adjacent to each other, each grid is surrounded by 4 grids, as shown in Fig. 3(a). If the edge and the top corner are adjacent, each grid is surrounded by 8 grids, as shown in Fig. 3(b). Therefore, the path planning search neighborhood is divided into two types, four-neighbor search and eight-neighbor search. The light green fill in the raster map indicates the neighborhood search range, and the red arrow indicates the planned path direction.

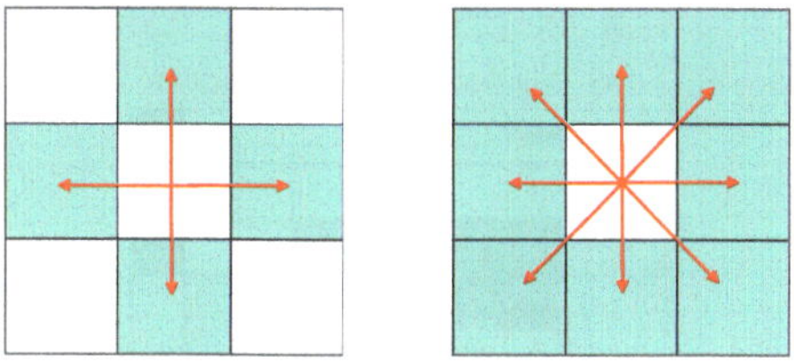

(a) Four neighborhood search (b) Eight neighborhood search

Fig. 3. Path planning search type

At present, four-neighbor search is widely applied to the path planning of the A* algorithm, but when the A* algorithm performs path planning search, the number of extended nodes affects the path search rate. The eight-neighbor search is doubled compared with the extended node of the four-neighbor search, which can improve the path planning search rate and ensure that the planned path is the optimal path.

4 Simulation Experiment and Result Analysis

In order to compare the optimization degree of the improved A* algorithm heuristic function distance algorithm, This paper constructs a 10 × 10 raster map environment model in which obstacle points are randomly distributed. Assuming the same starting point and target point, the heuristic function distance algorithm performs path planning simulation according to Manhattan distance, Euclidean distance and improved complex diagonal distance. The simulation results are shown in Fig. 4. The light green fill in the raster map represents the search node for the path plan.

From the simulation data of Table 1, it can be seen that the complex diagonal distance algorithm reduces the number of search nodes and path nodes by 30%, the path length by 20%, and the calculation time by 45%. Therefore, the mobile robot uses the complex diagonal distance algorithm for path planning, which can achieve the ideal effect of narrowing the search range, reducing the number of path nodes, shortening the path length and increasing the calculation rate.

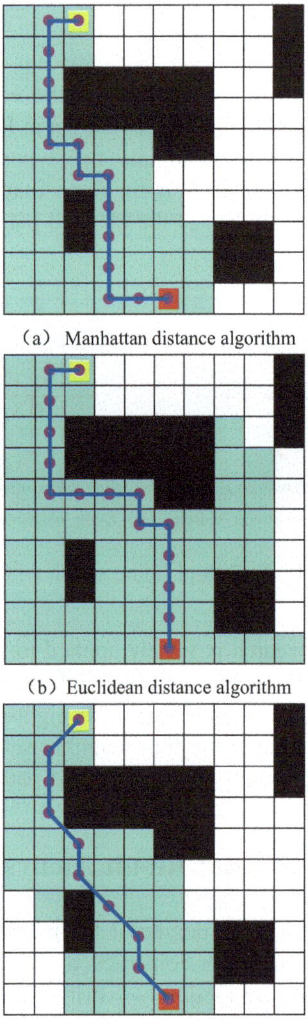

(a) Manhattan distance algorithm

(b) Euclidean distance algorithm

(c) Complex diagonal distance algorithm

Fig. 4. Mobile robot path planning simulation diagram

Table 1. Mobile robot path planning simulation data

Distance algorithm	Search node	Path node	Path length	Calculating time
Manhattan distance	45	15	14	18.94
Euclidean distance	60	15	14	26.71
Complex diagonal distance	39	10	11	12.48

In order to compare the superiority of A* algorithm in path planning under four-neighbor search and eight-neighbor search, select the same starting point and target point, use the improved complex diagonal distance algorithm to calculate, and set the accessibility environment as a comparison. To make the simulation comparison result of path planning more accurate, the simulation results are shown in Figs. 5 and 6. The light green fill portion of the raster map represents the search node for the path plan.

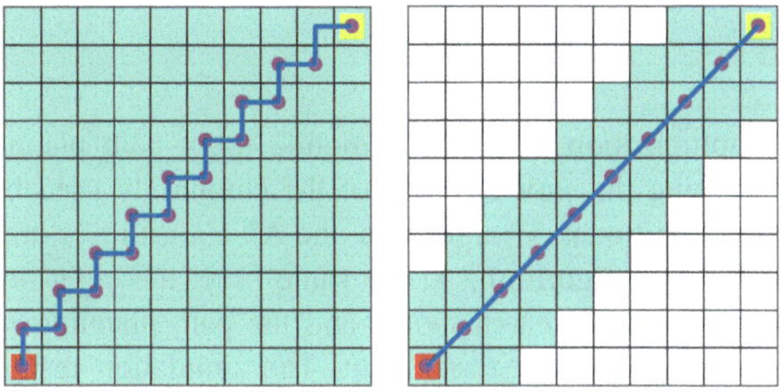

(a) Four neighborhood search (b) Eight neighborhood search

Fig. 5. Path search in an accessible environment

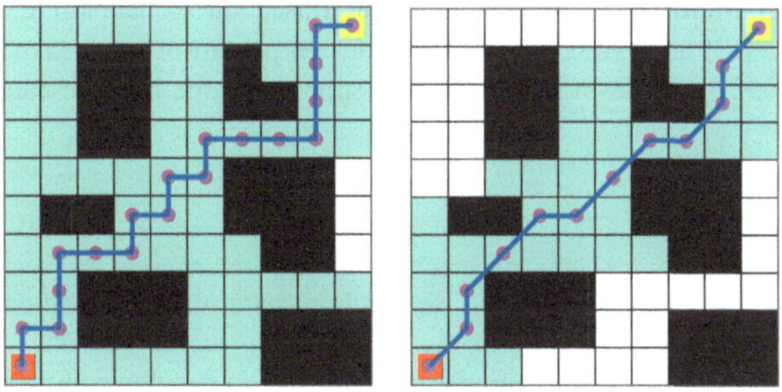

(a) Four neighborhood search (b) Eight neighborhood search

Fig. 6. Path search in an obstacle environment

From the simulation data of Table 2, it can be seen that in the barrier-free environment, the number of nodes and the path length of the eight-neighbor search are better than the four-neighbor search, in which the number of search nodes is reduced by nearly 1/2 and the path length is shortened by nearly 1/3. Therefore, the eight-neighbor search for path planning can obtain an optimal path with a small search range and a short path length.

Table 2. Four-neighbor and eight-neighbor search of A* algorithm

Obstacle environment	Search node		Path length	
	Four neighborhood	Eight neighborhood	Four neighborhood	Eight neighborhood
No	100	44	18	12.73
Yes	66	40	18	13.9

5 Conclusion

In this paper, the optimization research of mobile robot path planning algorithm is carried out. By analyzing the shortcomings of the commonly used heuristic function distance algorithm in traditional A* algorithm, the A* algorithm is improved to get the complex diagonal distance algorithm. At the same time, the eight-neighbor search is proposed to optimize the path search mode, and the path planning simulation of the mobile robot is carried out in the raster map. The simulation results show that the improved A* algorithm can narrow the search range, reduce the number of path nodes, shorten the path length, solve the problem of too many search nodes and long calculation time in the path planning process, and the planned path is the optimal path.

Acknowledgment. Thanks to the Key Research Projects of Henan Higher Education Institutions "Research on Multi-UAV Mission Planning Based on Collaborative and Autonomous Decision" (19A413012) and "Research on scene design pattern of interactive data driven UAV product" (20A630035) provides research funding support for the basic research of this paper.

References

1. Qu, D.K., Du, Z.J., Xu, D.G., Xu, F.: Research on path planning method of mobile robot. Robot **30**(2), 97–101+106 (2008)
2. Dai, B., Xiao, X.M., Cai, Z.X.: Current status and future development of mobile robot path planning technology. Control. Eng. **12**(3), 198–202 (2005)
3. Wang, S.X.: The improved Dijkstra's shortest path algorithm and its application. Procedia Eng. **29**, 1186–1190 (2012)
4. Duchoň, F., Babinec, A.: Path planning with modified a star algorithm for a mobile robot. Procedia Eng. **96**, 59–69 (2014)
5. Yu, Z.Z., Yan, J.H., Zhao, J.: Mobile robot path planning based on improved artificial potential field method. J. Harbin Inst. Technol. **43**(1), 50–55 (2011)
6. Su, Z.B., Lu, J.L.: A study on the path planning of mobile robot with the fuzzy logic method. Trans. Beijing Inst. Technol. **23**(3), 290–293 (2003)
7. Wang, G.L., Wang, H.C., Li, Z.Y., Li, J.P.: Path planning for mobile robots based on optimized genetic algorithm. Mach. Tool Hydraul. **47**(3), 37–40 (2008)
8. Yang, G.H., Fu, S.B., Liu, Z.Y., Wang, J.L.: Path planning of indoor mobile robot based on improved ant colony algorithm. Sci. Technol. Eng. **19**(19), 1671–1815 (2019)
9. Wang, D.J.: Indoor mobile-robot path planning base on an improved A* algorithm. J. Tsinghua Univ. (Sci. Technol.) **52**(8), 1085–1089 (2012)

10. Chen, R.N., Wen, C.C., Peng, L., You, Z.C.: Application of improved A* algorithm in indoor path planning for mobile robot. J. Comput. Appl. **39**(4), 1006–1011 (2019)
11. Lu, W.L., Lei, J.S., Shao, Y.: Path planning for mobile robot based on an improved A* algorithm. J. Ordnance Equip. Eng. **40**(4), 197–201 (2019)
12. Hart, P.E., Nilsson, N.J., Raphael, B.: A formal basis for the heuristic determination of minimum cost paths in graphs. IEEE Trans. Syst. Sci. Cybern. **4**(2), 100–107 (1968)
13. Fox, D., Burgard, W., Thrun, S.: The dynamic window approach to collision avoidance. IEEE Robot. Autom. Mag. **4**(1), 23–33 (1997)
14. Wang, Z.Z.: Automatic robot path planning under complicit dynamic environment. Modul. Mach. Tool Autom. Manuf. Tech. (1) 64–68 (2018)
15. Zhou, F.F., Gao, S.Y., Liu, K.: Research on optimal path planning of mobile robot based on simulation. J. QINGDAO Univ. (E&T) **33**(3), 47–51 (2008)

Supply Chain and Scheduling

Construction of Emergency Material Disturbance Management Model Considering Demand Change

Cheng-wu Fang, Jing-jing Chen(✉), and Fu-yu Wang

School of Management Science and Engineering,
Anhui University of Technology, Maanshan, China
1943787189@qq.com

Abstract. Emergency material dispatching plays an important role in rescue operations. Because of many uncertainties in the objective world, emergency materials will be disturbed by many uncertain events in the distribution process, and the initial distribution plan will be affected. Based on the interference management of emergency material dispatching, this paper combs and discusses the forecast of material demand, the classification of material demand points, and the interference management of emergency material dispatching. And an emergency material disturbance management model considering the change of demand is constructed. Finally, the future research is prospected.

Keywords: Disruption management · Model analysis · Material demand · Emergency material dispatching · Optimal analysis

1 Introduction

In recent years, unexpected disasters have occurred frequently in the world, which has brought great property losses and casualties to people. After disasters occur, emergency materials need to be quickly and effectively distributed to the disaster site. Due to the uncertainty of the objective world, emergency materials are often disturbed by many uncertain events in the distribution process, which makes the original distribution plan affected or even impossible. With the progress of technology and the improvement in human abilities, people begin to take active measures to reduce and eliminate the impact on uncertain events, which results in interference management.

According to the definition of interference management according to Yu and Qi [1], Interference management needs to establish corresponding optimization models and effective solving methods for various practical problems and the nature of interference events, and quickly give an optimal adjustment plan for dealing with interference events. The adjustment plans are based on this state to quickly generate system disturbances. The smallest adjustment plan, although also considering cost savings, is often not the most cost-effective option. How to make full use of the idea of interference management to minimize the disturbance to the original scheme becomes very important.

C.-F. Chien et al. (Eds.): *IE&EM 2019*, pp. 165–171, 2020.
https://doi.org/10.1007/978-981-15-4530-6_17

2 Research Status of Emergency Material Scheduling and Disturbance Management Considering Material Demand

2.1 Research Status of Emergency Material Scheduling Considering Material Demand Forecasting

After natural disasters occur, it is of great significance to study the post-disaster emergency material dispatch considering material demand forecasting. Xu et al. [2] adopt a hybrid forecasting method which combines empirical mode decomposition with autoregressive integral moving average. Sun et al. [3] considered two fields of emergency material demand to forecast under the framework of rough set theory, and established two fields of fuzzy rough set model. Wang [4] describes the attribute characteristics of post-earthquake secondary disaster events by using multi-case analysis scheme, and establishes a Petri model combined with Markov chain.

2.2 Research Status of Emergency Material Scheduling Considering Material Demand Classification

Disaster itself has great uncertainty and unexpectedness, so considering the classification of material needs is very helpful to improve the efficiency of emergency rescue. Ling et al. [5] establish an emergency medical resource allocation model with the objective function of minimizing the total transportation distance. Wang et al. [6] apply fuzzy comprehensive evaluation method to the corresponding demand urgency level and demand urgency value for different emergency materials. In order to improve the efficiency of emergency material transportation, Cheng [7] takes the first transportation time and the total twice transportation time as two objective functions.

2.3 Research Status of Emergency Material Dispatching Disturbance Management

(1) Analysis of Single-objective Model
The shortest time consuming usually refers to the shortest time consumed in the whole material distribution process, which is also the primary factor for many interference management considerations. Zhao et al. [8] constructed a disturbance management model on the minimum generalized total deviation cost as the objective function. Yang et al. [9] established an interference management model with the objective function of the minimum of the customer time window violation, the total vehicle travel time and the new vehicle preparation time. Li et al. [10] established an operational model that minimizes the sum of operating costs, scheduling interference costs, trip cancellation costs, and delay penalty costs as objective functions.

(2) Analysis of Multi-objective Model
In addition to the shortest time-consuming and the lowest cost, some literatures also consider the shortest route and the least time deviation, which constitute a

multi-objective model. Ruan [11] and others put forward the method of helicopter and vehicle joint transportation for material transportation. Wang et al. [12] identified and analyzed the VRPB problem of changing demand of picking-up customer points. Wang et al. [13] established a mathematical model of the interference problem with the goal of minimizing the deviation from the customer's time window and minimizing the distribution cost. Zhang et al. [14] established a multi-objective model to measure the original problem target and time window deviation from the delay problem.

(3) Model considering demand satisfaction and disaster victim satisfaction
Demand satisfaction and disaster victims satisfaction are very important indicators in the research of emergency material dispatching interference management. They are taken into account in the objective function in the following literature. Ding [15] constructed a dictionary order multi-objective interference management model with the change of customer time window as the interference factor. Liu [17] combines human behavioral science to analyze the interference events from three aspects: the needs of victims, emergency decision-making departments and emergency logistics executives. Zhu [16] and others set up a two-stage relief material dispatching model based on the dynamic characteristics of emergency material demand. Cao et al. [18] established an interference management model considering customer value in order to effectively allocate resources to customers in logistics enterprises.

(4) Model optimization analysis

(1) Heuristic algorithm
Heuristic algorithms widely used in the problem of disturbance management of vehicle scheduling in emergency supplies distribution mainly include genetic algorithm, particle swarm algorithm, ant colony algorithm and so on. After considering the complexity of the model, Yang [9] proposed a knowledge-based heuristic algorithm to solve the model. Wang [12] and others proposed a heuristic algorithm based on two strategies to solve the interference problem. Ruan et al. [11] designed an improved genetic algorithm for the interference model of joint transport of emergency medical supplies aiming at the change of transit point. Cao [18] and others designed an improved particle swarm optimization (PSO) algorithm to solve the logistics distribution problem considering different values. Ding [15] and others considered the impact of interference events on the whole logistics distribution system, and proposed an improved ant colony algorithm - hybrid ant colony algorithm. Liu [17] and others proposed a hybrid ant colony algorithm to solve the interference model in order to deal with the interference events effectively. Li et al. [10] proposed a relaxation algorithm based on Lagrange.

(2) Multi-stage and hybrid algorithms
Zhu et al. [16] proposed a two-stage emergency resource scheduling model. In the first stage, branch and bound algorithm was adopted, and in the second stage, fast non-dominant sequencing genetic algorithm with elite strategy was adopted. Wang et al. [13] proposed the method of route enumeration and route selection to solve the distribution vehicle scheduling interference problem with

travel time delay. Zhang et al. [14] designed a hybrid algorithm combining ant colony algorithm with discrete search algorithm for multi-objective model.

3 Construction of Disturbance Management Model Considering the Change of Demand

3.1 Model Assumptions

(1) In the early post-earthquake period, the local emergency material distribution center is an important supply point of emergency rescue, without considering the transfer from other places, and the total amount of materials in the local emergency material distribution center are in short supply.
(2) An emergency material distribution center is allowed to supply materials to multiple demand points, and one demand point can receive materials from multiple distribution centers.
(3) The number of rescue vehicles for the local emergency distribution center is limited.
(4) Vehicles carry the same cargo and the same load capacity.
(5) The demand point where the vehicle is serving for the time of interference is regarded as the completed point.
(6) The supply of emergency materials distribution centers is known, and the demand of each disaster site may change from the early post-earthquake period.

3.2 Problem Description

When an earthquake occurs to a certain area, several disaster spots occur. How to distribute emergency materials reasonably in the early period after the disaster without external material assistance, so that the disturbance of path and cost is minimal under the constraints on shortage of supply and demand, the change of demand of the disaster site and a certain time.

3.3 Definitions of Parameters and Variables

K: remaining rescue vehicles assemble when interference occurs, $K = \{k \mid k = 1, 2, \ldots\ldots, n\}$; N: number of unserved disaster sites when interference occurs; L: number of vehicles in transit when interference occurs; R: local emergency material distribution center, $R = \{r \mid r = 1, 2, \ldots a\}$; RQ_r: remaining material quantity of distribution center after interference occurs; Q_k: the amount of residual materials loaded by vehicles k after interference occurs; Q_{max}: maximum load of distribution vehicle; C_1: fixed cost of a new car dispatched from the distribution center; C_2: vehicle driving cost per unit path length; D_{ij}: the distance from disaster site i to disaster site j; P: the position of vehicle k in the course of driving when disturbance occurs; D_{pj}: distance of vehicle k from current location to disaster site j after disturbance occurs; D_k: the total driving distance of the remaining unserved disaster areas after the interference occurs; q_j: demand for materials not serving the disaster site j after the interference occurs; T_{ik}: the time when the vehicle k arrives at the disaster site i; LT_i: the latest time for emergency materials to arrive at the disaster site i; T_{ijk}: the travel time of vehicle k from disaster site i to disaster site j; x_{ijk}: when the vehicle k is from the disaster point i to the disaster point j, $x_{ijk} = 1$, otherwise $x_{ijk} = 0$; y_{ir}: when the distribution center r provides material for disaster site i, $y_{ir} = 1$, otherwise $y_{ir} = 0$.

3.4 Model Construction

$$MinZ = \sum_{k=1}^{K} C_1 x_{0jk} + \sum_{k=1}^{L} C_2\left(\sum_{j=0}^{N} x_{pjk} \mathrm{d}_{\mathrm{pj}} + \sum_{i=1}^{N}\sum_{j=0}^{N} x_{ijk} \mathrm{d}_{\mathrm{ij}} - \mathrm{d}_{\mathrm{k}}\right) \quad (1)$$

$$\sum_{i=1}^{N} q_j y_{\mathrm{ir}} \geq \sum_{r=1}^{R} RQ_r \quad (2)$$

$$\sum_{i=1}^{N}\sum_{j=1}^{N} q_j x_{ijk} \geq Q_k \quad (3)$$

$$\sum_{i=1}^{N}\sum_{j=1}^{N} q_j x_{ijk} + \sum_{j=1}^{N} q_j x_{0jk} \geq Q_{max} \quad (4)$$

$$\sum_{k=1}^{K}\sum_{j=1}^{N} x_{0jk} \leq K \quad (5)$$

$$\sum_{r=1}^{R} y_{ir} \geq 1 \quad (6)$$

$$\lambda_i = \frac{\sum_{i=1}^{N} q_j}{\sum_{r=1}^{R} RQ_r} \quad (7)$$

$$T_{jk} = T_{ik} + t_{ijk} \quad (8)$$

$$T_{ik} \leq LT_j \quad (9)$$

$$x_{ijk} = \{1,0\}, y_{ir} = \{1,0\}, \lambda_i \geq \mathrm{m} \quad (10)$$

The objective function represents the minimum disturbance to the path cost. Constraints (2) indicates that the total capacity of the distribution center is not greater than the total demand of the disaster site. Constraints (3) indicates that the amount of vehicle surplus in transit is not greater than the demand for the next service disaster site. Constraints (4) denotes the constraints on newly dispatched vehicle transport materials. Constraints (5) indicates that the number of additional rescue vehicles should not exceed the number of remaining vehicles in the distribution center. Constraints (6) means ensuring that each disaster site has at least one emergency material distribution center to serve it. Constraints (7) defines the overall satisfaction function of material demand in all disaster areas. Constraints (8) represents the time window constraint of material demand in disaster areas. Constraints (9) denotes the time constraints of vehicle transportation. Constraints (10) denotes the constraints of decision variables and m is the lowest satisfaction.

4 Conclusion

According to the literature review of emergency material dispatch about demand change, a disturbance management model about demand change is constructed, and the satisfaction factor of disaster victims is added. The next step is to solve the model algorithm. In recent years, many new algorithms have emerged, which can be combined with existing algorithms to form a hybrid algorithm to solve the model. In the future, the interference in simultaneous changes of multiple factors of emergency material dispatching can be comprehensively considered. In addition, the combination of human behavior factors and interest pursuit of interference management is also a problem worth studying.

Acknowledgment. This research is supported by the National Natural Science Foundation of China (No. 71872002), and the Project of Humanities and Social Sciences of the Education Department of Anhui Province (No. SK2018A0073).

References

1. Yu, G., Qi, X.: Disruption Management: Framework, Models and Applications. World Scientific Publishing Co., Pte. Ltd., Singapore (2004)
2. Xu, X., Qi, Y., Hua, Z.: Forecasting demand of commodities after natural disasters. Expert Syst. Appl. **37**(6), 4313–4317 (2010)
3. Sun, B., Ma, W., Zhao, H.: A fuzzy rough set approach to emergency material demand prediction over two universes. Appl. Math. Model. **37**(10–11), 7062–7070 (2013)
4. Wang, X.: Research on evolution prediction and emergency decision-making of seismic secondary disasters with stochastic Petri nets (Chinese). Chin. J. Manage. Sci. **22**(Special Issue), 158–165 (2014)
5. Ling, S., Yang, B., Sun, S.: Emergency medical resources allocation based on demand hierarchy (Chinese). J. Guangxi Univ. Nat. Sci. Ed. **39**(2), 358–364 (2014)
6. Wang, J., Wang, H.: Gradation for demand urgency of emergency materials in emergency relief (Chinese). J. Catastrophol. **49**(5), 4–7 (2013)
7. Cheng, X.: Study on twice emergent resources distribution based on demand classification (Chinese). J. Harbin Univ. Commerce (Nat. Sci. Edn.) **10**(10), 2556–2559 (2011)
8. Zhao, L., Liang, X., Yang, H., Wang, Z.: Disruption management optimization for VRPSDP with changes of customer distribution requirement (Chinese). J. Transp. Syst. Eng. Inf. Technol. **18**(03), 222–228 (2018)
9. Yang, W., Hu, X., Wang, Z.: Research on management method for logistics distribution system under disruptions of customer time window changing (Chinese). J. Dalian Uni. Technol. **52**(2), 290–296 (2012)
10. Li, J.Q., Mirchandani, P.B., Borenstein, D.: A Lagrangian heuristic for the real-time vehicle rescheduling problem. Transp. Res. Part E Log. Transp. Rev. **45**(3), 419–433 (2009)

11. Ruan, J., Wang, X.: Disruption management of emergency medical supplies intermodal transportation with updated transit centers (Chinese). Oper. Res. Manage. Sci. **25**(4), 114–124 (2016)
12. Wang, X., Ruan, J., Sun, Z., Cao, H.: Disruption recovery model for vehicle routing problem with backhaul (Chinese). J. Syst. Eng. **28**(5), 608–616 (2013)
13. Wang, Z., Hu, X., Wang, X.: Disruption management model and algorithm for distribution vehicle scheduling problems under accidental travel time delay (Chinese). Syst. Eng.-Theory Pract. **33**(2), 378–387 (2013)
14. Zhang, X., Tang, L.: Disruption management for the vehicle routing problem with time windows. In: International Conference on Intelligent Computing, pp. 225–234. Springer, Heidelberg (2007)
15. Ding, Q.: Model of disruption management for the change of time window based on human behavior in logistic distribution (Chinese). Chin. J. Manage. Sci. **23**(5), 89–97 (2015)
16. Zhu, H., Zhou, H., Kong, J., Gao, D.: Problems of two stage emergency resources scheduling with demand interference (Chinese). J. Safety Sci. Technol. **14**(05), 69–76 (2018)
17. Liu, C., Zhu, Z., Liu, L.: Disruption management of Location-Routing Problem (LRP) for emergency logistics system in early stage after earthquake (Chinese). Comput. Eng. Appl. **53** (20), 224–230 (2017)
18. Cao, Q., Zhao, L., Ren, X., Liang, W.: Disruption management for vehicle scheduling with considering customer value (Chinese). J. Shenyang Univ. Technol. **40**(2), 185–191 (2018)

Using Simulation as a Tool for Multiple-Product Production and Logistics Planning Design in the Japanese Rental Housing Unit Supply Chain

Haixia Sang(✉) and Shingo Takahashi

Department of Industrial and Management Systems Engineering, Waseda University, Tokyo, Japan
sanghaixia7861@hotmail.co.jp, shingo@waseda.jp

Abstract. The purpose of this research is to introduce a new perspective on the framework for designing a seasonal demand multiple-product production and logistics planning project in the Japanese rental housing unit supply chain. All aspects of manufacturing and logistics (i.e., forecast, lead time, initial inventory level, and inventory fill rate) are considered. However, how the production quantity and transportation volume of multiple products should be set is constantly under investigation. Simulation studies that include opportunity loss and average inventory level help to identify the most suitable quantity within a supply chain. By utilizing simulation studies, designers can make reliable decisions regarding suitable production and logistics planning faster than conventional methods based on a manager's experience. To understand the framework of supply chain design, a project to design a seasonal demand multiple-product production and logistics planning was investigated. The proposed framework is both practical and powerful in assisting managers with their continuous and efficient decision-making efforts.

Keywords: Demand seasonal · Production planning · Logistics planning · Supply chain management · Rental housing unit

1 Introduction

Multiple-product production and logistics planning design play an important role in supply chain management. Developing efficient production and logistics planning and bringing it to realization is the main task of supply chain designers. After conducting research with the managers of Japanese supply chain companies, it was revealed that supply chain production and logistics planning were traditionally designed based on the manager's experience and simple calculations of opportunity loss and average inventory level by processing previous data. Simulations were not used in the supply chain investigations.

The purpose of this study is to introduce a new perspective on the framework of designing production and logistics planning in the rental housing unit supply chain. In addition, this research introduces the methodology regarding how these simulations were used for investigating design processes via case studies covering design of

C.-F. Chien et al. (Eds.): *IE&EM 2019*, pp. 172–182, 2020.
https://doi.org/10.1007/978-981-15-4530-6_18

multiple-product production and logistics planning of "Company S" for a rental housing unit inventory project.

The rest of this paper is organized as follows: Sect. 2 proposes the framework for all design process flows for carrying out a multiple-product production and logistics planning design, Sect. 3 provides a brief literature review, Sect. 4 provides results from a case study in designing production and logistics planning, and Sect. 5 concludes the paper.

2 Framework for Multiple-Product Production and Logistics Planning Design

The general concept and process for multiple-product production and logistics planning design projects are introduced in Fig. 1.

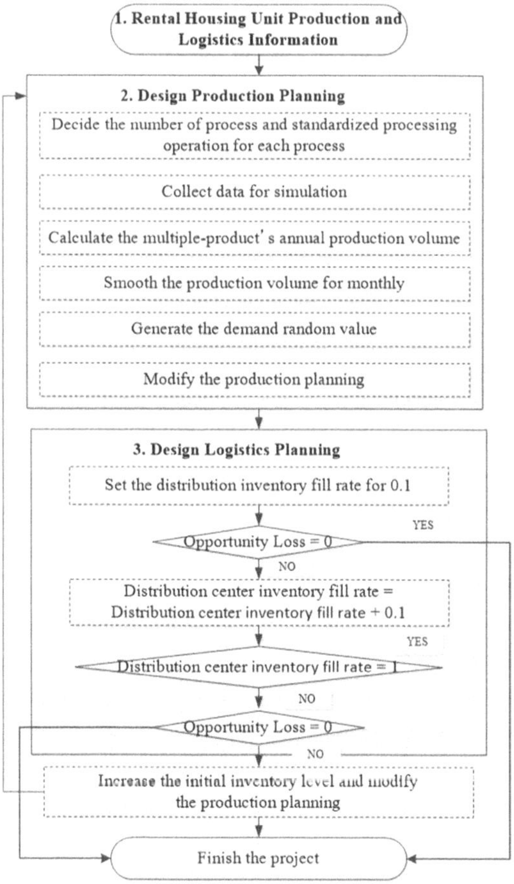

Fig. 1. Framework for multiple-product production and logistics planning design

2.1 Rental Housing Unit Production and Logistics Information

Unlike most products, the rental housing unit is a "circulation type" product; it is shipped from a distribution center and returned by the customer at their convenience. Rental housing units have durable life; the company may maximize the profit by selling the product to a pre-sale market. If a distribution center's inventory level cannot cover the customer demand, the factory produces more units and send the new product to a distribution center based on the inventory fill rate. Figure 2 shows the supply chain of a rental housing unit's business model.

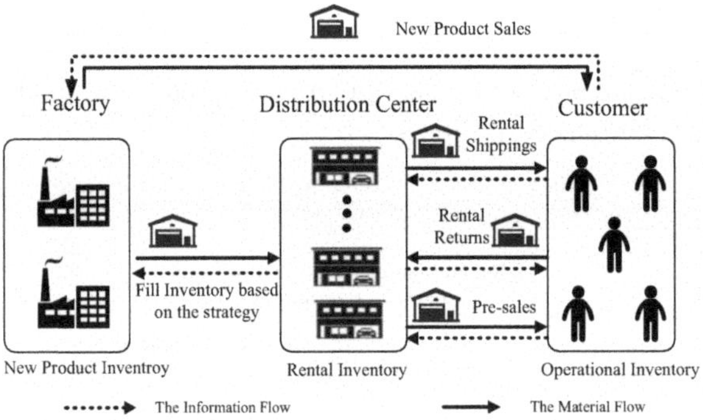

Fig. 2. The supply chain of a rental housing unit's business model

The following information on the project is given to designers:

1. Multiple-product monthly rental shipping volume.
2. Multiple-product monthly rental return volume.
3. Multiple-product monthly pre-sales volume.
4. Factory forecasting method.
5. Factory manufacturing capacity, production lead time.

With this information, the rental housing unit supply chain's production and logistics planning can begin.

2.2 Design Production Planning

Rental housing units have a seasonal demand, and demand in peak seasons exceeds the factory's production capacity. In this case, the demand can be satisfied by producing stock. Based on the smoothing production strategy, we consider not only regular production, but also overtime production to reduce both opportunity loss and average inventory level.

2.3 Design Logistics Planning

If the distribution center cannot meet the customers' demand, the product would be produced by the factory based on the rental housing unit company's fill rate strategy. Logistics planning design is developed through a simulation process. Opportunity loss and average inventory level are considered in order to develop the most efficient logistics planning. By increasing the fill rate from 0.1 to 1, the opportunity loss volume and average inventory level would be output. Even if we change the fill rate to 1, if the opportunity loss is inevitable, we then increase the initial production volume until opportunity loss does not occur.

3 Brief Literature Review

Many authors have long recognized that simulation is a valuable tool for the analysis of complex dynamics in supply chain planning [1, 2]. Simulation can analyze large and complex supply chain systems, including stochastic elements and complex relationships between system components [3, 4]. Especially, discrete event simulation are widely applied decision support tools in supply chain management [5–7]. However, the use of rental housing unit supply chain design is scarce.

Papier [8] studied the amount of appropriate holding inventory for railway freight rental businesses. Despite the seasonal nature of rental demand and return, the inventory was analyzed under constant conditions that eliminated seasonal fluctuations. Endo [9] proposed an approach that considered demand fluctuations in the rental business, but instead of the stock itself that actually existed in the company, the objective function was to maximize the expected profit. Because the rental housing unit business is more complex than the traditional one, planning and management tools must capture this emerging complexity. Sang, Kotakei and Takahashi [10] focus on rental unit production and identify the relationship between system components, such as forecasting method, lead time, and initial inventory level. However, the study was based on a simple product and did not consider the manufacturer's production capacity.

The first contribution of our research was to present the new framework for a design process throughout the whole supply chain for a Japanese rental housing unit business (Company S) from a practical point of view. The second was to introduce an empirical design project of a multiple-product supply chain to ensure the framework and understand the important role of simulation tools in making decisions regarding efficiency.

4 Case Study of Multiple-Product Production and Logistics Planning Design in a Japanese Rental Housing Unit Supply Chain

In this section, an example of designing a multiple-product production and logistics supply chain in Company S will be introduced to enhance understanding of how the simulation is utilized in the whole design process.

4.1 Rental Housing Unit Information

A significant amount of information occurs in the whole supply chain. For example, in the distribution center, a simple product's forecast inventory level can be calculated at the end of the month using the following equation:

$$FIL = FRS + FPS - FRR \tag{1}$$

Where:

FIL: Forecast inventory level.
FRS: Forecast rental shipping volume.
FPS: Forecast pre-sales volume.
FRR: Forecast rental return volume.

If the forecast inventory level is less than 0, then the product must be produced by the factory to avoid the opportunity loss. Figure 3 shows Product A and Product B's inventory level change over then last years. Based on Company S's forecasting method, next year's demand will be forecast, and the factory's production plan, which is shown in the green zone, will be created.

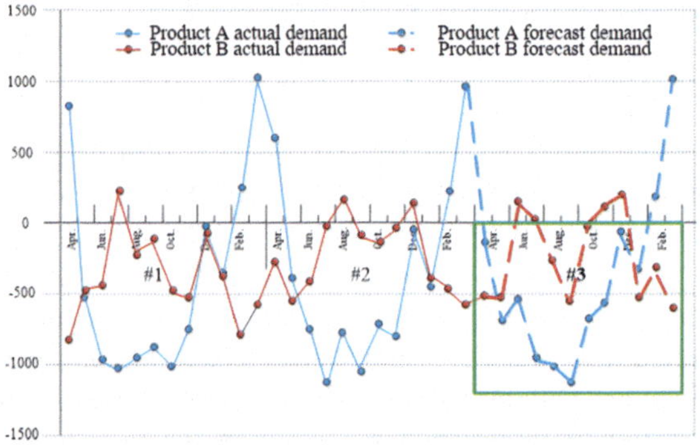

Fig. 3. The actual demand and forecast demand change for Product A and Product B in the distribution center.

4.2 Create a Basic Production Plan

Based on the forecast demand results, the factory calculates the monthly production volume for Product A and Product B, and levels off the production scheduling to stabilize employment of the workforce, as shown in Fig. 4 and Fig. 5.

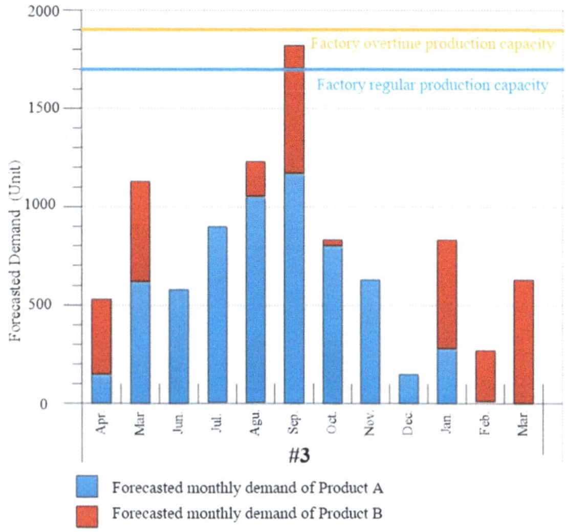

Fig. 4. The forecast monthly demand change of Production A and Production B.

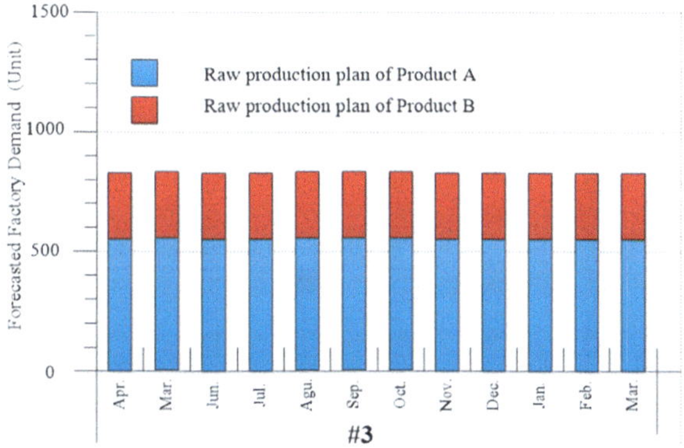

Fig. 5. The basic production plan of Production A and Production B.

4.3 Modify the Basic Production Plan to a Detailed Production Plan

The basic production plan is based on the forecast demand results. However, the demand is uncertain and the production plan must be modified based on the change in demand. Figure 6 shows how a basic production plan is converted into a detailed plan. This is based on the condition of real demand exceeding the forecast demand, as shown in the Fig. 6(a). For example, at the end of April, the real demand became clear, however, production had completed based on the basic production plan for April, and because the production lead time is one month, the gap between the raw production and real demand will be added to the basic production plan for June. That is, the detailed production for June will be decided at the end of April. If the detailed plan exceeds the production capacity, the excess volume will be shifted to the next month. On the contrary, as shown in Fig. 6(b), if the basic plan exceeds actual demand, the excess production volume must be modified in June's production plan to avoid excess inventory in the future.

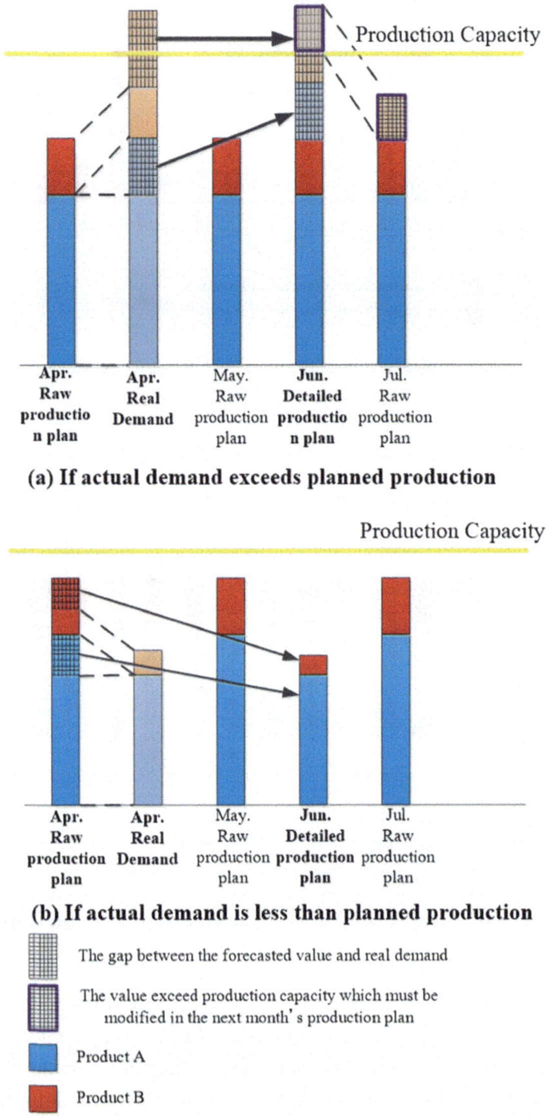

Fig. 6. Modifying the basic plan to a detailed plan.

4.4 Simulate the Distribution Center's Fill Rate to Avoid Opportunity Loss

The factory produces units to replenish stock. However, as shown in Fig. 2, the new product is not only to meet the distribution center's rental demand, but also for new product sales. Therefore, how to set the fill rate to avoid rental demand's opportunity loss is an important issue that must be considered.

The fill rate is an index, that is based on the distribution center's outgoing volume. If the filling indicator is set as 1, this means all of the monthly rental shipping volume and pre-sale volume will be replenished by the factory. Considering the monthly rental return volume, a fill rate value of 1 is extremely large. We set the distribution center's fill rate from 0.1 to 1 to simulate the opportunity change and average inventory change so that we could identify the optimal value that can reduce both opportunity loss and the average of the distribution center's inventory level. If the opportunity loss cannot be avoided by setting the fill rate as 1, this indicates that the product has seasonality; the factory must produce stock, and production capacity and the initial inventory level must therefore also be considered.

4.5 Simulation Model

Simulation models were constructed using Excel VBA software. Simulation input data and simulation run condition were set up based on information extracted from interviews or real business data.

The main factors for deciding alternatives for the production and logistics planning are the initial inventory level, the production capacity, and the distribution center's fill rate. For all alternative models, the simulation run time was set to working days in three years with 100 replications.

4.6 Simulation Results

The objective of this simulation was to reveal which alternative satisfies the design requirements (satisfied customer's demand and maintain a lower inventory level). The following results were obtained.

The relationship between the fill rate and the initial inventory level at regular production of Product B is shown in Fig. 7. The measure for evaluation is inventory level and opportunity loss, which must be considered simultaneously. If the fill rate is set as 0.1, although the alternative initial inventory level is low, the opportunity loss is 60 times greater, which is not good. By simultaneously increasing the fill rate and the initial inventory level, the inventory level is increased, and at the same time, the opportunity loss is reduced. If we set the fill rate as 0.5 and the initial inventory level for 628, the opportunity loss is 0 and the average monthly inventory level is 1604, which is the optimal value.

A simulation of inventory location was also examined. For example, Fig. 8 presents the regular output of Product B's relationship between the inventory level and opportunity loss if the initial inventory level is 628. By increasing the fill rate, the rental inventory level increases, and at the point of 0.5, the opportunity loss is 0.

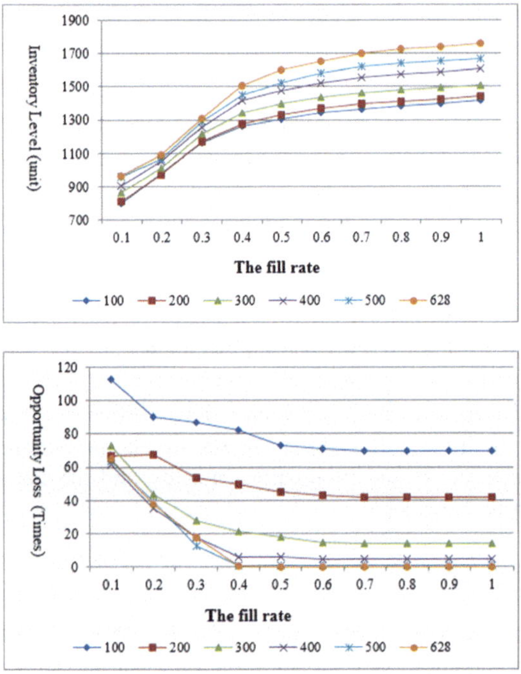

Fig. 7. The relationship between the fill rate and the initial inventory level at regular production of Product B.

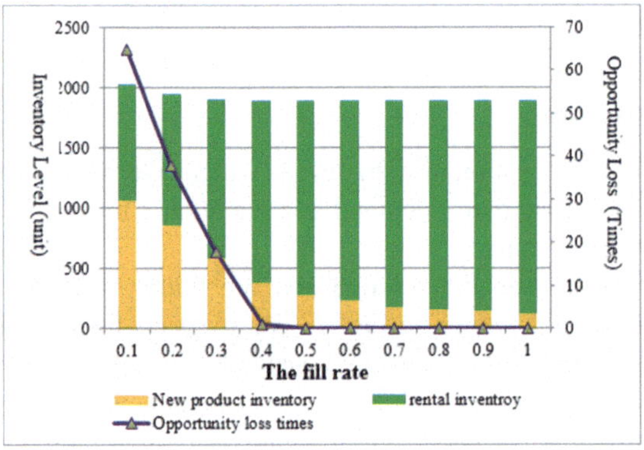

Fig. 8. The relationship between the inventory level in the factory and distribution center, and opportunity loss for regular production of product B.

We perform the same experiment for Product A and Product B under the conditions of regular and overtime production, the optimal results of each experiment are presented in Table 1.

Table 1. Optimal simulation results.

Product	Production capacity	Initial inventory level (Unit)	The filling indicator	Averege inventory level (Unit)	Opportunity loss (Times)	Inventory capacity utilization (%)
A	Regular production	3279	0.2	2257	0	88%
	Overtime production	3101	0.2	2193	0	88%
B	Regular production	628	0.5	1604	0	93%
	Overtime production	607	0.7	1766	0	93%

5 Conclusions

This study constructed a production and logistics inventory-level framework for the Japanese rental housing unit supply chain to demonstrate the contribution of a simulation study in the design process.

This research also introduced some advantages in utilizing simulations for designing supply chains. The simulation study can help managers identify better options for developing efficient indicators, but can also save time regarding mistakes in redesign compared to conventional methods that were mainly based on manager's experience.

Acknowledgment. This work was supported by JSPS KAKENHI Grant Number JP118K13954. The sample data related to the rental housing unit supply chain operation is fictitious.

Clarification. Interview are not conducted in our study. And the data used in our study is not interview data.

References

1. Min, H., Zhou, G.: Supply chain modeling: past, present and future. Comput. Ind. Eng. **43**(1), 231–249 (2002)
2. Terzi, S., Cavalieri, S.: Simulation in the supply chain context: a survey. Comput. Ind. **53**(1), 3–16 (2004)
3. Manuj, I., Mentzer, J.T., Bowers, M.R.: Improving the rigor of discrete-event simulation in logistics and supply chain research. Int. J. Phys. Distrib. Log. Manage. **39**(3), 172–201 (2009)
4. Minn, D.N., Takakuwa, S.: Emergence of simulations for manufacturing line designs in Japanese automobile manufacturing plants. In: Proceedings of the 2008 Winter Simulation Conference, pp. 1847–1855 (2008)

5. Semini, M., Fauske, H., Strandhagen, J.O.: Applications of discrete-event simulation to support manufacturing logistics decision-making: a survey. In: Proceedings of the 2006 Winter Simulation Conference, pp. 1946–1953 (2006)
6. Chik, M., Rahim, A., Rejab, A.M., Ibrahim, K., Hashim, U.: Discrete event simulation modeling for semiconductor fabrication operation. In: International Conference on Semiconductor Electronics (ICSE), Kuala Lumpur, Malaysia. Institute of Electrical and Electronics Engineers, Inc., pp. 325–328 (2004)
7. Caridi, M., Cavalieri, S.: Multi-agent systems in production planning and control: an overview. Prod. Plan. Control **15**(2), 106–118 (2004)
8. Wangphanich, P., Kara, S.S., Kayis, B.: Analysis of the bullwhip effect in multi-product, multi-stage supply chain systems–a simulation approach. Int. J. Prod. Res. **48**, 4501–4517 (2010)
9. Endo, U., Morito, S.: An approach to determine the amount of ownership in the rental business. In: Proceeding of the 2014 Japan Industrial Management Association Autumn Conference, pp. 194–195 (2014)
10. Sang, H.X., Kotake, Y., Takahashi, S.: A dynamic modeling simulation for supply chain management inventory service: a case study on a rental housing unit manufacturing and logistics company. In: Proceeding of 2018 International Conference on Information Management and Management Sciences (IMMS 2018), pp. 125–132 (2018)

Research on Resource Scheduling Optimization of Prefabricated Building Components Based on Improved Particle Swarm Optimization

Fu-yu Wang and Jie Ding(✉)

School of Management Science and Engineering,
Anhui University of Technology, Maanshan, China
xiaowang96@163.com, 1741022031@qq.com

Abstract. In order to effectively allocate resources and reduce project cost, this paper takes the scheduling of prefabricated building components as the research object, constructs the corresponding resource scheduling model, and solves the problem of the reasonable configuration of PC components with improved particle swarm optimization algorithm, so as to reduce construction cost. Through the simulation experiment and the result analysis, the rationality of the model and the feasibility and effectiveness of the algorithm are verified.

Keywords: Prefabricated building · Particle swarm optimization · PC component · Resource scheduling

1 Introduction

Reference [1] analysis and discussion of different types of buildings, found that China meets the basic elements of building industrial buildings. In [2], it is proved that the implementation of prefabricated buildings is the direction of the continuous development of the construction industry from four aspects: policy norms, technical system, industrial management and economic cost.

Assembly construction has many advantages. Its components are prefabricated at the factory to reduce the time required to wait for components on the construction site [3]. Through [4], it is found that the prefabricated construction can reduce the number of scaffolding or formwork and improve the construction efficiency of the project. Taking the prefabricated residential buildings in Shanghai as an example, it was found that prefabricated buildings can add value to construction projects [5].

In order to promote the prefabricated building, various provinces and municipalities have adopted different levels of subsidies for the use of prefabricated buildings. For example, Beijing Municipality provides financial incentives for non-government investment projects with a prefabrication rate of more than 50% and an assembly rate of over 70%. Chongqing Municipality subsidizes 350 (yuan/m^3) for the pilot project of

C.-F. Chien et al. (Eds.): *IE&EM 2019*, pp. 183–192, 2020.
https://doi.org/10.1007/978-981-15-4530-6_19

modern building construction in the construction industry. We can see that although the production cost of prefabricated buildings is relatively high, various regions are promoting the development of prefabricated construction projects through different financial subsidy policies. Of course, the rational use of fabricated building resources will also effectively reduce project costs.

Through [6], the criteria for evaluating the resource flow network are defined. When researching resource planning and allocation, new heuristic methods can be adopted, and the results show that the efficiency of resource scheduling can be effectively improved form [7–9]. Scholars also study emergency resource scheduling by [10]. Similarly, vehicle optimization problems in time windows can also be combined with the distribution of prefabricated buildings [11]. Through [12] and [13], we can see that the intelligent optimization algorithm also reflects the production scheduling of prefabricated components.

After reviewing the literature, it is not difficult to find that there are many types of research on resource scheduling problems. However, the research on resource scheduling of prefabricated buildings is not deep enough. Therefore, this article will learn and expand in this aspect, and provide ideas for future scholars to carry out related research. The purpose of this paper is to reduce the cost of the project by rationally allocating the resources of the components. Therefore, this paper constructs the resource scheduling model of the prefabricated building and solves it with the improved particle swarm optimization algorithm (PSO). The validity of the algorithm is verified by the case.

2 Problem Statement

There are many resources for prefabricated buildings. This article is only optimized for the single component of PC components. Components can be broadly divided into a prefabricated exterior siding, columns, laminated slabs and laminated beams. Different costs are generated when choosing different manufacturers, and under the premise of determining the component demand, this paper proposes the demand for resources for each manufacturer, thus achieving the goal of cost-saving.

At the same time, because such projects and China's energy-saving emission reduction policies complement each other. Therefore, when building a prefabricated construction project, it will have the support of the government, mainly in terms of financial subsidies. Based on comprehensive consideration of relevant policies and actual conditions, this paper stipulates that the prefabrication rate of construction projects exceeds 45%, and 300 (yuan/m^3) is subsidized according to the construction volume.

3 Model Construction

A. *Model assumption*

Since many complicated parameters are involved in calculating the cost of fabricated building components, the following assumptions are made for the convenience of this research topic:

(1) The total demand for each component is known;
(2) Relevant data of various manufacturers can be ascertained through field research;
(3) The government financial subsidy price is known;
(4) The prefabrication rate of each component is higher than 30% of its total;
(5) The prefabricated components are still charged in a cast-in-place manner.

B. *Established model*

The symbols of the model built are defined as follows:

n Number of factories
N_{nl} The unit price of the prefabricated beam of the nth factory
N_{nq} The unit price of the prefabricated wall of the nth factory
N_{nz} The unit price of the prefabricated column of the nth factory
N_{nb} The unit price of the prefabricated board of the nth factory
N_{nd} The distance from the nth factory to the construction project
N_{nk} The transportation cost of the nth factory unit
N_{nc} The carrying capacity of the nth factory unit vehicle
V_l The total volume of the beam
V_q The total volume of the wall
V_z The total volume of the column
V_b The total volume of the plate
N_x The unit price of the cast-in-place method
X_0 The total cost of the beam when combined in two ways
Y_0 The total cost of the wall when combined in two ways
W_0 The total cost of the column when combining the two methods
B_0 The total cost of the board when combining the two methods
σ The total cost of combining the two approaches
Z The price of government subsidies, the prefabrication rate is more than 45% is 300 (yuan/m^3), otherwise, it is 0.
K Building volume

Decision variables:

V_{nl0} The resources of the beam allocated by the nth factory
V_{nq0} The resources of the wall allocated by the nth factory
V_{nz0} The resources of the column allocated by the nth factory
V_{nb0} The resources of the board allocated by the nth factory

$$\min(\sigma) = \sum_{i=1}^{n}(X_0 + Y_0 + W_0 + B_0) - Z * K \tag{1}$$

$$\begin{aligned}\sum_{i=1}^{n} X_0 = & \sum_{i=1}^{n}(N_{nl} * V_{nl0} + N_{nd} * N_{nk} * \frac{V_{nl0}}{N_{nc}}) \\ & + \left(V_l - \sum_{i=1}^{n} V_{nl0}\right) * N_x\end{aligned} \tag{2}$$

$$\begin{aligned}\sum_{i=1}^{n} Y_0 = & \sum_{i=1}^{n}(N_{nq} * V_{nq0} + N_{nd} * N_{nk} * \frac{V_{nq0}}{N_{nc}}) \\ & + \left(V_q - \sum_{i=1}^{n} V_{nq0}\right) * N_x\end{aligned} \tag{3}$$

$$\begin{aligned}\sum_{i=1}^{n} W_0 = & \sum_{i=1}^{n}(N_{nz} * V_{nz0} + N_{nd} * N_{nk} * \frac{V_{nz0}}{N_{nc}}) \\ & + \left(V_z - \sum_{i=1}^{n} V_{nz0}\right) * N_x\end{aligned} \tag{4}$$

$$\begin{aligned}\sum_{i=1}^{n} B_0 = & \sum_{i=1}^{n}(N_{nb} * V_{nb0} + N_{nd} * N_{nk} * \frac{V_{nb0}}{N_{nc}}) \\ & + \left(V_b - \sum_{i=1}^{n} V_{nb0}\right) * N_x\end{aligned} \tag{5}$$

$$\sum_{i=1}^{N} V_{nl0} \geq V_l * 30\% \tag{6}$$

$$\sum_{i=1}^{N} V_{nq0} \geq V_q * 30\% \tag{7}$$

$$\sum_{i=1}^{N} V_{nz0} \geq V_z * 30\% \tag{8}$$

$$\sum_{i=1}^{N} V_{nb0} \geq V_b * 30\% \tag{9}$$

$$V_{nl0}, V_{nq0}, V_{nz0}, V_{nb0} \geq 0 \tag{10}$$

Use (1) to calculate the total cost of the combination of the two. Use (2), (3), (4) and (5) to calculate the total cost of beams, walls, columns, and plates combined with them respectively. (6), (7), (8) and (9) are the constraint conditions of the prefabrication rate of each component. (10) is a non-negative constraint on the number of resources allocated to each plant.

4 An Improved Algorithm for PSO

Particle swarm optimization has been applied in resource scheduling [14–16].

The flow of the traditional PSO is as follows:

1. Randomly initialize the particle swarm in the initialization range, including random position and velocity;
2. Calculate the fitness value of each particle;
3. Update the historical optimal position of the individual particle;
4. Updating the historical optimal position of the particle population;
5. Updating the velocity and position of the particle;
6. If the termination condition is not reached, then step 2.

A. *PSO based on adaptive weight*

In the adjustable parameters of the particle swarm algorithm, the inertia weight W is a very important parameter. A larger W is beneficial to improve the global searchability of the algorithm, while a smaller W enhances the local searchability of the algorithm. Therefore, in order to balance the global searchability and local improvement ability of the particle swarm algorithm, a nonlinear dynamic inertia weight coefficient is adopted, and the expression is as follows:

$$W = \begin{cases} w_{min} - \frac{(w_{max} - w_{min}) * (f - f_{min})}{(f_{avg} - f_{min})}, & f \leq f_{avg} \\ w_{max}, & f > f_{avg} \end{cases} \tag{11}$$

w_{max}, w_{min} are the maximum and minimum values of W respectively, f is the current target value of the particle, f_{avg} and f_{min} represent the average target value and the minimum target value of all current particles, respectively. Using (11), we can make the inertia weight automatically change with the particle's objective function value.

The improved particle swarm algorithm flow is shown in Fig. 1.

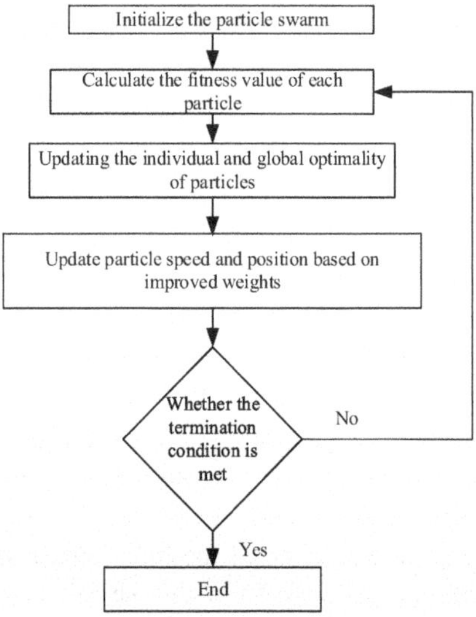

Fig. 1. Improved particle swarm algorithm flow

When the target values of the particles tend to be consistent or tend to be locally optimal, the inertia weights will be increased, and when the target values of the particles are dispersed, the inertia weight will be reduced, and the objective function value is better than the average target value. The particle has a corresponding inertia weighting factor that is small, thereby protecting the particle. Conversely, for a particle whose objective function value is worse than the average target value, the corresponding inertia weighting factor is larger, so that the particle is closer to a better search area.

5 Computational Experiments

In response to the call of the national policy, a project uses the cast-in-place and prefabricated construction of the four members of the column, beams, wall, and board. After consulting the relevant personnel of the project, it is known that different cast-in-place concrete has different information prices, and the prices of different grades of concrete are also different. At the same time, labor costs and material costs must be considered during construction. The factors involved in the actual situation are numerous and complicated. In order to simplify the model and improve efficiency, this paper sets the price of the cast-in-place to 492 (yuan/m^3) after combining the "Nanjing Project Cost Management" and the relevant data provided by the staff. At the same time, after comprehensively referring to the requirements of each province and city for the prefabrication rate of construction projects, the minimum prefabrication rate is set at 30%.

To simplify the calculation, the sum of the volume of the components is the volume of the construction project. The relevant data of each factory is shown in Table 1, in which the volume of columns, beams, walls, and plates is fixed and known.

A. *Experimental data*

In this paper, the improved particle swarm optimization algorithm is compared with the standard particle swarm optimization algorithm. The specific parameters of the algorithm are set as follows:

The range of iteration speed is from -1 to 1, the population number $popsize = 20$, maximum iteration number $iterations = 1000$, the inertia weight $w = 0.8$, $w_{max} = 0.9$, $w_{min} = 0.6$, the learning factor $C_1 = C_2 = 1.5$.

Table 1. Factory related data

C	F	P (yuan/m³)	Td (km)	Tp (yuan/km)	Vc (m³/car)	Tv (m³)
Column	A	3554	15	22	15	371.506
	B	2652	14.5	20	16	
	C	3426	15.2	23	14	
Beam	A	2824	15	23	14	298.657
	B	2647	14.5	22	15	
	C	2888	15.2	24	13	
Wall	A	2322	15	21	15	331.897
	B	2299	14.5	20	14	
	C	1976	15.2	23	16	
Plate	A	2057	15	20	14	482.128
	B	1932	14.5	18	13	
	C	2100	15.2	22	16	

In Table 1, C denotes component, F denotes factory, P denotes assembly type unit price, Td denotes transportation distance, Tp denotes transportation unit price, Vc denotes vehicle capacity, and Tv denotes total volume.

B. *Analysis of results*

Both algorithms are implemented using MATLAB 2018a programming. The prefabricated component resource scheduling scheme, cost, and prefabrication rate solved by the standard particle swarm optimization algorithm are shown in Table 2.

Table 2. Component allocation table of the standard particle swarm algorithm

F	C			
	Column (m³)	Beam (m³)	Wall (m³)	Plate (m³)
A	9.00	15.83	129.84	48.61
B	101.13	17.40	19.51	168.82
C	16.33	61.84	74.82	8.20
Total cost (yuan)	1532047.30			
Prefabrication rate (%)	45.23			

The pre-built component resource scheduling scheme, cost and pre-production rate improved by the improved particle swarm optimization algorithm are shown in Table 3.

Table 3. Component allocation table for improved particle swarm optimization

F	C			
	Column (m^3)	Beam (m^3)	Wall (m^3)	Plate (m^3)
A	3.09	39.99	51.80	93.59
B	120.59	11.50	19.08	105.30
C	5.84	39.75	126.57	58.79
Total cost (yuan)	1509396.12			
Prefabrication rate (%)	45.54			

It can be seen from the comparison of Table 2 and Table 3 that the particle swarm optimization algorithm is feasible and effective for the model built in this paper. The optimal cost of the improved particle swarm optimization algorithm is smaller than that of the standard algorithm, and its corresponding prefabrication. The rate is greater than the standard algorithm, which not only satisfies the owner's requirements for cost reduction, but also enables the project to enjoy the financial subsidies provided by the government to the greatest extent, and has a positive effect on the future evaluation of the construction project.

C. *Algorithm performance analysis*

The relationship between the fitness value change and the iteration number of the two algorithms is shown in Fig. 2.

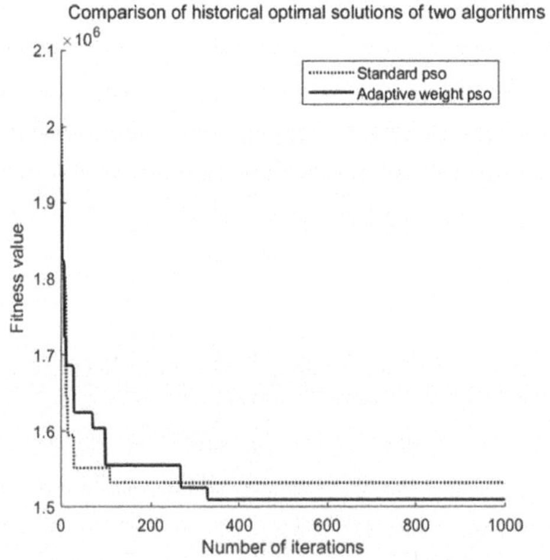

Fig. 2. The relationship between the fitness value of two algorithms and the number of iterations

It can be seen from the observation that the convergence rate of the two algorithms is fast at the beginning of the iteration, but around 150 generations, the standard particle swarm optimization algorithm falls into the local optimal solution and has not jumped out of the solution in the subsequent iteration. After the particle swarm optimization algorithm with adaptive weight briefly falls into the local optimal solution, the inertia weight increases, which enables the algorithm to jump out of the local optimal solution and carry out the next iterative optimization process. In the end, the curve continues to decline, the convergence speed is accelerated, and the accuracy of the optimal solution is also higher than the standard algorithm.

6 Conclusion

In the context of China's vigorous promotion of prefabricated building projects, this paper solves the resource scheduling problem of different components of prefabricated buildings under the condition of prefabrication rate and constructs a mathematical model with the lowest construction cost as the optimization goal. At the same time, according to the model of this paper, the improved particle swarm optimization algorithm is used to solve the problem. Through simulation analysis and comparison, it is further verified that the improved particle swarm optimization algorithm is superior to the standard particle swarm optimization algorithm in accuracy and efficiency. The experimental results not only meet the needs of the development of the construction industry but also enhance the value of the enterprise itself.

As the country is gradually promoting the assembly-type construction project, the research results of this paper are difficult to implement under the existing circumstances. Although the resource scheduling problem of prefabricated building projects has not yet been fully taken into account, scholars will certainly have more and more research in this area. Therefore, this paper can provide some reference value for future scholars to study this aspect.

Acknowledgment. This research is supported by the National Natural Science Foundation of China (No. 71872002), and the Project of Humanities and Social Sciences of the Education Department of Anhui Province (No. SK2018A0073).

References

1. Yang, X., Qian, Q., Tang, Z.: Preliminary study on the characteristics and adaptability of prefabricated construction in background of domestic architectural industrialization. Value Eng. **34**(27), 78–82 (2015)
2. Qi, B., Zhang, Y.: Prefabricated construction development bottleneck and countermeasures research. J. Shenyang Jianzhu Univ. (Soc. Sci.) **17**(02), 156–159 (2017)
3. Zhao, X., Yu, Y., Chen, X.: A versatile microwave plethysmograph for the monitoring of physiological parameters (Periodical style). Technol. Wind **06**, 91 (2018)
4. Li, H., Ji, Z., Sun, Y.: Development and prospect of prefabricated construction. China Build. Mater. Sci. Technol. **26**(03), 72–75 (2017)

5. Jiang, W., Li, L., Hou, J., Zhang, Z., Zhi, Z.: Research on integration of design and construction of precast residence. Constr. Technol. **46**(22), 72–74 (2017)
6. Klimek, M., Łebkowski, P.: Resource allocation for robust project scheduling. Bull. Pol. Acad. Sci.: Tech. Sci. **59**(01), 51–55 (2011)
7. Roubila, L., Boctor, F.: An efficient genetic algorithm to solve the resource-constrained project scheduling problem with transfer times: the single mode case. Eur. J. Oper. Res. **265**(02), 454–462 (2018)
8. Liu, X., Qi, M., Cheng, C.: Green vehicle routing problem with path flexibility. In: IEEE International Conference on Industrial Engineering and Engineering Management, pp. 1037–1041 (2017)
9. Mahendra, B.G., Subhash, K.S.: Task scheduling and resource allocation in cloud computing using a heuristic approach. J. Cloud Comput. **7**(01), 1–16 (2018)
10. Chi, H., Li, J., Shao, X., Gao, M.: Timeliness evaluation of emergency resource scheduling. Eur. J. Oper. Res. **258**(03), 1022–1032 (2017)
11. Li, J., Song, M., Deng, J., Han, Y.: Research on VRP with time window for delivery problem in prefabricated building. J. Liaocheng Univ. (Nat. Sci. Ed.) **32**(04), 101–110 (2019)
12. Xie, S., Chen, X.: Two phase job-shop scheduling optimization for prefabricated construction projects based on multilayer coding genetic algorithm. J. Eng. Manage. **32**(01), 12–18 (2018)
13. Ma, G., Wang, G.: Optimization method of prefabricated building project procedure based on QPSO algorithm. Mach. Tool Hydraul. **46**(12), 111–115 (2018)
14. Weng, S., Sun, M., Li, J.: Optimal operation of Hanjiang river basin based on particle swarm optimization. Pearl River **39**(02), 82–85 (2018)
15. Zhang, Y., Ma, J.: Emergency resources scheduling based on chaotic particle swarm optimization. J. Catastrophology **32**(02), 185–189 (2017)
16. Li, S., Zhang, C., Zhou, Z.: Flexible job-shop scheduling optimization of discrete manufacturing based on the improved particle swarm optimization algorithm **11**, 150–152 (2018)

Research on Optimization of Intelligent Express Terminal Considering User Experience

Tian-rui Zhang(✉), Ming-qi Wei, Chuan-sheng Qu, and Jia-nan Xu

Department of Industrial Engineering, School of Mechanical Engineering, Shenyang University, Shenyang, China
tianjiangruixue@126.com

Abstract. Aiming at the problems existing in the operation of intelligent express terminal, based on the factors of user experience satisfaction and cluster analysis, the size and quantity of intelligent express terminal were optimized and improved. Based on the theory of customer satisfaction, the index system of influencing factors of user satisfaction of intelligent express terminal was constructed in this paper. The key factors affecting user satisfaction were identified by collecting relevant data through questionnaires and using SPSS22.0 software for validity, reliability, descriptive statistical analysis and regression analysis. On the basis of this result, the existing data were further clustered by system and K-means clustering, and the size ratio and number of express terminals suitable for four different districts were obtained, which provided a foundation for the good development of express terminals in the future.

Keywords: Cluster analysis · Intelligent express terminal · User satisfaction

1 Introduction

Online shopping has become an inseparable part of people's daily life. Express delivery points are all over the streets. China's express industry ushered in a high-speed development stage. By December 28, 2018, China's express business totaled more than 50 billion yuan, far exceeding other countries. The rapid development of express industry in China is accompanied by many problems. According to the statistics of the National Post Office, the satisfaction degree of users to express experience is not very high, and the delivery mode of terminal express is mainly manual delivery, the direct reception of the receivers is the main way [1]. In 2012, the emergence of new express terminals, but the diversity of user needs led to the gradual exposure of other aspects of the size and number of express terminals, seriously affecting the satisfaction of users.

User experience refers to the user's subjective feeling after enjoying a certain service or product, which directly affects whether the user will continue to purchase this service. Fu Zhongnan established KQI calculation formula based on user experience to evaluate the operation of school infinite network [2].Chen Mengyue established the mathematical function model of system module and user experience according to the characteristics of user experience [3]. Ta-Sheng Lo took an operator in Taiwan as a

C.-F. Chien et al. (Eds.): *IE&EM 2019*, pp. 193–202, 2020.
https://doi.org/10.1007/978-981-15-4530-6_20

specific research object, and expounds the user experience-oriented service development model [4]. Yamakami, T compared the relationship between user experience and social experience, and expounds the importance of both relationships [5]. Edward Luca makes an analysis of how libraries used design thinking to improve user experience in physical library space [6]. Li Junjun divided user experience into different stages [7].

In recent years, the rapid growth of e-commerce has led to a sharp increase in the volume of delivery business, so scholars continue to deepen their research on the issue of express delivery. Zhang Jin used quantitative method to optimize the robustness of express delivery system [8]. Considering the time-constrained window problem, Wang Shuai constructed a distribution route decision-making model under uncertain time conditions [9]. Li Dongmei designed the terminal of the system considering the dislocation of delivery time in express business [10]. Jie Du built the index system of distribution center layout based on the characteristics of express delivery [11].

With the development of express industry, intelligent express terminal received extensive attention. Wang Lijuan analyzed the causes of intelligent express terminal and the problems encountered in the development of intelligent express [12]. Zou Hengtao used smart210 hardware platform to design the intelligent express terminal system [13]. Wang Haihua used Lingo11 to solve the shortest path of students'pick-up according to the situation of campus express delivery [14].

The above research does not involve the design of the number and size of express terminals with satisfactory user experience. Therefore, this paper puts forward specific improvement schemes for the practical problems in the development of intelligent express terminals.

2 Research on the Influencing Factor of User Satisfaction

A. Design of Index System

According to the current status, development and expert research results of intelligent express terminal, combined with user experience satisfaction and logistics service quality, 3 levels of hierarchical indicators are established, and the influencing factors based on user experience satisfaction are calculated as shown in Table 1.

B. Design and Distribution of Questionnaires

The purpose of social questionnaires is to find out the problems by understanding people's use of smart terminals. The contents of this questionnaire include: introduction of research content, survey of basic information of users, survey of satisfaction factors of user experience, and suggestions for the development of intelligent terminals. 600 questionnaires were distributed and 458 valid papers were collected.

C. Data Analysis of Questionnaire

The data analysis of the questionnaire includes the following aspects: basic information of sample data, descriptive statistical analysis, reliability analysis and regression analysis.

Table 1. Index system based on user satisfaction

Level 1	Level 2	Level 3	Source
Influencing Factors of Intelligent Express Terminal Service Based on User Satisfaction	Service model	V01 Notification Forms (e.g. SMS, WeChat, QQ, Telephone)	SERVQUAL
		V02 Payment methods (e.g. WeChat, Alipay, bank card, cash)	Logistics Scholar
		V03 Various types of services (e.g. special mail)	network
		V04 Additional functional categories (e.g. express inquiries)	network
	Failure response rate	V05 Attitude and Professional Ability to Handle Problems	SERVQUAL
		V06 Timeliness in dealing with problems	LSQ
		V07 Handling Speed of Express Terminal Failure	Logistics Scholar
		V08 Ways of handling complaints	Logistics Scholar
	Service reliability	V09 Reasonable Appearance Design of Express Terminal	Logistics Scholar
		V10 The size of express terminal meets user's requirement	network
		V11 Reasonable number of express terminals	This study
	Convenience of service	V12 Coverage Area of Express Terminal	LSQ
		V13 Easy for users to pick up parts everyday	SERVQUAL
		V14 Express terminal operation process is simple and convenient	network
	Overall Satisfaction	User Experience Satisfaction	This study

(1) *Basic Information of Sample Data*
In order to get more objective results, the subjects involved in this survey are different gender, age, occupation, income and different educational background. According to the survey, there are problems in express terminal as shown in Fig. 1.

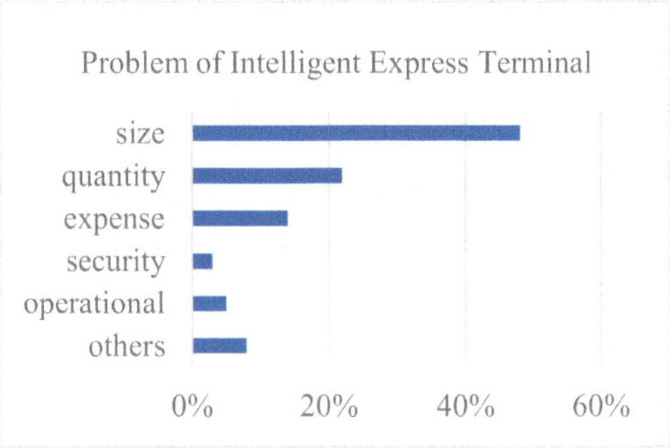

Fig. 1. Possible Problems of Intelligent Express

(2) *Descriptive Statistics*
Descriptive statistical analysis is used to analyze the characteristics of the data and to understand comprehensively the comprehensive perception of the respondents in order to make a clear data interpretation of the final results. The result of descriptive statistical analysis of 458 valid papers is shown in Table 2.

Table 2. Descriptive statistical analysis of factors affecting user satisfaction

Factor	Mean Value	Standard Deviation
V01	4.28	0.572
V02	4.14	0.458
V03	4.04	0.543
V04	3.27	0.610
V05	4.1	0.621
V06	4.02	0.564
V07	3.85	0.793
V08	2.98	0.745
V09	2.89	0.633
V10	2.57	0.752
V11	2.64	0.808
V12	3.12	0.601
V13	3.69	0.406
V14	3.37	0.561
Overall Satisfaction	3.49	0.62

The user satisfaction is divided into 5 points. From Table 2, we can find that the overall satisfaction is 3.49. The satisfaction of v01, V02 and V03 is higher, while the satisfaction of V10 and V11 is lower. The results are consistent with the statistical results of the basic information survey.

(3) *Reliability Analysis*
The reliability score is used to verify the reliability of the questionnaire. When α is greater than 0.6, it shows that the feasibility of the questionnaire meets the requirements [15]. This paper uses SPSS22.0 to calculate the reliability weight coefficient of the questionnaire, and the results are shown in Table 3.

Table 3. Reliability checklist

Variable	Cronbzch'α	No. of questions
Service model	0.847	4
Failure response rate	0.861	4
Service reliability	0.781	3
Convenience of service	0.803	3

In Table 3, the Cronbzch' αcoefficients of all variables are greater than 0.7, which indicates that the reliability of the questionnaire survey is high and the data is available.

(4) *Regression Analysis*
Through the analysis, it is found that service diversity, fault handling ability, service reliability and service convenience are the reasons that affect user experience satisfaction. Therefore, the four factors are regarded as common factors, and regression is used to find out the key reasons. Firstly, ANOVA variance analysis is obtained by software SPSS, in which F = 75.468 indicates that at least one of the four factors has a greater impact on user satisfaction. Then the overall satisfaction was analyzed by regression, as shown in Table 4.

Table 4. Regression analysis of overall satisfaction

Common factor	Non-standardization coefficient		Standardization coefficient	*t*	Sig.	Collinearity statistics	
	B	Standard error	*β*			Allow	VIF
(Constant)	−1.435	0.355		−6.464	0.000		
Service model	0.210	0.061	0.103	3.437	0.035	0.898	1.114
Failure response rate	0.468	0.054	0.368	8.126	0.000	0.799	1.251
Service reliability	0.670	0.052	0.380	10.324	0.000	0.857	1.166
Convenience of service	0.369	0.055	0.303	6.162	0.000	0.940	1.064

According to the data in Table 4, sig of each variable is less than 0.05, which indicates that these variables have a certain impact on user satisfaction. According to the standardized regression coefficient, the impact of satisfaction on the overall user experience from large to small is service reliability, fault handling capacity, service convenience, service mode.

Based on the above conclusions, this paper will focus on the size limitation and service reliability of express terminals.

3 Size and Quantity Optimazation and Improvement of Intelligent Express Terminal

A. SPSS22.0 System Clustering

(1) *Descriptive Statistical Analysis*

Considering the actual situation, this paper divides Shenyang's residential districts into four categories, and uses data statistics to reflect the overall characteristics of four different types of residential districts. 1-middle and high-end areas, 2-urban school districts, 3-old urban districts, 4-ordinary districts.

The description of district document express is shown in the Table 5. From Table 5, we can see that the maximum number of document express delivery in middle and high-grade residential and urban school districts reaches 12 pieces per day. Because of the majority of the elderly in the old urban area, they basically do not have document express delivery.

Table 5. Description of district document express

	Sample	Min	Max	Mean	SD
1	25	3	12	7.12	2.587
2	25	3	12	8.04	2.638
3	25	0	3	1.36	1.036
4	25	0	9	5.64	2.215
Effective N	25				

The description of small express is shown in the Table 6. For Table 6, the daily express delivery volume of ordinary community users is as high as 29 pieces, the lowest 10 pieces. From the average point of view, the express delivery of middle and high-grade residential areas will be slightly higher than that of urban school districts. For the old city, there is less express delivery.

Table 6. Description of small express

	N	Min	Max	Mean	SD
1	25	9	28	16.96	5.232
2	25	9	28	16.56	4.9504
3	25	3	12	7.96	2.6375
4	25	10	29	17.64	5.057
Effective N	25				

Medium-sized express delivery refers to articles such as wearable items. Large-sized items are usually small household appliances or assembly parts of some furniture. The conclusions drawn by subdividing market districts are shown in Tables 7 and 8 respectively.

Table 7. Description of middle express

	N	Min	Max	Mean	SD
1	25	9.0	23.0	14.800	3.4034
2	25	9.0	28.0	16.560	4.9504
3	25	0.0	5.0	1.920	1.4119
4	25	8.0	26.0	14.400	3.9264
Effective N	25				

Table 8. Description of large express

	N	Min	Max	Mean	SD
1	25	0	6	2.32	1.574
2	25	0	4	1.56	1.261
3	25	0	9	4.20	2.327
4	25	0	5	2.08	1.441
Effective N	25				

(2) *Systematic Clustering*

Based on the above analysis, systematic clustering first needs to standardize the data, and then uses different distance measures and clustering methods to cluster the data.

Data standardization is based on −1 to 1 standardization: $x'_{ij} = \frac{x_{ij} - \overline{x}_j}{R_j}$, selected distance formula: $d_{ik} = \sum_{j=1}^{k} \left(x_{ij} - y_{kj}\right)^2$.

The single cluster in SPSS system is used to get the number of specific types of each cluster, and the results are summarized, as shown in Table 9.

Table 9. Number of cells included in different cluster number

	Community type	Category I	Category II		
Grouped into two categories	1	25	0		
	2	25	0		
	3	0	25		
	4	25	0		
	Community type	Category I	Category II	Category III	
Grouped into three categories	1	21	4	0	
	2	22	3	0	
	3	0	0	25	
	4	21	4	0	
	Community type	Category I	Category II	Category III	Category IV
Grouped into four categories	1	19	4	2	0
	2	20	3	2	0
	3	0	0	0	25
	4	20	4	1	0

	Community type	Category I	Category II	Category III	Category IV	Category V
Grouped into five categories	1	16	4	3	2	0
	2	15	3	5	2	0
	3	0	0	0	0	25
	4	17	4	2	2	0

From the analysis, it can be concluded that there are differences in the proportion of express delivery size in the subdivided areas, but how can the specific differences not give a clear result? But it can be seen that clustering into four categories is the most reasonable, and the following use of SPSS K-means clustering function, the specific discussion of the proportion problem.

B. K-means Clustering

According to the analysis above, we classify it into four categories. Using K-means function, we input the prepared data into SPSS22.0 and K = 4. After software calculation, we get the start center and the final center. The results are shown in Table 10.

Table 10. Initial cluster and final cluster center

Initial Cluster Center					Final Cluster Center			
Clustering	1	2	3	4	1	2	3	4
File	5	9	4	1	7	7	6	1
Small express	15	18	29	3	15	16	27	8
Middle express	10	26	15	1	13	22	16	2
Large express	3	1	1	3	2	3	2	4

C. Design of Number and Size of Express Terminal

According to the results of Table 10, the common residential areas are clustered into the first category, and the final proportion of cluster centers is 7:15:13:2. The middle and high-grade residential areas are clustered into two categories, with the final proportion being 7:16:22:3.

The urban school districts are clustered into the third category, with the proportion being 6:27:16:2. And the last type is the old urban districts. At the same time, there are obvious differences in proportion of express delivery size between different regions, as shown in Table 11.

Table 11. Comparisons of express size in different types of cells

Community type	File	Salle	Middle	Large
1	7	16	22	3
2	6	27	16	2
3	1	8	2	4
4	7	15	13	2

Express terminals are generally divided into three types: small, medium and large, in which documents can be regarded as small express delivery. Therefore, the final conclusion is that the proportion of middle and high-grade residential areas is about 8:7:1; the proportion of urban school areas is 17:8:1; the proportion of old urban areas is 4:1:2; and the proportion of ordinary residential areas is 11:7:1.

4 Conclusion

With the rapid development of express industry and the increasing volume of express business, it is necessary to study the intelligent terminal of express delivery. Combined with the specific situation, this paper optimized and improved the number and size of intelligent express terminals in different regions based on the factors affecting user experience satisfaction and cluster analysis.

(1) By compiling reasonable social questionnaires and collecting data, descriptive statistical analysis, reliability analysis and regression analysis are carried out to ensure the reliability of the data.
(2) The existing data are processed by SPSS software, and the main factors affecting user satisfaction are determined as follows: service mode, fault processing speed, service reliability and service convenience.
(3) Classify the district and express, and use the system clustering method and K-means clustering analysis to get the reasonable size and quantity proportion of express terminal grid.

Acknowledgment. This work is supported by the Liaoning Province Natural Science Fund Project (Grant No.: 20180551001).

References

1. Yubing, S.: Research on size ratio optimization and revenue distribution of intelligent express terminal based on customer satisfaction. Thesis, Zhejiang University of Technology (2017). (in Chinese)
2. Shangqun, F.Z., Xiao, G., Qingjie, C., Changling, Z.: Evaluation of campus wireless network operation based on user experience quality. J. China Univ. Sci. Technol. **46**(1), 143–147 (2016)
3. Mengyue, C., Dongping, C., Xuening, Z.: User experience-based redesign module recognition for product service systems. Comput. Integr. Manuf. Syst. **22**(2), 2522–2529 (2016). (in Chinese)
4. Lu, T.S., Lu, H.S., Wei, H.L.: New service development model: a user experience-oriented design (Conference Paper). In: 18th Conference on Computer Supported Cooperative Work in Design, pp. 127–131 (2014)
5. Yamakami, T.: Relationship models of social experience design and user experience design (Conference Paper). In: 2014 Conference on Computing, Management and Telecommunications, pp. 531–540 (2014)
6. Luca, E., Narayan, B.: Signage by design: a design-thinking approach to library user experience. Weave: J. Libr. User Exp. **20**(5), 67–89 (2016)
7. Junjun, L., Fengyun, Y., Yuanyuan, C.: Dynamic behavior model and empirical study of mobile digital reading user experience. Mod. Inf. **39**(3), 24–25 (2019). (in Chinese)
8. Qindong, Z.J.: Research on robustness and promotion strategy of distribution network of express enterprises. J. Transp. Eng. Inf. **16**(3), 7–9 (2018). (in Chinese)
9. Shuai, W.: Delivery route decision-making of ecommerce express with uncertain time window. Thesis, Hebei University, pp. 1–3 (2018). (in Chinese)
10. Dongmei, L., Suidong, W.S.: Application of intelligent express terminal based on internet of things. Mod. Comput. **15**(8), 67–71 (2017). (in Chinese)
11. Du, J., Lu, W., Yinsheng, L.: Research on the layout of distribution centers of express enterprise for E-ecommerce (Conference Paper). In: 2010 Conference on Internet Technology and Applications, pp. 112–118 (2010)
12. Lijuan, W., Ni, W., Haiyan, L., Ling, Z.: Based on the analysis of the development status and causes of intelligent express terminal in the era of Internet of Things. Natl. Circ. Econ. **22**(3), 14–15 (2018). (in Chinese)
13. Hengtao, Z.: Design and implementation of smart express terminal system based on smart 210. Thesis, Anhui University, pp. 3–5 (2016). (in Chinese)
14. Wang Haihua, W., Yinan, L.X.: Research on the location of campus intelligent express terminal considering service radius. Logist. Eng. Manage. **40**(8), 90–92 (2018). (in Chinese)
15. Shengliang, Z., Xiaodong, L.: Study on the influencing factors of online shopping customer satisfaction. J. Tianjin Univs. **15**(2), 109–115 (2013). (in Chinese)

Study on Influencing Factors of Cold Chain Logistics Supply Chain for Fresh Agricultural Products

Di Liang[1], Kai Zhang[1](✉), Yu-cheng Pan[2], and Yue Liu[1]

[1] Department of Mechanical Engineering, Shenyang University, Shenyang, China
rafal120@163.com
[2] Yantai Zhenghai Magnetic Material Co., Ltd., Yantai, China

Abstract. Due to the perishable and fragile characteristics of agricultural products, cold chain logistics of agricultural products is more specific than general logistics activities. This paper proposes the combination of the interval gray number DEMATEL method, which solves the problem that the traditional DEMATEL method is too certain when scoring and ignores the differences of different enterprises. At the same time, the flexibility of interval gray number is used to solve the ambiguity of factor selection in evaluation. Deterministic, build a more flexible and realistic Grey-DEMATEL method decision model. Based on the interval gray number theory, the relationship matrix of the influencing factors for fresh agricultural products cold chain logistics supply chain is established. The comprehensive impact matrix of the cold chain logistics supply chain of fresh agricultural products was solved by MATLAB, and the cold chain logistics supply chain were quantitatively analyzed. The key influencing factors of the supply chain have laid the foundation for further optimizing the cold chain logistics supply chain of fresh produce.

Keywords: Cold chain logistics supply chain · Fresh agricultural products · Influencing factors

1 Introduction

Modern Industrial Engineering is the project which is gradually formed and developed on the basis of Production Manufacturing Management and System Engineering. It can be applied to services, manufacturing, food, medical, logistics etc. [1].

With the development of economy and the improvement of human living standard, increasingly attention has focused on the freshness and safety of food.

Low fresh food of loss rate is the foundation to ensure the freshness and quality safety of fresh food [2, 3]. However, the construction of fresh agricultural products cold chain logistics supply chain system has just started. It has faced the various problems like backward infrastructure, incomplete cold chain logistics supply chain system, low industry level, low marketization and different standards [4–6]. Although scholars have put forward many corresponding to the solutions, the lack of prioritization of these problems has greatly reduced the practical significance of lots of measures. Therefore,

C.-F. Chien et al. (Eds.): *IE&EM 2019*, pp. 203–212, 2020.
https://doi.org/10.1007/978-981-15-4530-6_21

this paper combines DEMATEL (Decision making Trial and Evaluation Laboratory) method with interval grey number to construct Grey-DEMATEL model to analyze the influencing factors of cold chain logistics supply chain for fresh agricultural products.

Fang Kai and Zhong Chengbao [7] has used three-stage data envelopment analysis model to research the efficiency of China's cold chain logistics enterprises. Finally, they found that the main obstacle to their development was scale inefficiency. They put forward five related improvement suggestions based on that. Yuan Xueguo and Zou Ping [7] analyzed the development situation of China's cold chain logistics industry. The study shows that there are many problems in its development and then put forward corresponding countermeasures and suggestions according to the problems which need to be solved. The factors affected the development of a certain events from various aspects and the respective influencing factors are interrelated and interacted with each other. The status and influencing degree of the respective influencing factors are quite different among them. Therefore, from the point of view of System Engineering, this paper uses Grey-DEMATEL method model to analyze fresh agricultural products cold logistics chain supply chain and determine the hierarchical position and influencing relationship of the influencing factors on it.

2 Model Construction

A. *Construction of Key Influencing Factor Index System of Cold Chain Logistics Supply Chain for Fresh Agricultural Products*

This paper uses brainstorming method to analyze influencing factors of fresh agricultural products with cold chain logistics supply chain [9]. Screen 17 factors from five aspects which are vulnerability factors, personnel factors, management factors, facility factors and cost factors. Grey-DEMATEL method was used to establish the index system of key factors for fresh agricultural products cold chain logistics supply chain. There are 5 first-level indexes and 17 second-level index (the indexes are represented by F_1, F_2, $\cdots$, F_{17}.) in the index system. The key influencing factors index system fresh agricultural products cold chain logistics supply chain established in this paper is shown in Fig. 1.

B. *Grey Number Research Method*

The grey number used in this paper is interval grey number [10, 11], which is recorded as $\otimes\lambda \in [\underline{\otimes}\lambda, \overline{\otimes}\lambda]$. The $\underline{\otimes}\lambda$ is lower limited of $\otimes\lambda$, $\overline{\otimes}\lambda$ is upper limited of $\otimes\lambda$. The detailed operation steps are as follows:

$$\otimes\lambda_1 + \otimes\lambda_2 = \left[\underline{\lambda_1} + \underline{\lambda_2}, \overline{\lambda_1} + \overline{\lambda_2}\right] \tag{1}$$

$$\otimes\lambda_1 - \otimes\lambda_2 = \left[\underline{\lambda_1} - \overline{\lambda_2}, \overline{\lambda_1} - \underline{\lambda_2}\right] \tag{2}$$

$$\begin{aligned}\otimes\lambda_1 \times \otimes\lambda_2 = [&min\left(\underline{\lambda_1\lambda_2}, \underline{\lambda_1}\overline{\lambda_2}, \overline{\lambda_1}\underline{\lambda_2}, \overline{\lambda_1\lambda_2}\right), \\ &max\left(\underline{\lambda_1\lambda_2}, \underline{\lambda_1}\overline{\lambda_2}, \overline{\lambda_1}\underline{\lambda_2}, \overline{\lambda_1\lambda_2}\right)]\end{aligned} \tag{3}$$

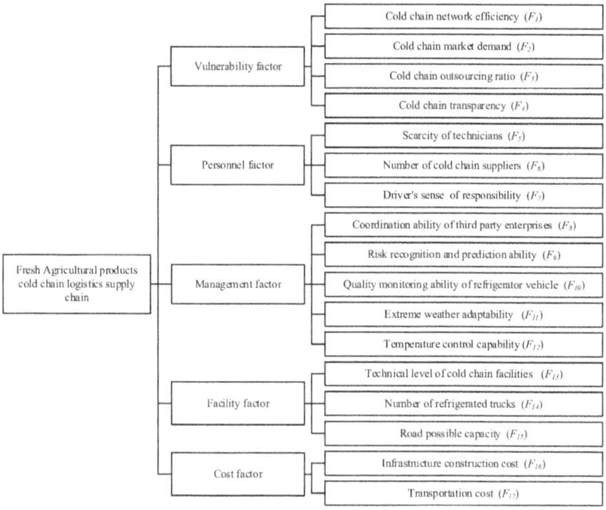

Fig. 1. Influencing factors index system of fresh agricultural products cold chain logistics supply chain

$$\otimes\lambda_1 \div \otimes\lambda_2 = \left[\underline{\lambda_1}, \overline{\lambda_1}\right] \times \left[\frac{1}{\underline{\lambda_2}}, \frac{1}{\overline{\lambda_2}}\right] \tag{4}$$

Considering that experts usually have a certain degree of fuzziness and uncertainty when they scored. The formula $\otimes\lambda_{ij}^k$ has been defined as the score of expert K for evaluating the influencing of influencing factor i on influencing factor j, and the $\otimes\lambda_{ij}^k \in \left[\underline{\otimes}\lambda_{ij}^k, \overline{\otimes}\lambda_{ij}^k\right]$. The clarificatory process for semantic evaluation variables for experts as follows:

(1) Standardize the upper and lower limit of grey number

$$\begin{aligned} \overline{\otimes}\tilde{\lambda}_{ij}^k &= \frac{\overline{\otimes}\lambda_{ij}^k - min\overline{\otimes}\lambda_{ij}^k}{\Delta_{min}^{max}} \\ \underline{\otimes}\tilde{\lambda}_{ij}^k &= \left(\underline{\otimes}\lambda_{ij}^k - min\underline{\otimes}\lambda_{ij}^k\right)/\Delta_{min}^{max} \\ \text{and, } \Delta_{min}^{max} &= max\overline{\otimes}\tilde{\lambda}_{ij}^k - min\underline{\otimes}\lambda_{ij}^k \end{aligned} \tag{5}$$

(2) Clear processing of standardized grey number

$$Y_{ij}^k = \frac{\left\{\underline{\otimes}\tilde{\lambda}_{ij}^k\left(1 \quad \underline{\otimes}\tilde{\lambda}_{ij}^k\right) + \left(\overline{\otimes}\lambda_{ij}^k \times \overline{\otimes}\lambda_{ij}^k\right)\right\}}{\left(1 - \underline{\otimes}\tilde{\lambda}_{ij}^k + \overline{\otimes}\lambda_{ij}^k\right)} \tag{6}$$

(3) Work out the clarity value

$$Z_{ij}^{k} = \underline{min} \otimes \lambda_{ij}^{k} + Y_{ij}^{k} \Delta_{min}^{max} \tag{7}$$

C. *DEMATEL Method*

This method is proposed by the Bottelle Institute of the United States to apply graph theory and matrix theory to effectively analyze the logical relationship between various factors. The specific operation steps are as follows:

(1) Construct influencing factor matrix. Use 0–5 to represent the relationships among the influencing factors. Detailed steps as follows:

$$a_{ij} = \begin{cases} 0 & \text{no influence between two} \\ 1 & \text{slight influence between two} \\ 2 & \text{low influence between two} \\ 3 & \text{obvious influence between two} \\ 4 & \text{great influence between two} \end{cases}$$

(2) Establish influencing factor matrix A. According to the expert's evaluation results to get directly influencing factor matrix A. Matrix A indicates the influencing degree of row factor i on column factor j.

$$A = \left(a_{ij}\right)_{n \times n} = \begin{bmatrix} a_{11} & a_{12} & \cdots & a_{1n} \\ a_{21} & a_{22} & \cdots & a_{2n} \\ \vdots & \vdots & \vdots & \vdots \\ a_{n1} & a_{n2} & \cdots & a_{nn} \end{bmatrix}$$

(3) Standardize matrix A. Sum of the rows of matrix A, *max* means the maximum of rows sum, let $X = A/max$.
(4) Calculate the comprehensive influencing matrix T. $T = X(I - X)^{-1}$
(5) Analyze influencing factors. According to the elements T_{ij} from the comprehensive influencing matrix to determine the relationships between the influencing factors. To get every element's influencing degree R_i, influenced degree D_j, centrality degree P_i and causality degree E_i [12, 13].

D. *Grey-DEMATEL Method*

In order to solve the uncertainty and fuzziness of factor selection in evaluation. This paper combines DEMATEL method with interval grey number to establish Grey-DETEMAL method decision model which is practical and flexible. Its operation steps are as follows:

(1) Establish fresh agricultural products cold chain logistics supply chain relationship matrix based on interval Grey Number Theory. The relationships between influencing factor i and influencing factor j are devided five kinds which are direct influence, weak influence, lower influence, obvious influence and great influence recoded as N, W, L, H and VH, the interval grey number in detail (see Table 1).

Table 1. Semantic variable of expert's evaluation

Semantic variable	Interval grey number
Direct influence (N)	[0, 0]
Weak influence (W)	[0, 0.25]
Lower influence (L)	[0.25, 0.5]
Obvious influence (H)	[0.5, 0.75]
Great influence (VH)	[0.75, 1]

The experts' evaluations would correspond change with experts' different attention on fresh agricultural products cold chain logistics supply chain. Therefore, the weight value is given the fuzziness characteristic of interval grey number (see Table 2).

Table 2. Semantic variable of expert's weight

Semantic variable	Interval grey number
Unimportant	[0, 0.3]
A little important	[0.3, 0.5]
Important	[0.4, 0.7]
More important	[0.5, 0.9]
Great important	[0.9,1]

(2) Establish the grey number matrix. Transformed the formed grey number matrix to get grey number matrix $\otimes\lambda$ according to Table 1.
(3) Clear processing of the grey number matrix according to formula (5), (6), (7), and used the formula (8) to calculate the weighted weight matrix Z. Z_{ij} represents the element which lies in column j and row i of the weight matrix.

$$Z_{ij} = \omega_1 Z_{ij}^1 + \omega_2 Z_{ij}^2 + \ldots + \omega_n Z_{ij}^n$$

ω_n represents the weight proportion of element Z_{ij}

$$\sum\nolimits_{i=1}^{n} \omega_i = 1 \tag{8}$$

(4) Gain the standardized matrix M according to formula (9), (10) to standardized Z. Then used formular (11) to calculate comprehensive influencing matrix T $(T = \left[t_{ij}\right]_{n\times n})$.

$$M = S \cdot Z \tag{9}$$

$$S = \frac{1}{\max\limits_{1\le i\le n} \sum\nolimits_{j=1}^{n} Z_{ij}} \quad i,j = 1, 2, \cdots n. \tag{10}$$

(5) Calculate the comprehensive influencing matrix.

$$T = M(I - M)^{-1} \tag{11}$$

(6) Calculate influencing degree R_i, influenced degree D_j, centrality degree P_i and causality degree E_i

$$R_i = \sum\nolimits_{i=1}^{n} t_{ij} \quad (i = 1, 2, \cdots, \mathrm{n}); \tag{12}$$

$$D_j = \sum\nolimits_{j=1}^{n} t_{ij} \quad (j = 1, 2, \cdots, \mathrm{n}); \tag{13}$$

$$P_i = R_i + D_j | i = j; \tag{14}$$

$$E_i = R_i - D_j | i = j \tag{15}$$

Through the above calculation, the influencing degree of each influencing factor on fresh agricultural products cold chain logistics supply chain could be judged by the influence degree and influenced degree. The importance of each index could be determined in the agricultural cold chain logistics' influencing factors index system by the centrality degree. It could further analyze the relationships between each index by the causality degree [12, 13]

3 Example Analysis

In order to better use Grey-DEMATEL method to analyze the key influencing factors of fresh agricultural products cold chain logistics supply chain, this paper carried out an example analysis of fresh agricultural cold chain logistics in M enterprise. Firstly, the questionnaire was distributed to seven experts who knew about the cold chain logistics supply chain of agricultural products or had been engaged in the cold chain logistics of fresh agricultural products for a long time to fill in, then recycled and analyzed. According to the results of investigation and interview, seven experts were given different weights which are processed by interval grey number method (see Table 3).

According to the above formulas and the analysis of the results of questionnaires, the Grey-DEMATEL method is used to construct a direct influencing matrix A.

The direct influencing matrix A is converted into grey number matrix X according to Table 3. Formula (5)–(7) are used to clarify the grey number matrix X. Finally, a comprehensive influencing table (see Table 4) is obtained. The influencing degree R_i, influenced degree D_j, centrality degree P_i and causality degree E_i of each factor are be worked out (see Table 5).

According to the cause index of each factor, the influencing factors of fresh agricultural products cold chain logistics supply chain are divided into two categories.

(1) Cause factors. The rank of cause factors ($E_i > 0$) by their size is: $F_1 > F_6 > F_9 > F_4 > F_2 > F_{14} > F_5 > F_{11} > F_3 > F_7$. These factors are positive factors to promote the development of cold chain logistics supply chain

Table 3. Expert weight values

Category of experts	Expert weight
Head of cold chain logistics project of M company	[0.3, 0.7]
Head of cold chain logistics development planning department of M company	[0.5, 0.8]
Cold chain logistics supply chain expert 1	[0.3, 0.6]
Cold chain logistics Supply chain expert 2	[0.3, 0.5]
Cold chain logistics supply chain expert 3	[0.4, 0.7]
Cold chain logistics supply chain expert 4	[0.6, 0.9]
Cold chain logistics supply chain expert 5	[0.7, 1]

due to they have a direct influencing on it. These factors should be strengthened. Advanced logistics equipment and technology such as GPS technology, RFID and wireless handheld terminal can be applied to realize the information management of all operation links for vehicle logistics and improve the efficiency of cold chain network. What's more, the strength of leading enterprises, changes in market supply and demand, construction of marketing channel and regional economic development should be paid attention to promote the circulation of the whole chain.

(2) Result factors. The rank of result factors ($E_i < 0$) by result factors' absolute value is: $F_{17} > F_{10} > F_{13} > F_8 > F_{12} > F_{15} > F_{16}$. The quality control capability and transportation cost are the factors which are easily changed by other results. Therefore, the relevant rules of fresh agricultural products cold chain logistics supply chain should be standardized. The organization and coordination of all links should be strengthened. Improve transport efficiency and reduce cost. The assessment and training of refrigerated vehicle drivers and professionals should be strengthened gradually.

According to the centrality degree ($P_i > 0$) index of each factor, the rank of fresh agricultural products cold chain logistics supply chain's influencing factors is: $F_8 > F_{12} > F_{17} > F_{11} > F_5 > F_{13} > F_{14} > F_{10} > F_2 > F_4 > F_3 > F_{15} > F_9 > F_{16} > F_6 > F_1 > F_7$. It shows that temperature control capability, coordination ability of third party enterprises and extreme weather adaptability would make enterprises actively face the problems caused by it and reduce can losses in maximum. Therefore, enterprises should constantly debug in the application process of information platform to determine the application environment and optimal technical parameters of each technology. And give full play to the role of information platform.

Table 4. Comprehensive influencing table

Factor	F_1	F_2	F_3	F_4	F_5	F_6	F_7	F_8	F_9	F_{10}	F_{11}	F_{12}	F_{13}	F_{14}	F_{15}	F_{16}	F_{17}
F_1	0.0242	0.1776	0.1180	0.0873	0.2170	0.0492	0.1913	0.3419	0.1570	0.2814	0.1630	0.2424	0.2886	0.1022	0.2575	0.1404	0.3425
F_2	0.0166	0.0462	0.1787	0.0645	0.1771	0.0354	0.0760	0.2559	0.0331	0.1315	0.0959	0.1658	0.2237	0.0559	0.2015	0.0932	0.2616
F_3	0.0352	0.1024	0.1346	0.1009	0.2199	0.0624	0.1083	0.3633	0.1630	0.2256	0.2679	0.3515	0.2945	0.2097	0.1774	0.2519	0.3728
F_4	0.0340	0.1812	0.1200	0.0967	0.1302	0.0520	0.0938	0.3427	0.0610	0.2984	0.2590	0.3405	0.2836	0.2128	0.2525	0.2526	0.2698
F_5	0.0360	0.1867	0.2318	0.1059	0.1497	0.1472	0.2125	0.3802	0.1640	0.2231	0.1801	0.3597	0.3191	0.1203	0.2763	0.1645	0.3817
F_6	0.0216	0.0295	0.0649	0.0514	0.0547	0.0222	0.0472	0.2145	0.0182	0.1915	0.0943	0.2163	0.0834	0.0541	0.1615	0.0688	0.2214
F_7	0.0238	0.0534	0.0714	0.0591	0.1632	0.0328	0.0527	0.1535	0.0428	0.1148	0.0819	0.2385	0.1992	0.1543	0.1030	0.1719	0.1571
F_8	0.1199	0.0846	0.2115	0.1701	0.1317	0.0658	0.1085	0.3193	0.0585	0.2024	0.2369	0.1986	0.1826	0.1835	0.1360	0.1400	0.3355
F_9	0.0328	0.0782	0.2209	0.1867	0.1382	0.0614	0.1967	0.2576	0.0409	0.3044	0.2564	0.3278	0.2819	0.0986	0.2501	0.2389	0.3680
F_{10}	0.0273	0.1037	0.2333	0.0987	0.2406	0.1543	0.2224	0.3923	0.1721	0.3320	0.1765	0.2735	0.3212	0.1948	0.1953	0.1567	0.3927
F_{11}	0.0291	0.0553	0.1130	0.0838	0.1843	0.0432	0.1814	0.3094	0.0380	0.2569	0.1352	0.2909	0.2375	0.0682	0.1340	0.1937	0.3182
F_{12}	0.0243	0.0490	0.0777	0.1608	0.0784	0.0385	0.0591	0.2512	0.0255	0.2232	0.1212	0.2427	0.1139	0.0651	0.1887	0.0939	0.2599
F_{13}	0.0367	0.1844	0.1278	0.1030	0.2180	0.0544	0.0993	0.3697	0.1614	0.3077	0.1865	0.3673	0.2929	0.1203	0.2671	0.2520	0.3720
F_{14}	0.0264	0.1946	0.1283	0.0936	0.2376	0.1627	0.1191	0.3667	0.0731	0.3171	0.2694	0.2640	0.3053	0.1099	0.2719	0.2508	0.3846
F_{15}	0.0263	0.1910	0.2191	0.0856	0.2413	0.1514	0.2089	0.2901	0.0800	0.3134	0.1612	0.2634	0.3024	0.1901	0.1789	0.2396	0.3884
F_{16}	0.0199	0.0393	0.0724	0.0492	0.0568	0.0310	0.0571	0.2074	0.0199	0.1012	0.1838	0.1989	0.0852	0.0525	0.0628	0.0687	0.2223
F_{17}	0.0199	0.0393	0.0724	0.0492	0.0568	0.0310	0.0571	0.2074	0.0199	0.1012	0.1838	0.1989	0.0852	0.0525	0.0628	0.0687	0.2223

Table 5. Causality degree and centrality degree of influencing factor

Index	F_1	F_2	F_3	F_4	F_5	F_6	F_7	F_8	F_9	F_{10}	F_{11}	F_{12}	F_{13}	F_{14}	F_{15}	F_{16}	F_{17}
R_i	3.6874	3.4413	2.6721	3.5751	3.6388	3.1815	2.1126	3.3395	3.2808	1.5284	3.5311	2.8854	2.0731	3.5205	1.8734	1.6155	1.5284
D_i	0.5540	1.7964	2.3958	1.6465	2.6955	1.1949	2.0914	5.0231	1.3284	3.9258	3.0530	4.5407	3.9002	2.0448	3.1773	2.8463	3.9258
P_i	4.2414	5.2377	5.0679	5.2216	6.3343	4.3764	4.2040	8.3626	4.6092	5.4542	6.5841	7.4261	5.9733	5.5653	5.0507	4.4618	5.4542
E_i	3.1334	1.6449	0.2763	1.9286	0.9433	1.9866	0.0212	−1.6836	1.9524	−2.3974	0.4781	−1.6553	−1.8271	1.4757	−1.3039	−1.2308	−2.3974

4 Conclusion

In this paper, the brainstorming method was used to find out the key factors which influenced fresh agricultural products cold chain logistics supply chain. Due to the complex relationships between influencing factors of fresh agricultural products clod chain logistics supply chain, an index system was established for its influencing factors. Through the analysis of the relationships among the influencing factors of it, the grey number matrix was used to describe quantitatively. In view of the fuzziness and uncertainty when experts scored for influencing factors, the interval grey number from grey number theory was used to clarify it. Then Grey-DEMATEL model was established. The validity and operability of fresh agricultural products cold chain logistics supply chain's influencing factors model based on Grey-DEMATEL have been verificated by the case. The conclusion could better fit the actual situation of enterprises, reflect the problems existed in cold chain logistics supply chain and provide reference measures for enterprises to improve.

Clarification. The reason of anonymous is that all the experts involved in the scoring work in the enterprise, involving some private information of the enterprise, so anonymous words are used to score.

The Authenticity of the questionnaire, the questionnaire was scored by colleagues and leaders of the internship department at that time, and was analyzed against the background of the internship company.

References

1. Liang, D., Meng, W.: Application of Taguchi method in optimal combination of parameters design of main spindle system. J. Shenyang Univ. (Nat. Sci.) **28**(05), 405–409 (2016)
2. Fan, L.N., Dong, D.Y., Li, J.Y., Liu, C., Ding, Y.: Route optimization of cold chain logistics based on fresh agricultural products. J. Shenyang Univ. (Nat. Sci. Ed.) **29**(2), 125–131 (2016)
3. Yang, F., Xie, R.H.: Constructing the interpretive structural model for the cold chain logistics systems of fresh farm produce. Syst. Eng. J. **30**(12), 99–104 (2012)
4. Ju, H.: Research on cold chain logistics system of fresh agricultural products based on synergetics. Logist. Technol. **32**(11), 25–27 (2013)
5. Wan, Y.F.: Study on Income Balance of Three-Level Fresh Agricultural Products Logistics Supply Chain. Qufu Normal University, Qufu (2015)
6. Wang, Y.Y.: Study on the Safety Risk Assessment of Agricultural Cold Chain Logistics. Changan University, Xian (2014)
7. Fang, K., Zhong, S., Wang, H.J., Lan, L.: Analysis of the efficiency of China's cold chain logistics enterprises based on green supply chain. Agric. Technol. & Econ. **30**(12), 99–104 (2012)
8. Yuan, X.G., Zou, P., Zhu, J., Wu, D.: Development trend, problems and Countermeasures of cold chain logistics industry in China. China Agric. Sci. Technol. Rep. **17**(1), 7–14 (2015)
9. Liu, H.C., Liu, R.: Risk analysis of cold-chain logistics distribution based on an improved FMEA method. Preserv. Process. **18**(4), 119–125 (2018)
10. Wu, C.Y., Chen, X.H., Kuang, B.H.: Influencing factors Identification of enterprise green growth pattern based on rough-DEMATEL. Manage. Rev. **26**(08), 74–81 (2018)

11. Qu, H.: Research on grey-AHP assessment method based on grey number judgment matrix and its applications. J. Hefei Univ. Technol. (Nat. Sci.) **39**(11), 1567–1570 (2016)
12. Wang, Y.H., Zhang, H.W.: Elasticity evaluation of agricultural cold chain logistics supply chain in Heilongjiang. Logist. Technol. **35**(05), 146–150 (2016)
13. Wu, S.: M Dessert Company Cold Chain Logistics System Optimization Design. Shenyang University, Shenyang (2016)
14. Govindan, K., Khodaverdi, R., Vafadarnikjoo, A.: A grey DEMATEL approach to develop third-party logistics provider selection criteria. Ind. Manage. Data Syst. **116**(04), 132–138 (2016)
15. Wang, S.: Research on the Influencing Factors of Port Green Development Based on Grey-DEMATEL Method. Dalian University of Technology, Dalian (2016)

Effect Factors Analyzing on Supply Chain Logistics Collaboration Based on Structural Equation Model

Rui-min Zhang and Xiao-ying Yang(✉)

School of Mechatronics Engineering,
Henan University of Science and Technology, Luoyang, China
lyyxy111@163.com

Abstract. For the supply chain logistics collaboration problem with manufacturers as the core, the questionnaire was designed through investigation and analysis. The index system of three-dimension collaborative influence factors about logistics standardization, process information and decision-making intelligent was determined. The Structural Equation Model of logistics collaboration influencing factors was constructed based on the relationship hypothesis of index factors affecting logistics collaboration. Combined with the results of the questionnaires, SPSS & AMOS was used to analysis reliability and validity. And structural equation model was verified and evaluated. Finally, the path relationship and the degree of influence between the various factors were determined. It provides a reference for management and control of manufacturing supply chain logistics collaboration in China.

Keywords: Influencing factor · Logistics collaboration · Supply chain · Structural Equation Model

1 Introduction

With the increasingly complex organizational structure of the supply chain, the factors affecting the supply chain collaboration have increased dramatically, which has aggravated the difficulty of supply chain collaboration. Logistics as the physical foundation throughout the supply chain, logistics collaboration is the core of supply chain collaboration, and is essential to improve the overall service quality of the supply chain [1, 2]. In order to improve the rapid response of supply chain logistics to market demand and production, and to solve the logistics collaboration problem among various links in the complex supply chain, the mining and analysis of the factors of supply chain logistics collaboration is to achieve on time, efficient, orderly and low inventory is the key to the supply chain logistics collaboration.

At present, domestic and foreign scholars' research on supply chain logistics collaboration mainly focuses on logistics collaborative management mode, collaboration mechanism and strategy, collaborative performance and so on. Rhea and Shrock first proposed the logistics utility measurement index to measure the contribution of logistics to corporate profits [3]. Sawik pointed out that logistics performance

C.-F. Chien et al. (Eds.): *IE&EM 2019*, pp. 213–223, 2020.
https://doi.org/10.1007/978-981-15-4530-6_22

evaluation can better reflect the logistics operation of the supply chain under certain customer service levels [4]. In China, Ma studied the optimization of supply logistics collaborative strategy under the Supply-Hub model from the perspective of transportation and inventory, VMI integration and JIT technology, production and distribution [5, 6]. Xie et al. studied the synergy of multiple suppliers' supply logistics and its impact on supply chain enterprise performance from the three dimensions of supplier-manufacturer collaboration, supplier collaboration and logistics service capability [7]. Representative research results of logistics collaboration factors: Qian and He studied the factors affecting the synergy between the two by constructing a grounded theoretical model of cross-border e-commerce and cross-border logistics coordination [8]. Liu and Li used the structural equation model to reveal the influence path and influence degree of the various factors affecting the resource allocation of coal production logistics system [9]. Fu and Li studied the hierarchical structure of the influencing factors of regional logistics and regional economic collaboration development by improving the Interpretative Structural Model [10]. Mi and Zeng used regression analysis to study the logistics links current status in the Yangtze River Economic Belt and its influencing factors [11]. In terms of manufacturing logistics, Huang and Chen constructed a PCA-GRA-based manufacturing and logistics industry collaboration evaluation model for the collaboration between Xiamen's manufacturing and logistics industries [12]. For the application of research methods, some scholars use factor analysis, principal component analysis, gray correlation, AHP, regression analysis and other methods to analyze the factors. These methods are more subjective to the measurement and evaluation of qualitative indicators, and it is difficult to achieve simultaneous estimation of the relationship between factor structure and factors. The Structural Equation Model (SEM) can make up for the shortcomings of the above methods. Zhang and Zhang used SEM to analyze the influencing factors of the selection behavior of the dilemma in the signalized intersection, and obtained the key influencing factors and their influence degree [13].

In summary, the existing research results mainly focus on logistics collaborative management mode, collaboration mechanism and strategy, and collaboration performance, which laid the foundation for this research. The factors of the logistics collaboration to analyse problem mainly concentrated in the fields of regional logistics and e-commerce logistics, but there are few researches on the factors of manufacturing supply chain logistics collaboration. Therefore, this paper focuses on the analysis of the influence factors for the supply chain logistics collaboration, taking supply chain logistics as the research object, and determining the index system of logistics collaboration factors by the investigation and analysis. On this basis, constructing multiple dimensions of supply chain logistics synergy influencing factors analysis model, and the SEM is used to measure the logistics collaboration factors, reveal the influence path and influence degree between each influencing factor, and provide the theory for the logistics association of supply chain related enterprises.

2 Determination of Synergistic Factors

2.1 Structural Equation Model

The structural equation model consists of two parts: the measurement equation and the structural equation. The measurement model describes the relationship between the latent variable and the explicit variable. The structural model describes the relationship between exogenous latent variables and endogenous latent variables, and this relationship is expressed in the form of a graph to form a path diagram.

Research on the Influence Factors upon Supply Chain Logistics Collaboration belongs to the uncertainty problem of combining qualitative and quantitative factors among multiple factors. For traditional statistical methods, it is impossible to deal with latent variables that cannot be directly and accurately measured. SEM can be indirectly reflected by the data of observation indicators, and at the same time estimate the relationship between variable structure and variables. Therefore, the structural equation model has been widely used in the fields of behavioral science and sociology [14]. In addition, SEM allows for some errors in the measured variables, which can avoid the problem that the evaluator subjectively cannot accurately measure. In summary, this paper uses SEM to study the influence path and influence degree of the factors affecting supply chain logistics collaboration.

2.2 Determining the Impact Factor Indicator

The research literature found that there is currently no direct application to the manufacturer-centric supply chain logistics synergy impact indicator system. In order to ensure the rationality of the indicators, on the basis of research on enterprise research and interviews with logistics experts, 26 influencing factors were selected from all relevant logistics contents in the whole process of supply chain purchase order, material transportation, receipt and storage, sorting and distribution, and production assembly and assembly.

According to the survey results, the key factors affecting supply chain logistics collaboration are identified through factor analysis and logistic regression analysis, namely 20 indicators in 4 dimensions. The identified four dimension influencing factors are used as the latent variables of the structural equation model, and the corresponding 20 influencing factor indicators are used as the observed variables of each latent variable. The analysis found that the content of the indicators contained in the first three latent variables coincides with the three core elements of logistics collaboration: logistics standardization, process informationization and decision-making intelligence matching, so the three latent variables are named. According to the index content of the 4th latent variable, the variable is named as supply chain logistics collaboration. In summary, the preliminary identified influencing factors indicator system of supply chain logistics collaboration is shown in Table 1.

Table 1. Factors affecting supply chain logistics collaboration

Key influencing factors (variable)	Item	Indicator code
Logistics standardization (BZ)	Material delivery punctuality rate	BZ1
	Material delivery order rate	BZ2
	Material delivery accuracy	BZ3
	Logistics operation standardization	BZ4
	Standardization of logistics equipment	BZ5
	Material data account matching rate	BZ6
Process information (XX)	Logistics management system information platform coverage (LMS, SCM, SRM, WMS, etc.)	XX1
	Application level of information acquisition technology (RFID, EDI, etc.)	XX2
	Real-time information transfer ratio	XX3
	Information transmission accuracy	XX4
	Information sharing depth	XX5
	Information sharing breadth	XX6
Intelligent decision making (ZN)	Information system integration level (SCM, ERP & SRM, ERP & CRM, MES & ERP, MES & PLM, etc.)	ZN1
	Intelligent level of logistics and storage equipment	ZN2
	Material dynamic tracking and quality traceability	ZN3
	The applicability of intelligent decision making (Automatically generate and dynamically adjust material distribution plans)	ZN4
	Intelligent control agility (Anti-error material, material shortage warning)	ZN5
Supply chain logistics collaboration (XT)	Order completion rate	XT1
	Order response speed	XT2
	Inventory turnover rate	XT3

3 Influential Factor Structural Model

From the analysis of the influencing factors of supply chain logistics collaboration, logistics standardization, process informationization, and intelligence decision making will have a direct impact on supply chain logistics collaboration. However, there may also be interactions between these factors, such that one factor has an indirect effect on logistics collaboration by acting on other factors. According to the analysis of the key influencing factors of supply chain logistics collaboration, the relationship hypothesis between each influencing factor is proposed. The content of the hypothesis is shown in Table 2. Then, based on the hypothesis, the structural model of logistics synergy path is constructed. As shown in Fig. 1, the ellipse represents the latent variable.

Table 2. Relationship hypothesis of logistics synergy factors

Serial number	Assumed content
H1	Logistics standardization has a positive impact on supply chain logistics collaboration
H2	Process informatization has a positive impact on supply chain logistics collaboration
H3	Intelligent decision-making has a positive impact on supply chain logistics collaboration

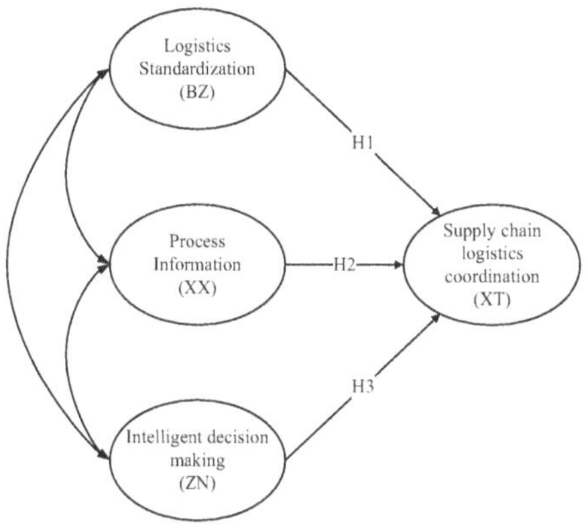

Fig. 1. Structural model of logistics synergy path

4 Model Analysis and Evaluation

4.1 Data Sources

This paper mainly uses questionnaires to conduct in-depth research and collect data. The investigation was carried out in 2018, mainly for the first-line employees and middle- and high-level managers of the supply chain logistics operations of the assembly enterprises. The survey includes the basic situation of the respondents and the impact of logistics standardization, informationization, and intelligence on collaboration. The design of the questionnaire is mainly based on the Likert 5 scale method. This questionnaire is mainly based on the questionnaire star push survey. The questionnaire design is a mandatory submission form, which effectively avoids the invalid questionnaire problem caused by too many blank answers, and a total of 389 questionnaires were collected. There were 351 valid questionnaires after eliminating the questionnaire with obvious errors. The effective rate was 90.3%.

4.2 Reliability and Validity Analysis

Before the structural equation model analysis, it needs to analyze the reliability and validity of the data used is required.

Reliability analysis is reliability analysis. The reliability of the questionnaire results is analyzed by SPSS 17.0. The consistency and stability of each latent variable were measured by Cronbach's alpha value. The reliability analysis results of each latent variable are shown in Table 3. From the survey analysis results, Cronbach's alpha values for each latent variable and all observed variables were greater than 0.7, with higher reliability and passed an internal consistency test. Validity analysis is the effectiveness analysis used to test the merits of the structural validity of the scale. This paper mainly tests by factor analysis. First judge the relevance of the scale. When the KMO is close to 1 and the P value (Sig.) of the Bartlett spherical test is less than 0.05, the correlation between the variables is better. The calculation results are shown in Table 3. The KMO value is 0.809. The Bartlett sphere test is significant at the level of P = 0.000. The data has good structural validity and can be used to analyze the structural equation model.

Table 3. Test results of scale reliability and validity

Variable	Cronbach's alpha	Standardized item based Cronbach's alpha	KMO	Number of items
BZ	0.816	0.831	0.768	6
XX	0.842	0.847	0.864	6
ZN	0.784	0.793	0.826	5
XT	0.813	0.815	0.795	3
Total amount	0.815	0.826	0.809	20

4.3 Model Fitting Test and Evaluation

Using AMOS23.0 to draw the initial path map of logistics synergy factors, the data through the reliability and validity test is imported into the initial model for rationality test and fitting analysis, and the rationality of the structural model is evaluated according to the fitting index. Correct the unreasonable model and then estimate the parameters.

According to the principle of "gradual rationality", the rationality of each measurement model is first tested. Through the convergence validity of the CFA test scale, the combination reliability (CR) of each variable is more than 0.7, and the average variance extraction value (AVE) is more than 0.5, indicating that each factor has good convergence validity. The results are shown in Table 4. The normalized loads of the measured indicators on the measured latent variables are all in the range of 0.5–0.95, all of which passed the rationality test.

Table 4. Rationality test results of measurement model

Variable	Item	Factor loadings	CR	AVE
BZ	BZ1	0.719	0.851	0.617
	BZ2	0.607		
	BZ3	0.801		
	BZ4	0.612		
	BZ5	0.549		
	BZ6	0.822		
XX	XX1	0.817	0.914	0.590
	XX2	0.839		
	XX3	0.735		
	XX4	0.741		
	XX5	0.619		
	XX6	0.554		
ZN	ZN1	0.819	0.873	0.578
	ZN2	0.705		
	ZN3	0.693		
	ZN4	0.782		
	ZN5	0.734		
XT	XT1	0.783	0.866	0.526
	XT2	0.717		
	XT3	0.689		

Then estimate the parameters of the structural model. By calculating the path coefficient and the degree of significance between the latent variables, the rationality of the path relationship between the latent variables is judged. As shown in Table 5, it can be seen from the parameter estimation results of the structural model that the standard error of the measured variables contained in each latent variable is less than 0.20, the CR value is greater than 2.58, and the path relationship between the latent variables is $P < 0.05$. The level is statistically significant and passed the test.

Table 5. Parameter estimation of structural model

Hypothesized paths	Estimate	S.E.	C.R.	P	Results
H1: XT ← BZ	0.529	0.103	4.761	***	Supported
H2: XT ← XX	0.614	0.125	5.590	***	Supported
H3: XT ← ZN	0.498	0.076	4.172	***	Supported
BZ ↔ XX	0.561	0.084	4.619	***	
XX ↔ ZN	0.473	0.054	3.826	**	
BZ ↔ ZN	0.607	0.109	4.803	***	

Note: ***indicates significant at 0.01; **indicates significant at 0.05.

Model fitting mainly uses Relative Fitting Index (RFI), Incremental Fit Index (IFI), Tucker-Lewis Index (TLI), Comparative Fit Index (CFI), Root Mean square Residual (RMR), Root Mean Square Error of Approximation (RMSEA), Goodness of Fit Index (GFI) and Normed Fit Index (NFI) are used to judge the fit of the model. After repeated fitting, the final fitting data is shown in Table 6, RFI and IFI and TLI are close to 1, well fitted, CFI more than 0.9, RMR and RMSEA less than 0.08, indicates that the fit is good. GFI more than 0.9, NFI less than 0.9, but the acceptance criteria of more than 0.8, considering the relationship between the latent variables and the number of measured indicators, indicate that the model fits well and can accept the model.

Table 6. Evaluation results of structural equation model

Fitting indicator	Actual value	Suggested value
χ^2/df	2.163	1–3
RFI	0.946	>0.9
NFI	0.894	>0.9
GFI	0.905	>0.9
IFI	0.951	>0.9
TLI	0.993	>0.9
CFI	0.962	>0.9
RMR	0.065	<0.08
RMSEA	0.061	<0.08

Using AMOS23.0 to run and identify the SEM model, the finalized standardized path of output and model fitting effect are shown in Fig. 2.

4.4 Result Analysis

It can be seen from Fig. 2 that the three external potential variables through the significance test have a positive effect on the supply chain logistics collaboration. The order of influence is as follows: process informationization (0.614) > logistics standardization (0.529) > decision-making intelligence (0.498), and there is a significant correlation between logistics standardization, logistics informationization and logistics intelligence.

Logistics standardization has a significant positive impact on supply chain logistics coordination, and the path coefficient is 0.529. The degree of synergy between the various factors of logistics standardization on supply chain logistics is: the material data account is consistent with 0.822; the average load of accurate delivery of materials on time and in order is 0.709; the standardization impact of logistics operation standardization and logistics equipment is relatively weak in order of 0.612 and 0.549. In addition, logistics standardization is the ablest one dimensionality to obtain high returns with low input in the three key influencing factors. Logistics standardization is the basis for achieving supply chain logistics coordination.

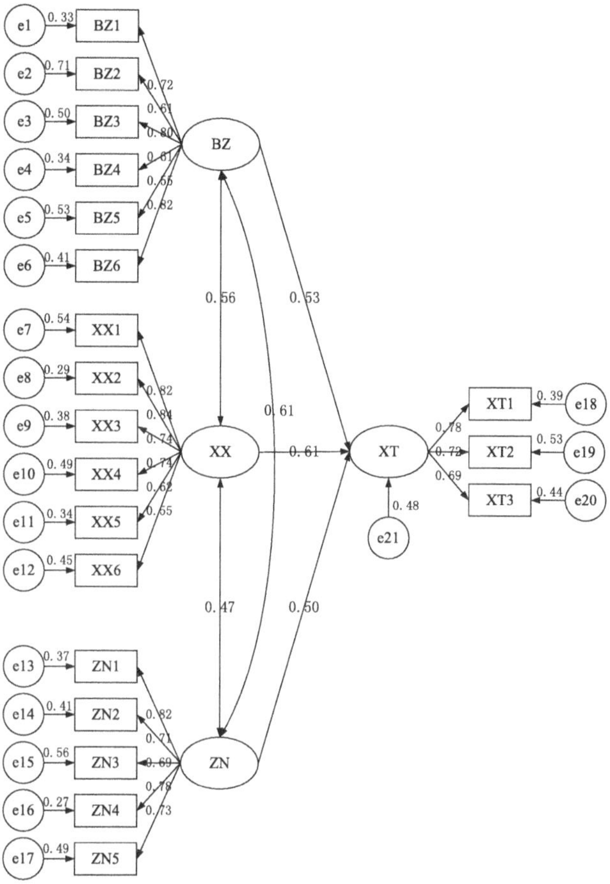

Fig. 2. Fitting results of structural equation model

Logistics process informatization has the highest impact on logistics synergy. The standardized path coefficient is 0.614, in which information platform, information acquisition technology infrastructure and so on are the most important, in which the influence factor loadings are 0.817 and 0.839 respectively. The enterprise should be on the basis of standardization of logistics operations, increase input and support of the information platform represented by SCM, SRM, ERP, CRM, LMS, WMS, MES and so on, expand the promotion and use of the information acquisition technology such as RFID, EDI, realize the information sharing of all platforms and units.

Decision-making intelligence has a significant positive impact on supply chain logistics coordination, of which information integration is the most critical 0.819, followed by intelligent decision-making fitness and intelligent control level of 0.782 and 0.734 respectively. Logistics equipment intelligence and logistics dynamic tracking impact factor coefficient are relative. The weaker reason may be that the advanced equipment investment is higher in the early stage, and most manufacturing enterprises have not yet realized it, but the intelligence is the inevitable future development of the manufacturing industry, which helps the enterprises to achieve better logistics coordination.

5 Conclusion

In order to achieve precise management and control of logistics coordination, this paper studies the synergistic factors of supply chain logistics with manufacturers as the core. First, the questionnaire for influencing factors was designed based on in-depth research. The index system of logistics synergy factors consisting of logistics standardization, process informationization and decision-making intelligence was determined. Then the structural equation model of logistics synergy factors is constructed. Combined with the results of the questionnaire, SPSS17.0 was used to analyze the reliability and validity and AMOS23.0 was used to verify and fit the model and data. The results showed that the data has high reliability and validity and the model data has a good fit. Finally, according to the factor load and path relationship coefficient of observed variables, the influence path and influence degree of supply chain logistics coordination key factors were determined, which lays a solid foundation for supply chain logistics collaborative evaluation. This research helps enterprises to further standardize logistics standards, promote the rapid development of logistics informationization and intelligence, achieve efficient collaborative operation in the supply chain, and at the same time has reference significance for collaborative management of supply chain logistics coordination.

Acknowledgment. This research is supported by the fund of China innovation method project (No. 2016IM030200) and doctoral research foundation of Henan University of Science and Technology.

Paper Survey Questionnaire Statement. I promise that the contents of the questionnaire involved in the paper Effect Factors Analyzing on Supply Chain Logistics Collaboration Based on the Structural Equation Model are true and effective, reasonable and reliable, and can be used for the next analysis and research work of this paper. In addition, the process of expert interview and questionnaire survey is all anonymous, which does not involve any personal privacy and work content information leakage of the respondents, and the survey data is only used for the subject research.

Promiser: ZHANG Ruimin, YANG Xiaoying.

References

1. Long, G.: Collaborative forecasting, inventory hedging and contract collaboration in dynamic supply risk management. Eur. J. Oper. Res. **245**(1), 133–145 (2015)
2. Qin, X., Liu, Z.: A review and future prospects on logistics service quality in supply chain. Chin. J. Manag. **15**(11), 1731–1738 (2018). (in Chinese)
3. Rhea, M.J., Shrock, D.L.: Measuring the effectiveness of physical distribution customer service programs. J. Bus. Logist. **8**(1), 31 (1987)

4. Sawik, T.: On the fair optimization of cost and customer service level in a supply chain under disruption risks. Omega **53**, 58–66 (2015)
5. Li, G., Ma, S., Gong, F., Wang, Z.: Research reviews and future prospective of collaborative operation in supply logistics based on Supply-Hub. J. Mech. Eng. **47**(20), 23–33 (2011). (in Chinese)
6. Ma, S., Lv, F.: Research on production and distribution collaborative mode based on Supply-Hub. Chin. J. Manag. Sci. **6**, 50–60 (2014). (in Chinese)
7. Xie, L., Ma, S., Gui, H., Huang, P.: Empirical analysis on influence mechanism of supply logistics synchronization. Sci. Res. Manag. **3**, 147–154 (2014). (in Chinese)
8. Qian, H., He, J.: Factors influencing synergy of cross-border e-commerce and cross-border logistics: an exploratory study based on ground theory. Ind. Econ. Rev. **8**(6), 110–122 (2017). (in Chinese)
9. Liu, Q., Li, X., Luo, X., Lv, F.: Measurement of influence factors on resources allocation in coal production logistics system from the perspective of safety and productiveness. Sci. Technol. Manag. Res. **38**(16), 225–233 (2018). (in Chinese)
10. Fu, W., Li, M.: Analysis of the influencing factors of regional logistics and regional economic cooperative development based on the improved Interpretative Structural Model. Mod. Manag. **36**(3), 23–25 (2016). (in Chinese)
11. Mi, Z., Zeng, G.: Logistics linkage pattern, characteristics and influencing factors of the Yangtze River economic belt under different scales. Sci. Geogr. Sin. **38**(7), 1079–1088 (2018). (in Chinese)
12. Huang, J., Chen, X.: Research on the cooperative development of manufacturing and logistics industry in Xiamen based on PCA-GRA. J. Fuzhou Univ. (Philos. Soc. Sci. Ed.) **32** (5), 57–63 (2018). (in Chinese)
13. Zhang, H., Zhang, Y.: Effect factors analyzing on the driver choices behavior on the dilemma zone at the signalized intersections based on structural equation model. Sci. Technol. Eng. **18**(28), 254–259 (2018). (in Chinese)
14. Xiong, B., Skitmore, M., Xia, B., Masrom, M.A., Ye, K., et al.: Examining the influence of participant performance factors on contractor satisfaction: a structural equation model. Int. J. Project Manag. **32**(3), 482–491 (2014)

Research on Emergency Mechanism of Social Emergency Management Based on Blockchain

Le Sun, Xiu-quan Deng(✉), Nueraili Aihemaiti, Zhi-yuan Dang, Yu-shan Xu, and Hong-wei Jia

School of Economics and Management, BeiHang University, Beijing, China
dengxiuquan@buaa.edu.cn

Abstract. Blockchain as a distributed and sharing ledger technology is in the rising application in various fields. Its characteristics of decentralization, transparency, traceability and non-tampering meet the demands of social emergency management and provide a new train of thought for solving the problems existing in the current social emergency management, such as insufficient information resource sharing, imperfect trust mechanism and high use cost of donations. This paper first introduces the core technologies of blockchain and analyzes the coupling between the characteristics of blockchain and the demands of social emergency management. Secondly, the emergency mechanism of social emergency management based on blockchain is proposed from three dimensions of object, property and function. Thirdly, the typical scenarios are given to illustrate the specific application of blockchain in the social emergency management.

Keywords: Blockchain · Sharing · Social emergency management · Trust · Value

1 Introduction

With the changes of international situation since the beginning of the 21st century, various domestic public emergencies have occurred from time to time. This not only brings great losses to people's lives and property, but also affects social stability and economic development. Social emergency management is referred to the emergency management in the face of emergent public events, with the participation of government at all levels and social organizations. In the new era, strengthening and improving social emergency management have become an important topic that governments at all levels and social organizations must face.

Various experts and scholars analyzed the social emergency management from different angles and pointed out the existing problems. Reference [1] showed that there has not yet been a nationwide or regional influential network in China's emergency management system and emergency management is inefficient. Reference [2] said that the information management system is not perfect, and information delay, concealment and omission often occur. Reference [3] revealed that in the era of big data, the formation of isolated information islands and severe information security situation contain new social risks. Some scholars also pointed out the lack of transparency in the

C.-F. Chien et al. (Eds.): *IE&EM 2019*, pp. 224–233, 2020.
https://doi.org/10.1007/978-981-15-4530-6_23

participation of charitable organizations in emergency management [4], the high use cost of charitable funds and low efficiency in the rescue process [5].

According to the higher requirements of social emergency management in the new era, scholars at home and abroad have published their own unique suggestions and countermeasures. Reference [6] proposed to monitor online data flow in social media to quickly and accurately identify the latest situation of disaster events. Reference [7] proposed to use social media to promote the sharing of emergency information. Reference [8] put forward the idea of building an open emergency management platform for sharing, integration and collaboration through the use of Internet technology and cloud platforms. Reference [9] proposed a new intelligent emergency management model- "cloud emergency".

Different countermeasures and solutions have promoted the improvement of emergency management. However, it is difficult to guarantee the authenticity and reliability of emergency information under the Internet technology, and centralized systems such as cloud platforms have problems of security risks and high costs. In addition, the problem of isolated data islands still exists, and the requirements of information resource sharing are still not fully met. Blockchain just meets the demands of social emergency management, and few people have set their hands on the research of the integration of blockchain and social emergency management. Therefore, this paper based on blockchain, combined with the coupling analysis of blockchain and social emergency management, proposes the emergency mechanism of social emergency management, and illustrates the specific application mode of blockchain through typical application scenarios of social emergency management, which provides a new train of thought and technical support for the development of social emergency management.

2 Blockchain Technology

A. Concepts and Characteristics of Blockchain Core Technologies

Blockchain is a decentralized sharing general ledger that combines data blocks into a specific data structure in a chain according to time sequence and ensures the ledger non-tampering and unforgeable by cryptography. It can safely store data that simple, sequential and can be verified in the system [10].

(1) *P2P network*

Blockchain network is a P2P network. Different from the traditional centralized organization, all nodes in the blockchain have the same status. Each node stores all data information in the whole blockchain. In this data framework, the data of each node is owned, managed and supervised by all participants. Each node can join or leave the network at will and ensure the stability of the network. What's more, the transparency and traceability of data are realized.

(2) *Consensus mechanism*

As the core technology of blockchain, consensus mechanism uses a set of consensus algorithm based on mathematics to ensure the authenticity of each record in a distrust environment. Each node will voluntarily and honestly judge the authenticity of each record according to the rules set in the protocol for the sake of maximizing its own interests, and finally add the true records into the blockchain.

(3) *Asymmetric encryption algorithm*

Blockchain uses asymmetric encryption algorithm to solve the privacy and trust problems of users between networks. Each participating user in the blockchain has an exclusive public key and a private key. The public key is published to all network users, whereas the private key is only mastered by the user himself. Users can use the private key to digitally sign the end of the data. Other users can verify the authenticity of the data source by decrypting with the public key.

(4) *Smart contract*

Smart contract is a piece of code deployed in a distributed sharing ledger. It is a computer program that can automatically execute the terms of the contract. The blockchain system automatically judges the contract execution conditions by specifying the obligations and conditions of the contract. When all the judgment conditions are met, the blockchain system will automatically enforce the contract terms.

B. Coupling Analysis of Blockchain and Social Emergency Management
The characteristics of blockchain are consistent with the demands of social emergency management (see Table 1), which is described below.

Table 1. Coupling analysis of blockchain characteristics and social emergency management demands

Characteristics	Blockchain	Social emergency management demands
Decentralization	All nodes store all information in a distributed ledger	Sharing of information and resources
Transparency, traceability and non-tampering	Distributed ledger maintained by all nodes	Information security, reliability and trust mechanism
Intelligent and efficient	Contracts that can be executed automatically	Reduce costs and improve benefit

(1) Blockchain is essentially a distributed ledger. In P2P network, each node has equal rights and obligations, and all information in the distributed ledger is stored. In the process of social emergency management, there are many channels for information dissemination of emergency information, effective collection is difficult, and information sharing is insufficient, resulting in poor coordination of emergency rescue, low rescue efficiency, idle waste of emergency resources, etc.
(2) The data cannot be tampered with once it is added to the block. In addition, consensus mechanism is used to realize authenticity and reliability of the data, and timestamp ensures its traceability and inquiry. The authenticity of emergency

information is crucial to command, decision-making and cooperative rescue in the emergency management network. Moreover, there are some problems in charitable donation, such as opaque use information and low public trust.

(3) Smart contract technology ensures the reliability and non-tampering of the contract, and can be automatically executed when the conditions are met. In the social emergency management, the donation of charity money and the transaction of idle emergency resources often require the third-party auditing and arbitration institutions to provide trust guarantee, which increases the cost and reduces the benefit of social emergency management.

3 Emergency Mechanism of Social Emergency Management Based on Blockchain

As a new generation of distributed information technology, blockchain based on P2P network, consensus mechanism, encryption algorithm, smart contract has the characteristics of decentralization, transparency, traceability and non-tampering. It provides a new train of thought to deal with the problems of insufficient information resource sharing, imperfect trust mechanism and high use cost of donations. The emergency mechanism of blockchain in social emergency management can be summarized and analyzed from three aspects: object dimension, property dimension and function dimension, as in Fig. 1.

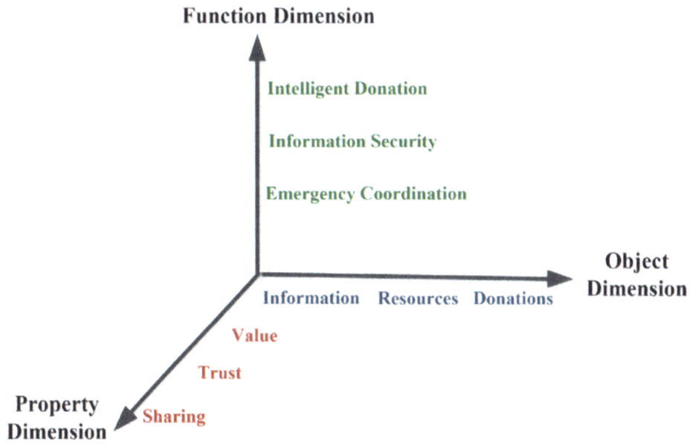

Fig. 1. Emergency mechanism model of social emergency management based on blockchain

A. Object Dimension

Emergency information, resources and donations play an important role in emergency management. Some scholars divide the emergency management involving social organizations into three parts: risk management, crisis management and recovery management, and summarize each part, including disaster information, rescue resources and donations [1]. Obviously, these three subjects are indispensable parts in social emergency management.

Information: One of the core mechanisms of emergency management is the communication and sharing of information. Emergency response strongly depends on the ability of information exchange and the ability of decision makers to formulate effective action policies based on collected information [11]. However, insufficient information sharing [12], lack of information linkage mechanism and mutual communication mechanism [13] have become a key factor restricting the efficiency of emergency response and rescue effect.

Resources: Emergency resources are playing an increasingly important role in large-scale emergency rescue. Big data can optimize the production, scheduling, storage, transportation, distribution, location and disposal of emergency resources [14]. However, problems such as waste of resources, insufficient emergency capacity and unreasonable use of emergency resources still exist [15].

Donations: Charitable donations initiated by charitable organizations as representatives are vital in saving lives and helping the reconstruction of disaster areas during the recovery process of emergency management. However, in recent years, problems such as incomplete accounting information, difficult access to financial information, lack of independent external supervision and trust mechanism, and high use cost of charitable donations have become bottlenecks restricting the development of charitable undertakings [5].

B. Property Dimension

Sharing: The distributed nature of the blockchain enables all parties involved in social emergency relief to collect and share data in the same network. As a node in the blockchain network, each participating social organization stores all the information, which can conveniently query the emergency information and resources information in the rescue process. It provides a technical solution to solve the problems of poor coordination, low efficiency, and idle waste of resources in emergency management.

Trust: Consensus mechanism ensures the authenticity of each record in a distrust environment and solves the trust problem between nodes. In addition, the value exchange based on mathematical principles on the blockchain is completely trustworthy, once the transaction is confirmed and recorded on the blockchain, it cannot be denied or tampered with, and every transaction recorded can be checked and traced by users, thus improving the public's trust in the donation of charity funds and eliminating the participation of third-party organizations such as auditing.

Value: Blockchain can reduce the cost, improve the efficiency and promote the development of social emergency management. P2P network and consensus mechanism have created conditions for full sharing of rescue resources. Through the way of digital token donation and the principle of consensus, the high handling fee in the process of inter-bank transfer is avoided, the participation of third-party organizations is eliminated, and the cost of donation is reduced. In addition, smart contract technology can ensure the automatic execution of donations, resource allocation and other transactions, which has the advantage of cost and efficiency.

C. Function Dimension

Intelligent donation is to use smart contract technology to realize the automation, high efficiency and safety of donation. And the security of the execution process and related transaction data can be ensured through encryption, which can better solve trust problem from the perspective of avoiding default and operational risks. What's more, it does not need an additional third party to supervise it. With the help of blockchain, the donation process can be effectively simplified, the use cost of donations can be reduced, and the public trust can be improved, as in Fig. 2.

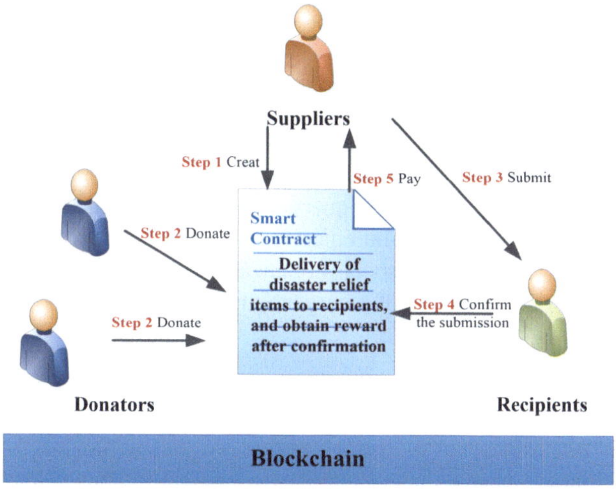

Fig. 2. Intelligent donation based on blockchain

Information security: Using encryption algorithm and digital signature can ensure the authenticity, security and reliability of information in emergency management, as in Fig. 3. Specifically, each social organization, as a node in the network, needs to add a digital signature encrypted with a private key every time it broadcasts data to the whole network, which makes it very difficult for attackers to fake data to cheat other nodes. Moreover, if an attacker attempts to tamper with the data in the blockchain database, due to the decentralized nature of system, the attacker must control at least 51% of the data nodes in the system to achieve data tampering, which greatly increases the cost and reduces the possibility of data tampering.

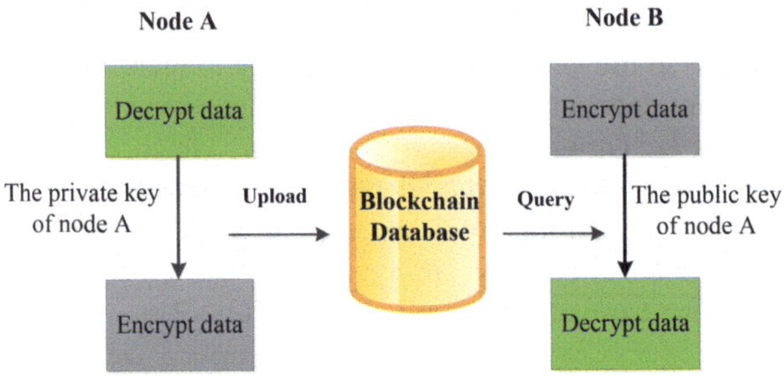

Fig. 3. Information security based on blockchain

Emergency coordination: Organizations participating in social emergency relief upload emergency information to the distributed network. Under consensus mechanism, real information will be added to the block and new block will be transmitted to the whole network. Therefore, each node in the network has written a complete backup of the blockchain data, and can freely query all of information, thus avoiding asymmetric and inconsistent information in emergency management, realizing data sharing and information exchange among social organizations, and promoting efficient coordination and orderly development of emergency rescue, as in Fig. 4.

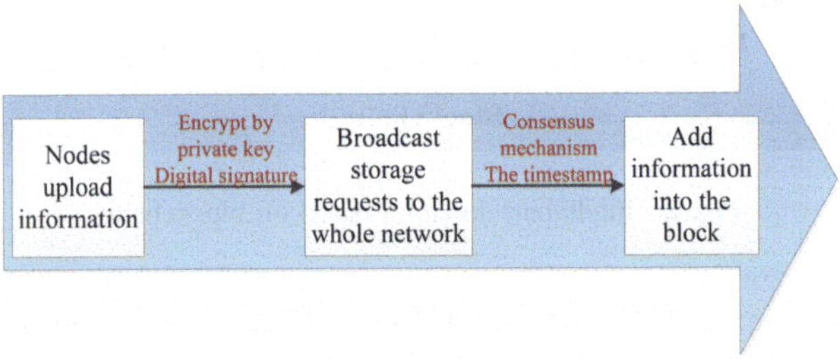

Fig. 4. Emergency coordination based on blockchain

4 Typical Application Scenarios of Social Emergency Management Based on Blockchain

A. Donation of Disaster Relief Funds

Charity donation, as one of the main ways for social organizations to participate in emergency management, plays an important role in the process of emergency relief and reconstruction of disaster areas. The application of blockchain can solve many problems of current charity donation, such as non-transparent donation accounting information, difficulty in ensuring authenticity, difficulty in accessing financial information, high cost of donation use, etc. As in Fig. 5, a donation model for social emergency management based on blockchain is presented [16].

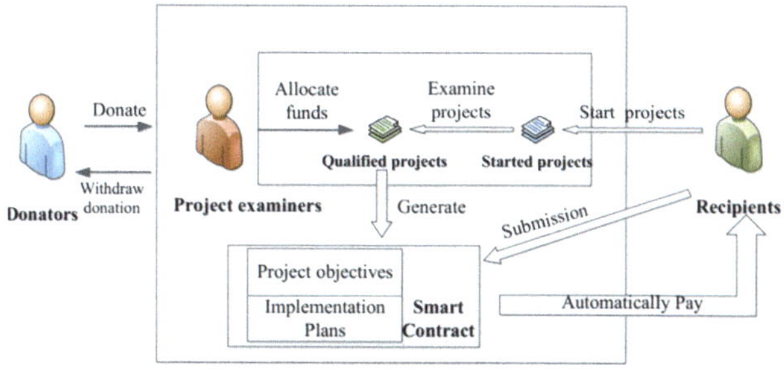

Fig. 5. Donation model for disaster relief funds based on blockchain

On the blockchain donation platform, recipients start fund-raising projects (such as rebuilding 100 houses for the disaster-stricken areas). After the projects have been examined by the project examiners, the qualified projects will generate smart contracts according to their project objectives and implementation plans, and write them into blockchain. At the same time the fund-raising project information will be transmitted to the whole network. Donors can choose projects that they are interested in to donate, and can withdraw donations before funds allocated to recipients. When fund-raising fund is raised entirely, the project will be started and smart contracts will automatically transfer corresponding fund to the recipient according to periodic results reports of the recipient.

B. Sharing of Emergency Resources

At present, all kinds of emergency rescue platforms lack effective integration of emergency resource information, which affects the effect of emergency rescue. Due to the particularity of emergency rescue, many emergency materials may be idle after emergency events, resulting in the waste of resources and increasing emergency management costs. Some scholars have proposed to use big data and the Internet of Things to build an emergency resource sharing platform [17], but the problems of information authenticity, public trust and privacy cannot be well solved.

Using blockchain to build an emergency resource sharing platform can solve the above problems, effectively avoid resource waste and reduce emergency management costs. As in Fig. 6, the operation process of the emergency resource sharing platform based on the blockchain is shown. The organizations participating in social emergency relief share emergency resource information to the platform. And the authenticity and security of the information transmitted by each node are guaranteed under consensus mechanism, encryption algorithm, etc. Each node can freely query all information of the platform, thus realizing the sharing of emergency resource information. In addition, blockchain is used to generate smart contract agreements according to the conditions of the resource supplier and store them in the block. When the resource demander presents an application and meets the required conditions for sharing resources, the contract will be automatically executed and the resource will be approved to the demander.

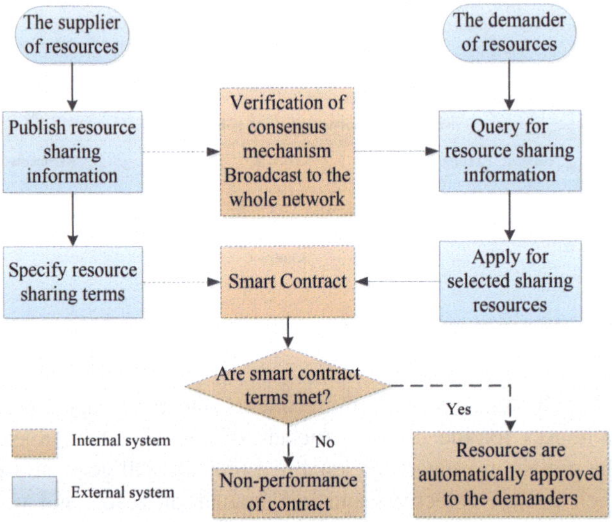

Fig. 6. Emergency resources sharing model based on blockchain

5 Conclusion

Social emergency management is directly related to social stability and is an important guarantee for social and economic development and people's quality of life. As a rapidly developing technology, blockchain has attracted much attention due to its characteristics of decentralization, transparency, traceability and non-tampering. By analyzing the coupling between blockchain and social emergency management, this paper studies from three dimensions of object, property and function, proposes emergency mechanism based on blockchain, and finally illustrates the specific application of blockchain in two typical scenarios of charity donation and resource sharing. It provides new theoretical guidance and technical solutions for social emergency management to meet the challenges in the new era.

References

1. Tao, P., Xue, L.: Discussing the establishment of cooperative partnership between government and social organizations in emergency management. J. Chin. Acad. Govern. **03**, 14–18 (2013). (in Chinese)
2. Xue, L., Liu, B.: New challenges of emergency management system and its top-level design. J. Chin. Acad. Govern. **01**, 10–14 + 129 (2013). (in Chinese)
3. Wu, S.J., He, L.B., Zhou, L.J.: Strengthening emergency management to deal with social risks – overview of 'risk society and emergency management international symposium'. Chin. Public Admin. **01**, 150–152 (2017). (in Chinese)
4. Ren, Q.F.: The problems and countermeasures of charity in China. Legal Econ. **04**, 118–119 (2009). (in Chinese)

5. Jia, H.W., Deng, X.Q.: A preliminary analysis of the application of blockchain in social emergency relief. In: Proeedings of 2nd Conference on Education, Management and Applied Social Science, EMASS2018, pp. 327–332, Hubei, China (2018)
6. Tsai, Y.F., et al.: Discovering disaster events from social media streams. Int. J. Ind. Eng. **25** (5), 685–701 (2018)
7. Xia, Z.J., Wu, Z., Luan, D.X.: The literature review of information sharing research for emergency respond based on social media. J. Intell. **32**(10), 125–130 (2013). (in Chinese)
8. Zhang, H., Liu, Y.: Key problems on fundamental science and technology integration in "scenario-response" based national emergency response platform system. Syst. Eng.-Theory Pract. **32**(05), 947–953 (2012). (in Chinese)
9. Li, C.D., Xie, T., Liu, Y.: Cloud emergency management – a new model of intelligent emergency management. China Emerg. Manage. **05**, 27–32 (2011). (in Chinese)
10. Yuan, Y., Wang, F.Y.: Blockchain: the state of the art and future trends. Acta Autom. Sin. **42** (4), 481–494 (2016). (in Chinese)
11. He, J.: The reform of emergency response model in the new media age. Chin. Public Admin. **4**, 20–24 (2012). (in Chinese)
12. Yang, L., Yang, S.H., Plotnick, L.: How the Internet of Things technology enhances emergency response operations. Technol. Forecast. Soc. Change **80**(9), 1854–1867 (2013)
13. Che, C.J.: Research on government emergency management in China, M.S. thesis, University of International Business and Economics, Beijing, China (2016). (in Chinese)
14. Zeng, Y.H.: Collaborative mechanism construction of government emergency management under the background of big data. Chin. Public Admin. **10**, 155–157 (2017). (in Chinese)
15. Xie, F., Chen, X.Y., Zhang, X.B.: Design and construction of emergency resource information sharing platform for unconventional emergencies. Occup. Health Emerg. Rescue **37**(3), 89–92 (2019). (in Chinese)
16. Jia, H.W.: Research on the blockchain based key mechanisms of social emergency relief, M. S. thesis, Beihang University, Beijing, China (2018). (in Chinese)
17. Ye, T.C., Zhao, D.J., Qian, H.W.: Design and exploration of emergency resource sharing platform based on big data and internet of things. Safety **39**(12), 20–26 (2018). (in Chinese)

A Study on the Government "Incentive–Supervision" Mechanism for Urban Terminal Joint Distribution Systems

Mei-Yan Li, Rui-Feng Qi(✉), Chuan Luo, and Jia-xin Chong

College of Mining and Safety Engineering, Shandong University of Science and Technology, Qingdao, China
qiruifengqd@163.com

Abstract. By constructing a single-principal multi-agent model involving governments, distributors and infrastructure operators, this paper studies the problem of government "incentive–supervision" for urban terminal joint distribution systems after logistics infrastructure operators are involved. The results show that the government can incent and supervise urban terminal joint distribution systems and improve their output efficiency by selecting participants, stabilizing the operating environment, guiding participants to establish a positive competition–cooperation relationship, encouraging participants to make active innovation, and encouraging enterprises to participate in joint distribution activities through appropriate policies. Incentive–supervision policies should aim at encouraging participants to upgrade their operation competence, minimize cost and maximize efficiency so as to create a favorable operating environment and promote positive competition among agents.

Keywords: Government "incentive–regulation" · Urban terminal joint distribution · Principal–agent · Use of infrastructure

1 Introduction

Development of urban logistics is disabling traditional terminal logistics distribution that mostly relies on own outlets of express companies to accommodate the "last-mile" and even the last 100-m distribution demanded by urban residents. Numerous express collection points, post stations, smart express cabinets and other outlets of non-express companies are finding their way into urban terminal distribution systems. To address this problem, the General Office of the State Council issued an "*Opinions on Promoting the Cooperative Development of Electronic Commerce and Express Logistics*" in 2018, which defined the public attribute of express terminal facilities for the first time and included smart express cabinets and express terminal service facilities into service infrastructure-related programs. In answer to the call of this policy, local governments began to include terminal service infrastructures such as public transit stations,

Funded by: Social Science Planning Project of Qingdao, Research on tripartite coordination of public facilities utilization in Qingdao joint distribution system (QDSKL1801117).

C.-F. Chien et al. (Eds.): *IE&EM 2019*, pp. 234–249, 2020.
https://doi.org/10.1007/978-981-15-4530-6_24

newsstands, convenience stores, and community property management companies into their urban distribution systems and have provided guidance and planning for the cooperative operation of terminal outlets of express companies.

At this point, urban terminal joint distribution systems have become a complex, multi-player system involving the government, industry associations, express enterprise alliances, logistics infrastructure operators, and urban residents; shared operation of urban terminal distribution outlets has changed from traditional enterprise alliance cooperation to shared utilization of public resources under government supervision and profit distribution in the presence of both competition and cooperation. David et al. investigated the problem of profit distribution when a public entity delegates the design, construction and maintenance of an infrastructure system to a private entity to solve such conflict: the goals of the public and private party do not coincide, there is information asymmetry between them and their interaction unfolds in environments under uncertainty [1]. Qiu proposed a "government principal–agent" governance structure dominated by the government from the perspective of the functions of the government in major infrastructures, constructed an "incentive–supervision" governance model on this basis and verified that intensive government involvement in the supervision of major infrastructures can effectively prevent potential risks and improve the social performances of infrastructures [2]. Wu et al. suggested obtaining the labor input of agents by introducing a supervisory mechanism to address the output of agents and their labor input during service and establish a full, scientific compensation incentive system [3]. Wen analyzed the task allocation of logistics supply chain management based on principal–agent theory and investigated the profit coordination mechanism between logistics service demanders and suppliers on this basis [4]. Zeng built a profit sharing game model for virtual R&D alliance based on principal–agent theory and yielded the optimal incentive base for distribution game incentives [5]. Liu incorporated the fairness preference concept into the original principal–agent model and built a profit distribution model for collaboration innovation of enterprise, college and institutes [6]. Dai incorporated a sharing mechanism between independent work teams and built a mutual incentive model among team members to investigate the active input among members. Nevertheless, these studies are mostly focused on profit sharing or incentive mechanisms between business partners or team members [7]. Few have looked at the roles played by the government in encouraging resource sharing in urban terminal distribution. In view of the structural characteristics of government-led urban terminal joint distribution systems, we constructed a single-principal, multiple-agent model involving the government, express distributors, and infrastructure operators to examine government incentive and supervision for urban terminal joint distribution systems when logistics infrastructure operators are involved.

2 Working Mechanism of an Urban Terminal Joint Distribution System

An urban terminal joint distribution system is a government-led complex system involving industry associations, express enterprise alliances, logistics infrastructure operators, and urban residents that work together and interact each other. Its structure with and interaction are illustrated by Fig. 1.

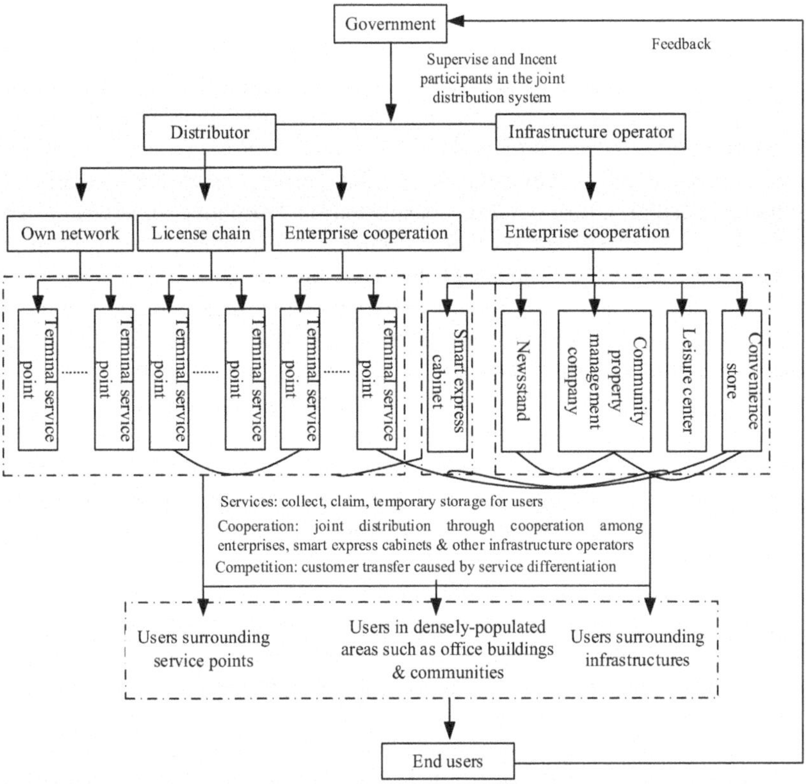

Fig. 1. Structure and interaction of an urban terminal joint distribution system

Within this system, the government incents and supervises the urban terminal joint distribution system as by the urban residents so that not only are urban residents provided with convenient, safe terminal distribution services, environmental pollution, traffic congestion and other social problems arising out of increasing delivery outlets are also limited to an acceptable level. On the one hand, the government will guide terminal distribution participants such as express companies, smart express cabinet operators, convenience stores, and newsstands to participate in this system to attain scale effect through incentive policies. On the other hand, it will also supervise how participants cooperate with other members within the system to ensure the safety and environmental friendliness of express services and the rationality of system operation.

Express companies, as the main implementers of terminal distribution services, try to extend the coverage of their service network and upgrade their distribution service level by building their own outlets, franchised outlets, business cooperation, and collection and procurement services. These operation activities, however, are affected by the construction and operation costs of outlets, requirements of government supervision, and strategic selection of competitors. Maximizing operation profit constitutes the core goal of these participants.

Infrastructure operators, such as convenience stores, newsstands, and community service centers, which are not part of a traditional terminal distribution system, try to and improve the organizational operation profit and customer satisfaction by cooperating with express companies or e-commerce enterprises and providing collection, claim, and temporary storage services for customers.

Smart express cabinets and other similar service facilities, as a new business form in the shared economy context, gain profit by providing shared terminal equipment for express companies and end users in densely populated areas such as communities, office buildings, industrial parks, schools, and hospitals. The working tools of these enterprises are both the means for them to gain profit and are subject to the supervision of the local operators, postal service administrations, public security bureaus and other related government authorities due to their attribute as public goods.

Among these participants, the government is the advocator and supervisor of joint distribution systems which delegates express companies and other infrastructure operators to provide terminal distribution services for urban residents. Logistics infrastructure operators, including the terminal distribution outlets and smart express cabinet of express companies, as well as convenience stores, newsstands and community property management companies, which are capable of providing terminal distribution services, are the implementers of terminal distribution services. Among these participants, there is both a momentum for cooperation and a competition transfer resulted from the similarity of services provided. Figure 2 shows the single-principal, multiple-agent relationship and governance among the government as the principal and express distributors and logistics infrastructure operators as the agents.

In order to provide appropriate incentive and supervision for express distributors and logistics infrastructure operators, it is necessary for the government to address uneven profit sharing between express distributors and logistics infrastructure operators arising as a result of cooperation degree, competition transfer, risk evasion, and operation cost, and encourage positive cooperation and competition between distributors and logistics infrastructure operators by defining regulatory indexes such as efficiency target and incentive coefficient so as to fulfill the ultimate goal of improving distribution efficiency and service quality and optimizing urban terminal distribution environment.

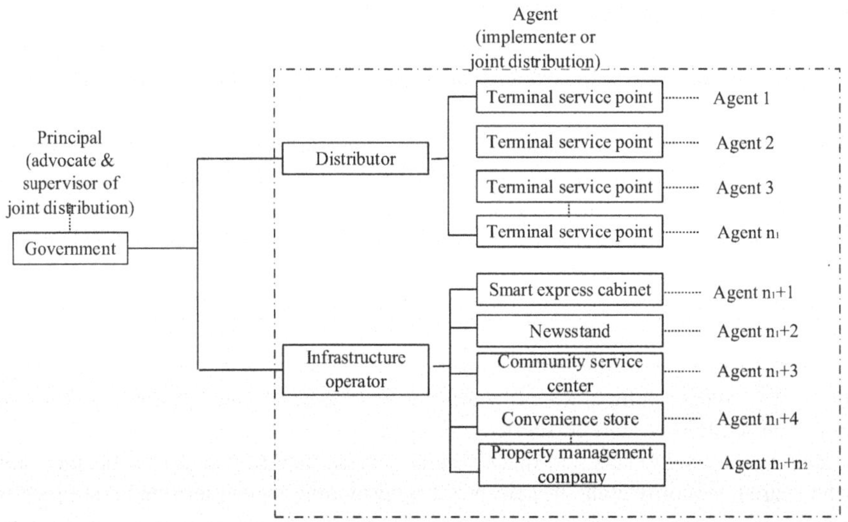

Fig. 2. Principal–agent relationship in an urban terminal joint distribution system

3 Model Construction and Hypotheses

In an urban terminal joint distribution system based on the government incentive–supervision mechanism, the government delegates distributors and infrastructure operators to provide terminal distribution services for urban residents. Information asymmetry exists between the government and distributors. This multiple principal–agent structure satisfies the two essential constraints of individual rationality (IR) and incentive compatibility (IC). IR means the conditions on which individual distributors and infrastructure operators are willing to accept government delegation, i.e., they will refuse to accept government delegation unless the profit from distributors and infrastructure operators' participating in terminal joint distribution is always greater than or equal to their own minimum profit $\overline{w}$. IC means to incent an agent to select its optimal effort level by maximizing the profit of a distributor or infrastructure operator. Other related hypotheses include [8–17]:

(1) Both the principal and the agent aim to maximize their income. Here, the income of the principal, i.e., the government, is expressed as w_g; the risk is neutral; the output of the agents, i.e., the distributors and infrastructure operators, is expressed as π_i; the income is expressed as $S(\pi_i)$; the profit is expressed as v_i and is risk aversive.
(2) According to the competitive agent output relationship proposed by Tian et al., the output of an agent is affected by the effort level of both itself and its competitor [18]. The output of a distributor or an infrastructure operator is typically composed of cooperation output and competition output, expressed as:

$$\pi_i = ka_i + \sum_{i,j \neq i}^{n} b(a_i - a_j) + \theta' \tag{1}$$
$$i - 1, 2, 3 \cdots, n; \; j = 1, 2, 3 \cdots, n \;\; i \neq j$$

Where: there are n_1 distributors and n_2 infrastructure operators, and $n_1 + n_2 = n$; $a_1, a_2 \cdots a_{n_1}$ is the effort level of the distributor, denoting the effort level of the distributor to cooperate with other gents when participating in terminal distribution; $a_{n_1+1}, a_{n_1+2} \cdots a_{n_1+n_2}$ is the effort level of the infrastructure operator, denoting the effort level of the infrastructure operator to cooperate with other agents when participating in terminal distribution; k is the economic efficiency coefficient of the distributor or infrastructure operator when participating in terminal distribution; b is the economic efficiency coefficient of customer transfer caused by competition between the distributor and infrastructure operator, between distributors, or between infrastructure operators; θ is the effect of external uncertainties on the output, which is a normally distributed random variable with a mean of 0 and a variance of σ^2.

(3) The income of an agent from participating in terminal distribution comprises two parts: the incentive offered by the government according to the output of the distributor or infrastructure operator and the fixed income provided by the government to the distributor or infrastructure operator. The income of a distributor or infrastructure operator is

$$S(\pi_i) = \alpha + \beta(\pi_i - \pi_0) \tag{2}$$

Where: α is the fixed income designed by the government for the distributor or infrastructure operator from participating in terminal distribution, which is independent from efficiency; β is the incentive coefficient defined by the government according to the output of the distributor or infrastructure operator; π_0 is the base output for government incentive. The expected government income can be expressed as:

$$\begin{aligned} Ew_g &= E[\pi - S(\pi_1) - S(\pi_2) - \cdots - S(\pi_n)] \\ &= -n\alpha + k(1 - \beta)(a_1 + a_2 + \cdots + a_n) + n\beta\pi_0 + n(1 - \beta)\theta \end{aligned} \tag{3}$$

(4) The cost of the distributor or infrastructure operator for participating in terminal distribution mainly includes distribution cost and risk cost. Hence the profit of a distributor or infrastructure operator can be expressed as:

$$v_i = \alpha + \beta(\pi_i - \pi_0) - da_i^2/2 - \rho\beta^2\sigma^2/2 \qquad i = 1,\ 2,\ 3,\ \cdots,\ n \tag{4}$$

Where: The joint distribution cost of a distributor or infrastructure operator is related to effort level, which can be expressed as: $c(a_i) = da_i^2/2$, $i = 1, 2, 3 \cdots, n$, where d is the cost coefficient, which is a constant and is greater than 0; the risk-aversive nature of the efficiency function of the distributor or infrastructure operator is a constant, i.e., $\mu = -e^{-\rho w}$, where ρ is the absolute risk aversion and the risk cost is $\rho\beta^2\sigma^2/2$.

Based on these analyses and hypotheses, we can build a single-principal, multiple-agent model as follows:

$$\max_{\alpha,\beta,a} w_g = -n\alpha + k(1-\beta)(a_1 + a_2 + \cdots + a_n) + n\beta\pi_0 + n(1-\beta)\theta$$

$$s.t.\ (IR)\ \alpha + \beta(\pi_1 - \pi_0) - da_1^2/2 - \rho\beta^2\sigma^2/2 \geq \overline{w}$$

$$(IR)\ \alpha + \beta(\pi_2 - \pi_0) - da_2^2/2 - \rho\beta^2\sigma^2/2 \geq \overline{w}$$

$$\vdots$$

$$(IR)\ \alpha + \beta(\pi_n - \pi_0) - da_n^2/2 - \rho\beta^2\sigma^2/2 \geq \overline{w} \quad (5)$$

$$(IC)\ a_1 \in \arg\max v_1 = \alpha + \beta(\pi_1 - \pi_0) - da_1^2/2 - \rho\beta^2\sigma^2/2$$

$$(IC)\ a_2 \in \arg\max v_2 = \alpha + \beta(\pi_2 - \pi_0) - da_2^2/2 - \rho\beta^2\sigma^2/2$$

$$\vdots$$

$$(IC)\ a_n \in \arg\max v_n = \alpha + \beta(\pi_n - \pi_0) - da_n^2/2 - \rho\beta^2\sigma^2/2$$

Equivalent to:

$$\max_{\alpha,\beta,a} w_g = -n\alpha + k(1-\beta)(a_1 + a_2 + \cdots + a_n) + n\beta\pi_0 + n(1-\beta)\theta$$

$$s.t.\ (IR)\ \alpha + \beta(\pi_1 - \pi_0) - da_1^2/2 - \rho\beta^2\sigma^2/2 \geq \overline{w}$$

$$(IR)\ \alpha + \beta(\pi_2 - \pi_0) - da_2^2/2 - \rho\beta^2\sigma^2/2 \geq \overline{w}$$

$$\vdots$$

$$(IR)\ \alpha + \beta(\pi_n - \pi_0) - da_n^2/2 - \rho\beta^2\sigma^2/2 \geq \overline{w} \quad (6)$$

$$(\mathrm{IC})\ a_1 = \frac{\beta[k + (\mathrm{n}-1)\mathrm{b}]}{d}$$

$$(\mathrm{IC})\ a_2 = \frac{\beta[k + (\mathrm{n}-1)\mathrm{b}]}{d}$$

$$\vdots$$

$$(\mathrm{IC})\ a_n = \frac{\beta[k + (\mathrm{n}-1)\mathrm{b}]}{d}$$

As the government needs to ensure that its delegation is justified, simply an equal symbol is needed for these constraints. By introducing the constraints into the target function, we yield that the expected government income is a function about β:

$$\max_{\alpha,\beta,a} w_g = n\beta k[k + (\mathrm{n}-1)\mathrm{b}]/d - n\beta^2[k + (\mathrm{n}-1)\mathrm{b}]^2/2d - n\rho\beta^2\sigma^2/2 - n\overline{w} + n\theta \tag{7}$$

Through first-order derivation of this target function, we have:

$$\frac{\partial w_g}{\partial \beta} = nk[k + (\mathrm{n}-1)\mathrm{b}]/d - n\beta[k + (\mathrm{n}-1)\mathrm{b}]^2/d - n\rho\beta\sigma^2 \tag{8}$$

Through second-order derivation, we have:

$$\frac{\partial^2 w_g}{\partial \beta^2} = -n[k + (\mathrm{n}-1)\mathrm{b}]^2/d - n\rho\sigma^2 < 0 \tag{9}$$

When the first-order derivative is 0 and the second-order derivative is smaller than 0, the function is maximized. Hence, let the first-order derivative be 0, and we get the optimal government incentive coefficient as:

$$\beta^* = \frac{k[k + (\mathrm{n}-1)\mathrm{b}]}{[k + (\mathrm{n}-1)\mathrm{b}]^2 + d\rho\sigma^2} \tag{10}$$

By introducing β^* into the model's incentive compatibility (IC), we get the optimal effort level of the distributor or infrastructure operator as:

$$a_1^* = a_2^* = \cdots = a_n^* = \frac{k[k + (\mathrm{n}-1)\mathrm{b}]^2}{d[k + (\mathrm{n}-1)\mathrm{b}]^2 + d^2\rho\sigma^2} \tag{11}$$

By introducing a_i^*, β^* into the single-principal, multiple-agent model, we obtain the optimal government income under information asymmetry as:

$$w_g^* = \frac{n[k(k + (\mathrm{n}-1)\mathrm{b})]^2}{2d[(k + (\mathrm{n}-1)\mathrm{b})^2 + d\rho\sigma^2]} - n\overline{w} + n\theta \tag{12}$$

The optima income of the distributor or infrastructure operator can be expressed:

$$S(\pi_1)^* = S(\pi_2)^* = \cdots = S(\pi_\mathrm{n})^* = \overline{w} + \frac{[k(k + (\mathrm{n}-1)\mathrm{b})]^2}{2d[(k + (\mathrm{n}-1)\mathrm{b})^? + d\rho\sigma^2]} \tag{13}$$

Table 1 summarizes the optimal decision variables for achieving Pareto efficiencies under information asymmetry:

Table 1. Optimal Decision Variables

Decision variable	Expression for optimal decision variable
a_i^*	$\frac{k(k+(n-1)b)^2}{d(k+(n-1)b)^2+d^2\rho\sigma^2}$
β^*	$\frac{k(k+(n-1)b)}{(k+(n-1)b)^2+d\rho\sigma^2}$
$S(\pi_i)^*$	$\overline{w}+\frac{[k(k+(n-1)b)]^2}{2d[(k+(n-1)b)^2+d\rho\sigma^2]}$
w_g^*	$\frac{n[k(k+(n-1)b)]^2}{2d[(k+(n-1)b)^2+d\rho\sigma^2]}-n\overline{w}+n\theta$

By analyzing the sensitiveness of parameters in the expressions for the optimal decision variables, we can obtain the distributions of the decision variables and the interaction between the parameters.

4 Multi-principal–Agent Modeling for Single-Parameter Variation

As it is hardly possible to directly identify the interaction between influencing factors in an information asymmetry context due to the relatively complex expressions and the many influencing factors for decision variables, in order to clarify the interactions of the effort level of distributors or infrastructure operators a_i^*, government incentive coefficient β^*, income of the distributor and infrastructure operator $S(\pi_i)^*$ and government income w_g^*, relative to output economic efficiency coefficient k, customer transfer economic efficiency coefficient b, cost coefficient d, risk aversion ρ, variance of external uncertainty σ^2, and number of participants n, we can yield the distributions of output economic efficiency coefficient k, customer transfer economic efficiency coefficient b, cost coefficient d, risk aversion p, variance of external uncertainty σ^2, and number of participants n, as shown in Table 2, by taking the derivatives of these parameters from the expressions for decision variables a_i^*, β^*, $S(\pi_i)^*$, and w_g^* in Table 1.

Table 2. Decision variable versus parameter

Analysis	Variable parameter	Decision variable			
		a_i^*	β^*	$S(\pi_i)^*$	w_g^*
(1)	$d\downarrow$	↑	↑	↑	↑
(2)	$\rho\downarrow$	↑	↑	↑	↑
(3)	$\sigma^2\downarrow$	↑	↑	↑	↑
(4)	$b\uparrow$	↑	First↑Later↓	↑	↑
(5)	$k\uparrow$	↑	↑	↑	↑
(6)	$n\uparrow$	↑	↓	↑	↑

From the effects of parameter variations on decision variables, we can see that:

(1) $\frac{\partial a_i^*}{\partial d} < 0, \frac{\partial \beta^*}{\partial d} < 0, \frac{\partial S(\pi_i)^*}{\partial d} < 0, \frac{\partial w_g^*}{\partial d} < 0$, i.e., a_i^*, β^*, $S(\pi_i)^*$ and w_g^* are negatively related to cost coefficient d. At a constant output, the lower the cost coefficient is, the higher the income of the government and the distributor/infrastructure operator, the greater the incentive offered by the government and the more willing the other distributors and infrastructure operators are to participate in the cooperation and provide their respective effort levels.

Conclusion 1: The government should select enterprises or infrastructure operators with high management competence to participate in joint distribution systems and incent participating enterprises or infrastructure operators to minimize operation cost and thereby improve the output of the entire joint distribution systems.

(2) $\frac{\partial a_i^*}{\partial \rho} < 0, \frac{\partial \beta^*}{\partial \rho} < 0, \frac{\partial S(\pi_i)^*}{\partial \rho} < 0, \frac{\partial w_g^*}{\partial \rho} < 0$, i.e., a_i^*, β^*, $S(\pi_i)^*$ and w_g^* are negatively related to risk aversion ρ. The greater the risk aversion level of the distributor or infrastructure operator is for participating in terminal distribution services, the more they are opposed to bearing the related risk cost, the less they are willing to make efforts, the lower the income they will make and the lower the government incentive level will become.

Conclusion 2: Enterprises or infrastructure operators with conservative risk preferences are less willing to participate in urban terminal joint distribution systems. The government may guide their participating in urban terminal joint distribution systems as an enterprise innovation project and incent more enterprises or infrastructure operators to participate by taking the advantage of innovation-oriented policies.

(3) $\frac{\partial a_i^*}{\partial \sigma^2} < 0, \frac{\partial \beta^*}{\partial \sigma^2} < 0, \frac{\partial S(\pi_i)^*}{\partial \sigma^2} < 0, \frac{\partial w_g^*}{\partial \sigma^2} < 0$, i.e., a_i^*, β^*, $S(\pi_i)^*$ and w_g^* are negatively related to variance of external uncertainty σ^2. Distributors and infrastructure operators are opposed to environmental fluctuations. The greater the environmental fluctuation is, the less the distributors and infrastructure operators are willing to make efforts, the lower the income they will make and the lower the government incentive level will become.

Conclusion 3: A stable working environment represents an important prerequisite for enterprises to participate in new businesses or make business innovations. As the leading player, the government must regard creating a stable, open, friendly working environment as an important part of nurturing the operation of urban terminal joint distribution systems.

(4) $\frac{\partial a_i^*}{\partial b} > 0, \frac{\partial S(\pi_i)^*}{\partial b} > 0, \frac{\partial w_g^*}{\partial b} > 0$, i.e., a_i^*, $S(\pi_i)^*$ and w_g^* are positively related to customer transfer coefficient b. Increasing customer transfer coefficient will lead to an increase in the effort level of the distributor and infrastructure operator and consequently the income of themselves and the government. However, as $\frac{\partial \beta^*}{\partial b} = \frac{k[d\rho\sigma^2 - (k+(n-1)b)^2]}{[(k+(n-1)b)^2 + d\rho\sigma^2]^2}$, when $\frac{\partial \beta^*}{\partial b} = 0$, $b = \sigma\sqrt{\rho d} - k$; when $\frac{\partial \beta^*}{\partial b} < 0$,

$b > \sigma\sqrt{\rho d} - k$; when $\frac{\partial \beta^*}{\partial b} > 0$, $b < \sigma\sqrt{\rho d} - k$. As customer transfer coefficient b increases, incentive coefficient β^* first increases and then decreases. Although increasing customer transfer coefficient will lead to an increase in the effort level of the agent, thereby stimulating distributors or infrastructure operators to improve their service level to attract more users, when competition reaches a given limit, the optimal government incentive coefficient will decrease.

Conclusion 4: The government may arouse the enthusiasm of agents for participating in joint distribution systems through incentive means but, instead of encouraging over-competition, it must determine an appropriate incentive level to guide participating enterprises or infrastructure operators to establish a positive competition–cooperation relationship.

(5) $\frac{\partial a_i^*}{\partial k} > 0, \frac{\partial \beta^*}{\partial k} > 0, \frac{\partial S(\pi_i)^*}{\partial k} > 0, \frac{\partial w_g^*}{\partial k} > 0$, i.e., a_i^*, β^*, $S(\pi_i)^*$ and w_g^* are positively related to economic efficiency coefficient k. The greater the economic efficiency coefficient is, the more effort the distributor and infrastructure operator will make to participate in urban terminal distribution, the greater the income they will obtain and the higher the government incentive level will become. This is obvious.

Conclusion 5: The government must not only encourage participants to minimize cost, it must also encourage enterprises or infrastructure operators to improve their distribution efficiency, make technical reform and increase the overall output of the distribution system through cost-reducing and efficiency-increasing approaches.

(6) $\frac{\partial a_i^*}{\partial n} > 0, \frac{\partial \beta^*}{\partial n} < 0, \frac{\partial S(\pi_i)^*}{\partial n} > 0, \frac{\partial w_g^*}{\partial n} > 0$, i.e., a_i^*, β^*, $S(\pi_i)^*$ and w_g^* are positively related to number n, β^* negatively related to number n, The larger the number of participants, the greater the profit the government and participants will make, the more effort the distributors and infrastructure operators will make in participating urban terminal distribution.

Conclusion 6: The government must disseminate and promote terminal distribution patterns, increase public recognition and acceptance of joint distribution and provide infrastructure operators with sufficient operation authority and appropriate participation income so as to involve more participants in joint distribution.

As discussed above, while a government-led "incentive–supervision" urban terminal joint distribution system can well resolve the low efficiency and environmental pollution confronting the current urban terminal distribution systems, a better result will be achieved by supporting innovation, guiding operating environment optimization, certifying participating enterprises, and supervising the operating process.

5 Modeling for Double-Parameter Variation

The model analysis above only considers the effects of single-parameter variation on decision variables. In the next section, we are going to examine how variation of any two of output economic efficiency coefficient k, customer transfer economic efficiency coefficient b, cost coefficient d and risk aversion ρ will affect decision variables so as to find out what must be highlighted during government incentive–supervision.

Let us assume that there is only one infrastructure operator and one distributor in the system, i.e., $n_1 = 1$, $n_2 = 2$, $n = 2$; external uncertainty θ is subject to standard normal distribution, i.e., $\theta \sim N(0, 1)$; output economic efficiency coefficient k, customer transfer economic efficiency coefficient b, cost coefficient d or risk aversion ρ is [0, 1] when variable and 1 when constant. On this basis, we can numerically simulate the distributions of the decision variables in Table 1 and produce surface plots indicating the effects of the parameter values on the decision variables. From Table 1, $S(\pi_i)^*$ and w_g^* vary in the same way, hence we only need to analyze the plot of $S(\pi_i)^*$. Figure 3 compares the effects of the parameters on the decision variables.

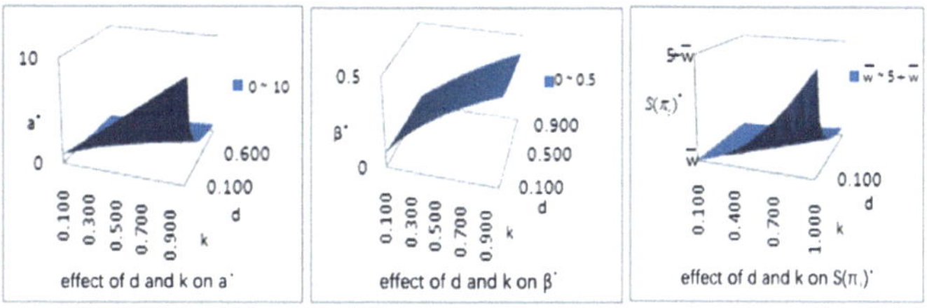

Fig. 3-1. Surface plots of decision variables as a function of d and k values

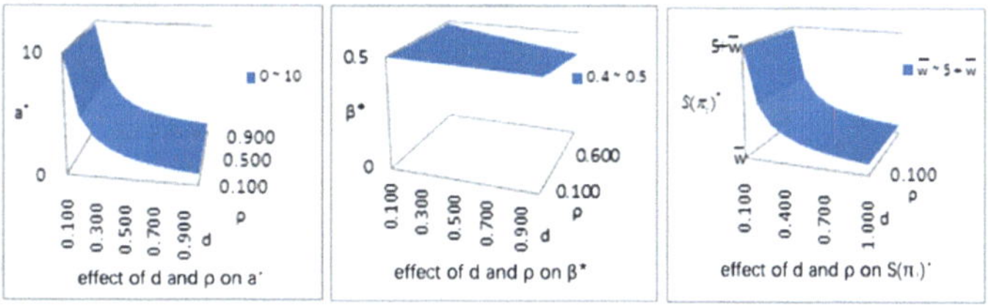

Fig. 3-2. Surface plots of decision variables as a function of d and ρ values

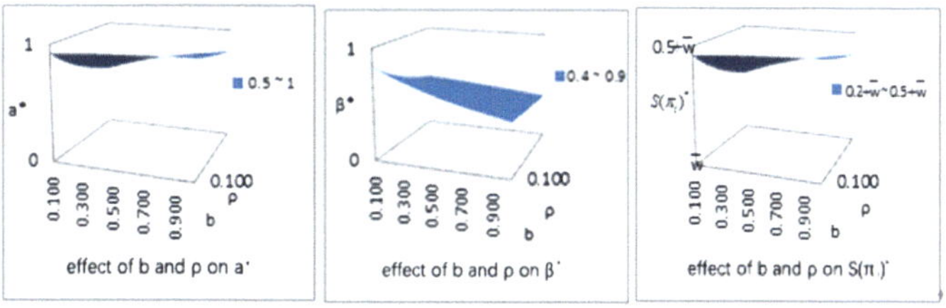

Fig. 3-3. Surface plots of decision variables as a function of b and ρ values

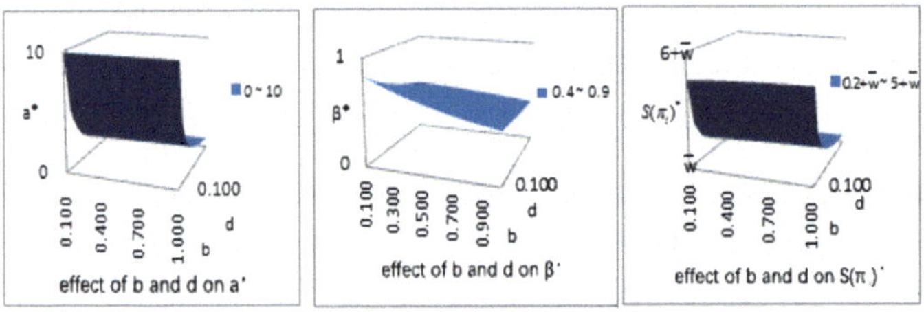

Fig. 3-4. Surface plots of decision variables as a function of b and d values

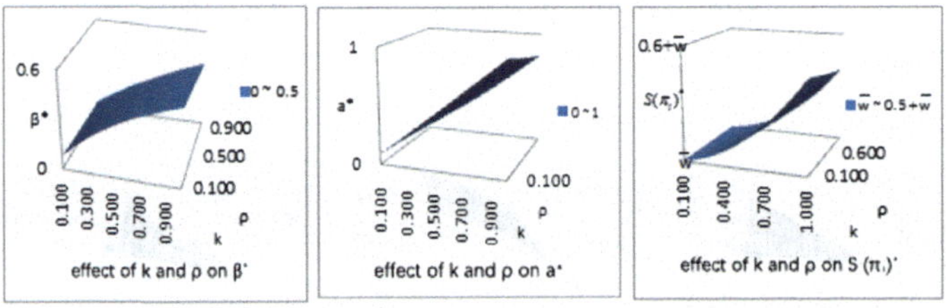

Fig. 3-5. Surface plots of decision variables as a function of k and ρ values

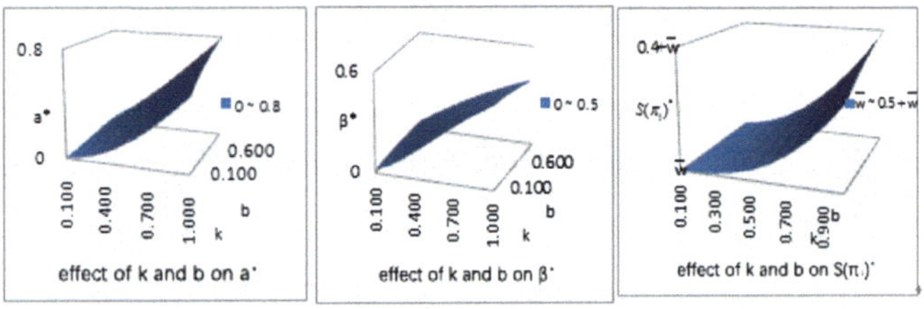

Fig. 3-6. Surface plots of decision variables as a function of k and b values

From these plots, β^* varies in a different manner from a_i^* and $S(\pi_i)^*$ (w_g^*) of the same group. Hence we can analyze the variation of β^* separately and tabulate the result into Table 3 and Table 4.

Table 3. Distributions of a_i^*, $S(\pi_i)^*$ and w_g^* values

Figure	Constant	Variable	a_i^*	$S(\pi_i)^*$	w_g^*
Figure 3-1	$\rho = 1,\ b = 1$	$d\downarrow,\ k\uparrow$	↑	↑	↑
Figure 3-2	$k = 1,\ b = 1$	$d\downarrow,\ \rho\downarrow$	↑	↑	↑
Figure 3-3	$k = 1,\ d = 1$	$b\uparrow,\ \rho\downarrow$	↑	↑	↑
Figure 3-4	$k = 1,\ \rho = 1$	$b\uparrow,\ d\downarrow$	↑	↑	↑
Figure 3-5	$b = 1,\ d = 1$	$k\uparrow,\ \rho\downarrow$	↑	↑	↑
Figure 3-6	$\rho = 1,\ d = 1$	$k\uparrow,\ b\uparrow$	↑	↑	↑

Table 4. Distribution of β^* values

Figure	Constant	Variable	β^*
Figure 3-1	$\rho = 1,\ b = 1$	$d\downarrow,\ k\uparrow$	$\uparrow$
Figure 3-2	$k = 1,\ b = 1$	$d\downarrow,\ \rho\downarrow$	$\uparrow$
Figure 3-3	$k = 1,\ d = 1$	$b\downarrow,\ \rho\downarrow$	$\uparrow$
Figure 3-4	$k = 1,\ \rho = 1$	$b\downarrow,\ d\downarrow$	$\uparrow$
Figure 3-5	$b = 1,\ d = 1$	$k\uparrow,\ \rho\downarrow$	$\uparrow$
Figure 3-6	$\rho = 1,\ d = 1$	$k\uparrow,\ b\uparrow$	First$\uparrow$Later$\downarrow$

From Fig. 3, Table 3 and Table 4, conclusion 1 through conclusion 6 holds true not only for single-parameter variation, but also for two-parameter variation.

In order to further investigate the variations of decision variables, let us introduce k=b=d=ρ=0 and k = b = d=ρ=1 into the optimal expressions for a_i^*, β^*, $S(\pi_i)^*$, and w_g^*, respectively and tabulate the result into Table 5:

Analysis of both Fig. 3 and Table 5 reveals that:

The government incentive coefficient β^* is the same between Fig. 3-1 and Fig. 3-5; the government incentive coefficient β^* is the same between Fig. 3-3 and Fig. 3-4; when cost coefficient d and risk aversion ρ are used together with customer transfer coefficient b and economic efficiency coefficient k to adjust government incentive coefficient β^*, cost efficient d has the same effect on government incentive coefficient β^* as risk aversion ρ.

Table 5. Distribution of a_i^*, β^*, $S(\pi_i)^*$ and w_g^* values

Figure	a_i^*	β^*	$S(\pi_i)^*$	w_g^*
Figure 3-1	(0, 10)	(0, 0.5)	$(\overline{w},\ 5+\overline{w})$	$(d = 1 - 2\overline{w},\ 10 - 2\overline{w})$
Figure 3-2	(0, 10)	(0.4, 0.5)	$(\overline{w},\ 5+\overline{w})$	$(0.8 - 2\overline{w},\ 10 - 2\overline{w})$
Figure 3-3	(0.5, 1)	(0.4, 0.9)	$(0.2+\overline{w},\ 0.5+\overline{w})$	$(0.4 - 2\overline{w},\ 1 - 2\overline{w})$
Figure 3-4	(0, 10)	(0.4, 0.9)	$(0.2+\overline{w},\ 5+\overline{w})$	$(0.4 - 2\overline{w},\ 10 - 2\overline{w})$
Figure 3-5	(0, 1)	(0, 0.5)	$(\overline{w},\ 0.5+\overline{w})$	$(-2\overline{w},\ 1 - 2\overline{w})$
Figure 3-6	(0, 0.8)	(0, 0.5)	$(\overline{w},\ 0.5+\overline{w})$	$(-2\overline{w},\ 1 - 2\overline{w})$

The effort level a_i^* and income $S(\pi_i)^*$ of distributors and infrastructure operators and government income w_g^* in Fig. 3-3, Fig. 3-5 and Fig. 3-6 are lower than in Fig. 3-1, Fig. 3-2 and Fig. 3-4. It is also obvious that the level of cost coefficient d has the greatest effect on the effort value a_i^* and income $S(\pi_i)^*$ of distributors and infrastructure operators and government income w_g^*; the values of customer transfer coefficient b and risk aversion ρ have the smallest effect on the effort level a_i^* and income $S(\pi_i)^*$ of distributors and infrastructure operators and government income w_g^*.

Conclusion 7: The government must perform incentive–supervision over urban terminal joint distribution systems by regulating cost coefficient and economic efficiency coefficient as the primary factors and risk aversion and customer transfer coefficient as the secondary ones. That is, the government must highlight advocating cost reduction and efficiency improvement as the main targets and stabilizing the market environment of terminal distribution and boosting positive competition among agents as the secondary goals. In the meantime, consideration may also be given to the ability of the government for cost control and risk prevention to achieve equivalent substitution of cost control and risk prevention for the incentive–supervision of joint distribution.

6 Conclusion

A single-principal, multiple-agent model involving the government, distributors and infrastructure operators is established after analyzing the structure of a government-led urban terminal joint distribution system and the interactions among participants, The problem of government incentive–supervision after public service facilities are involved in terminal distribution systems is analyzed. The results show that the government can perform incentive–supervision over the formation and operation of urban terminal joint distribution systems by selecting participants, stabilizing the working environment, guiding participants to establish a positive competition–cooperation relationship, encouraging participants to make active innovation and encouraging more enterprises to participate in terminal joint distribution systems through appropriate policies. This participation can effectively improve the output efficiency of urban terminal distribution systems. Incentive–supervision policies must aim at encouraging participants to improve their operation competition, minimize cost and maximize efficiency as the main purpose and creating a positive operating environment and promoting a positive competition among agents as the secondary goal. The findings of our research offer useful reference for the government in selecting the general regulation direction and developing guiding policies. The study doesn't include for changes in government risk attitude, which may be further examined in subsequent studies.

References

1. Davidpaez-perez, M.: A dynamic principal-agent framework for modeling the performance of infrastructure. Eur. J. Oper. Res. **254**(2), 576–594 (2016)
2. Qiu, Y.M., Cheng, S.P.: Governance model of major infrastructure projects: from a perspective of multiple functions of the government. J. Syst. Manage. **27**(1), 129–136 (2018). (Chinese)
3. Wu, K., Zhang, H.Y.: Supervisory intensity and incentive mechanism based on principal-agency relationship. System **07**, 68–72 (2016). (Chinese)
4. Wen, L.G.: Task allocation model of logistics supply chain and benefit coordination mechanism based on the principal-agent model. Econ. Prob. **04**, 62–66 (2011). (Chinese)
5. Zeng, Z.: Research on the profit distribution games of virtual R&D alliance. Modern. Manage. **02**, 101–103 (2013). (Chinese)
6. Liu, S.T., Cai, Q.M., Zhan, M.S.: Profit distribution in collaboration innovation of enterprise, college and institutes under the leading of enterprise based on fairness preference theory. Sci. Technol. Manage. Res. **36**(13), 205–209 (2016). (Chinese)

7. Dai, J.S., Meng, W.D.: A study on profit-sharing mechanism in team production. Chin. J. Manage. Sci. **18**(01), 120–127 (2010). (Chinese)
8. Holmstrom, B.: Aggregation and linearity in the provision of intertemporal incentives. Ecomometrica **55**, 303–328 (1987)
9. Ren, S.M., Sun, F., Wang, H.J.: Double principal-agent, shareholder superiority and incentive contract design-empirical study based on food listed companies. J. Dalian Univ. Technol. (Soc. Sci.) **35**(02), 44–49 (2014). (Chinese)
10. Corgent, B., Gomez-minambres, J.: Goal setting in the principal–agent model: weak incentives for strong performance. Games Econ. Behav. **109**, 311–326 (2018)
11. Guo, H.D., Hao, H., Zhang, Y.X.: Analysis of agency chain and research on multi-level incentive mechanism of project quality government supervision. Chin. J. Manage. Sci. **25**(6), 82–90 (2017). (Chinese)
12. Wangab, Y., Liu, J.: Evaluation of the excess revenue sharing ratio in PPP projects using principal–agent models. Int. J. Project Manage. **33**(6), 317–1324 (2015)
13. Chang, H., Cvatanic, J.: Optimal contracting with moral hazard and behavioral preferences. J. Math. Anal. Appl. **428**, 959–981 (2015)
14. Jia, C.R.: Research on the regulation policy of network car market based on principal–agent theory. Dongbei University of Finance and Economics (2017). (Chinese)
15. Garcia, D.: Optimal contracts with privately informed agents and active principals. J. Corp. Finance **29**, 695–709 (2014). (Chinese)
16. Shi, J.F., Qin, X.Y., Liu, J.C.: Study on double principle–agent model of "investors–operators" dual-role party's in PPP projects. Soft Sci. **32**(09), 124–128+137 (2018). (Chinese)
17. Chen, K.G., Huang, M.: Incentive Contract Design of Virtual Enterprises—a Study Based on Asymmetric Information. Economy & Management Publishing House (2017). (Chinese)
18. Tian, H.P., Liu, C.X., Guo, Y.J.: Pay scheme designing method with multiple agents' competition. J. Industr. Eng. Eng. Manage. **04**, 153–156 (2007). (Chinese)

High-Dimensional Objective Flexible Sand Foundry Scheduling Under Green Manufacturing

Hong-tao Tang(✉), Huan Zhang, Xiang-yi Li, and Yue Feng

Hubei Key Laboratory of Digital Manufacturing,
School of Mechanical and Electronic Engineering,
Wuhan University of Technology, Wuhan 430070, China
tanghongtaozc@163.com

Abstract. With the increasing social awareness of green manufacturing, the sand casting, as one of the major production methods of carbon emission in the manufacturing, faces a major challenge in the coordination between production and sustainable development of green. In view of the carbon emission caused by electrical energy consumption in sand mold casting production, for the sake of achieving sustainable scheduling, a high-dimensional target flexible sand mold casting job shop scheduling model with carbon emission constraint is established, aiming to minimize the maximum completion time, minimize carbon emission, minimize the maximum machine load and minimize the processing cost. In order to solve the model, the PCA in machine learning is introduced into the optimization scheduling of high-dimensional objectives, and an improved PCA-NSGAII algorithm is adopted to reduce the dimension of the target. Finally, a practical scheduling problem is taken as an example to verify the feasibility of the algorithm. By comparison with traditional NSGAII DPSO PESAII algorithm, the effectiveness of the algorithm is demonstrated.

Keywords: Green manufacturing · Sand casting production · Optimization scheduling of high-dimensional objective · PCA-NSGAII algorithm

1 Introduction

With the continuous promotion of made in China 2025, the relationship between production and environment is increasingly inseparable, and the impact on the environment and effective utilization of resources in the production process have become the target factors that cannot be ignored [1]. As one of the main sources of carbon emission in the manufacturing industry [2], through appropriate scheduling schemes [1], sand casting can greatly reduce the carbon emission in the production process and realize the green manufacturing with energy saving and emission reduction [3]. The job-shop scheduling problem (JSP) is considered NP-hard [4], which can be described as the allocation of work on the machine [5]. The method of energy intensity decomposition is proposed to analyze carbon emission, and the calculation formula of carbon dioxide emission in the casting process is constructed [6]. Moreover a method for estimating and evaluating the boundary of carbon emission in casting production

C.-F. Chien et al. (Eds.): *IE&EM 2019*, pp. 250–260, 2020.
https://doi.org/10.1007/978-981-15-4530-6_25

system is established, and the carbon emission in sand casting production process is calculated by carbon emission evaluation function [7]. A multiobjective particle swarm optimization algorithm based on preference for multiobjective flexible sand casting problem avoids the problem that decision makers are difficult to choose in many noninferior solutions [8]. Then, a hybrid fruit fly algorithm (HFFA) selects the Pareto frontier solution from existing solutions through fuzzy decision tools [9]. Actually aiming at the problem of high-dimensional objective optimization, PCA dominant mechanism based on weight coefficient is proposed, through the new selection mechanism to make up the disadvantage of Pareto sorting that the number of non-dominant solutions increases greatly due to the small selection pressure [10, 11].

2 Problem Description

2.1 Introduction of Sand Casting Process

Sand casting production is characterized by a small batch of individual pieces, and each casting has to go through a variety of complex technological processes from the material to the finished product. Process can be roughly classified into three major parts: modelling and core, melting and pouring, shake out and quality inspection. In addition, according to the machining features Process can be divided into two categories, machining and nonmachining of the technological process. Owing to the existence of nonmachining, sand casting job-shop scheduling problem is not entirely flexible job-shop scheduling problem (FJSP). In order to transform it into flexible sand casting job shop scheduling problem, the nonmachining process is virtualized into machine processing. Starting from the stage of pouring, the research on pouring, shake out, shot blasting, polishing and finishing is mainly carried out.

2.2 Process Buffer Time Interval

For the traditional FJSP, when a process of the job is completed, as long as the optional machine of next process is idle, the process can be processed. But in the field of sand casting, owing to its special processing technology characteristics, to meet the quality requirements of the castings, as soon as a process is complete, no matter whether the corresponding machine is idle or not, the next process can be processed after a certain process buffer the time interval.

2.3 Calculation Method of Carbon Emission in Processing

In the process of sand casting, the energy demand is different according to the process flow. For measuring the impact of electricity consumption on carbon emission, the standard coal conversion coefficient and carbon emission coefficient of electricity are used to convert the energy consumption value into carbon emission.

3 Mathematical Model

3.1 Shop Scheduling Model Analysis of Flexible Sand Casting

The classic FJSP can be described as: there are m machines and n job and each job has n_i processes. What is more, every process of each job can be processed on multiple machines, but the working procedure sequence of each process cannot be changed due to the limitation of technological process. Although each machine can process multiple operations, only one can be processed at a time. For the sake of studying the flexible sand casting shop scheduling model under the constraint of carbon emission, the following assumptions are made: (1) At the beginning of processing, each machine is idle. (2) Each machine can only process at most one process at any time. (3) Any process of any job can only be processed continuously in one machine. (4) The preparation time before the first process and the cleaning time of the last process are not considered. (5) Transport time between one process and the next is negligible. (6) The machine is only switched on and off once before starting and after finishing. The time can be ignored. By default, the machine is turned on before the job starts processing and always on during process. (7) The carbon emission mentioned here only considers the carbon generated by the power consumption of the processing equipment and dust removal equipment. The cost only considers the processing cost. (8) In the processing process, the worker's processing link is simulated as virtual process.

4 Notation Definition

(1) Because of the problem requires Table 1 to be some description of the notation used.

Table 1. Notation and description of FJSP Problem

Notation	Describe
n	casting total
m	total number of jobs
s	total dust removal equipment
n_i	total number of process for job i
i	serial number of job
j	serial number of process
h	serial number of processing equipment
c	serial number of dust removal equipment
O_{ij}	the process j of job i
C_{ijh}	the processing time of process O_{ij} on the machine h
T_{ijc}	the dust removal time of the process O_{ij} on the dust removal equipment c
J_{ij}	the buffer time interval between the process i and the process i + 1
y_{ijh}	the processing cost of process O_{ij} on the machine h
t_{ijh}	the finishing time of process O_{ij} on the machine h
P_h	rate operating power of the processing equipment h

(*continued*)

Table 1. (*continued*)

P_c	rate operating power of the dust removal equipment c
B	power standard coal conversion coefficient, 0.1229kgce/(KW · h)
EF	carbon emission coefficient of electric energy, 4.035 kgCO—2e/kgce
C_{max}	Maximum completion time
E	Maximum carbon emission
W_m	Maximum machine load
FC	The processing cost
A_{ij}	The starting time of process O_{ij}
B_{ij}	The ending time of process O_{ij}

(2) Decision variables

$$x_{ijh} = \begin{cases} 1, & I_{ij} \text{ is processed on machine } h \\ 0, & I_{ij} \text{ is not processed on machine } h \end{cases}$$

$$ST_{ijh} = \begin{cases} 1, & \text{process } O_{ij} \text{ is the first process on machine } h \\ max(t_{i'j'h}, t_{i(j-1)h} + j_{i(j-1)}), & \\ \text{process } O_{i'j'} \text{ is the previous process of process } O_{ij} \text{ on machine } h & \end{cases}$$

5 Building Model

For purpose of meeting the actual processing requirements, a mathematical model is established aiming at minimizing the maximum processing time, minimizing the maximum carbon emission, minimizing the maximum machine load and minimizing the minimum processing cost. The specific mathematical formula is as follows:

$$object \begin{cases} minC_{max} = min(\sum_{i=1}^{n}\sum_{j=1}^{n_i} C_{ijh} + \sum_{i=1}^{n}\sum_{j=1}^{n_i} J_{ij}) \\ minE = min[(\sum_{h=1}^{m}\sum_{i=1}^{n}\sum_{j=1}^{n_i} P_h C_{ijh} + \sum_{c=1}^{s}\sum_{i=1}^{n}\sum_{j=1}^{n_i} P_c T_{ijc}) \times B \times EF] \\ minW_m = min(max\sum_{i=1}^{n}\sum_{j=1}^{n_i} C_{ijh} x_{ijh}) \\ minFC = min(\sum_{i=1}^{n}\sum_{j=1}^{n_i} y_{ijh}) \end{cases}$$

$B_{ij} + J_{ij} \leq A_{ij+1}$, it limits the sequence of casting process. $\sum_{h=1}^{m} x_{ijh} = 1$, it limits each casting process to one machine.

$B_{ij} = A_{ij} + C_{ijh}$, the end time of the process j of the casting i is equal to the sum of its starting processing time and processing time, which restricts the continuous and uninterrupted processing of the process.

$ST_{ijh} \geq t_{i'j'h}$, It restricts each machine to only one process at any one time.

6 Design of Algorithm

6.1 The Introduction to the Improved NSGAII Algorithm

The general NSGAII algorithm mainly uses the Pareto algorithm to perform non-dominated sorting for scheduling problems with multiobjective. However, as the goal increases to high-dimensional (four or more), solves the nonlinear scheduling model using the nondominated sorting optimization algorithm. The difficulty of selection increases, and the number of nondominated solutions increases exponentially, which greatly increases the difficulty of solving the problem. For the purpose of solving this problem, principal component analysis (PCA) algorithm is adopted for noise reduction of targets. Then PCA dominant mechanism is used to replace the nondominated sorting in population evolution, which increases the selection pressure of the algorithm on the nondominated solutions. On behalf of increasing the population diversity, the non-dominated solution set obtained from the first stage of population evolution and the parent solution set generated from the second stage are combined into a new population.

6.2 Population Initialization and Crossover Operations

The two-stage coding method is adopted, as the population is initialized, that is, the process sequencing and the machine selection [12]. The purpose of crossover and mutation in the population is to obtain more optimized chromosomes through some changes while retaining the characteristics of excellent individuals in the population. So as to improve the overall quality of the population, the selection strategy of the tournament is adopted. Select a part of chromosomes from the population for crossover and mutation, in which the process part is POX crossover [13]. Manipulating multiple genes on a pair of parent chromosomes enables the children chromosome to better inherit the gene of the parent chromosome. For example, a pair of process genes are Parent1 = [2,4,1,1,3,2,3,4] and Parent2 = [4,2,2,1,4,1,3,3]. Randomly select the processes 2 and 3 for crossover operation to obtain the child chromosomes Children = [2,4,1,4,3,2,3,1]. As in Fig. 1.

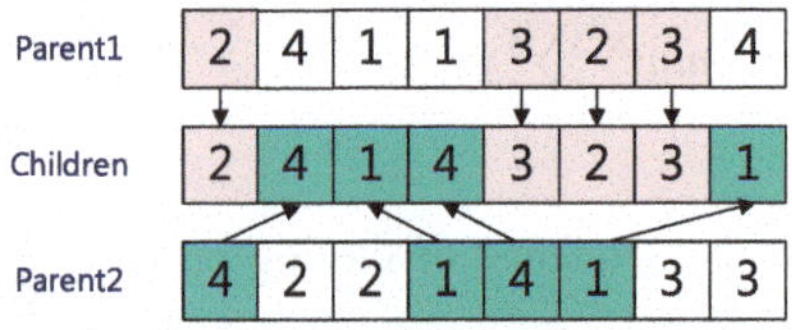

Fig. 1. Schematic diagram of POX intersection.

In addition, machine parts for the RPX crossover [14]. Randomly generate a matrix whose element size between [0, 1] and length is the same as the machine segment. Assuming that pf = 0.7, marking the machine code gene index of a member whose size is less than 0.7, the parent Parent1 is copied to the children, and delete the gene of the marked machine code. Then copy the machine gene in Parent2 to the children in the light of the correspondence between the marked gene and the process of Parent1. As in Fig. 2. The crossover weight pf is defined as follows:

$$pf = pf_{max} - \frac{pf_{max} - pf_{min}}{MaxIt} \times It$$

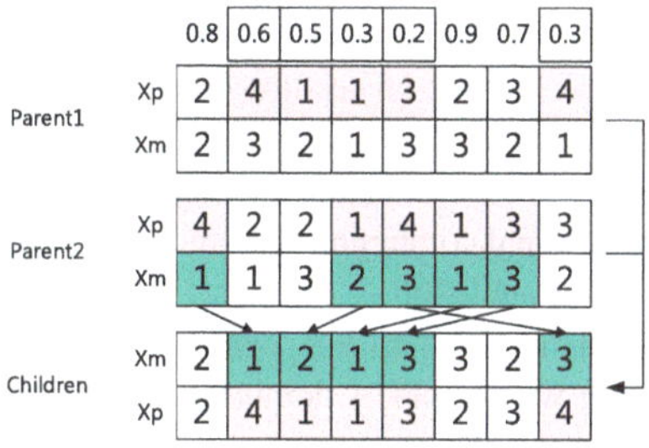

Fig. 2. Schematic diagram of RPX intersection.

6.3 PCA Dominant Mechanism

In the PCA analysis, in order to not change the target characteristics of the problem as much as possible, the discarding strategy is not adopted for the obtained redundant target vector. What is said here is that the noise reduction of the redundant target is to give it a weight according to its corresponding principal component, and then the redundant target will be fitted as a virtual target. If the target f_i is a nonredundant target determined by principal component N(N = 1,2…m), the weight of the target f_i is the proportion of the corresponding eigenvalue of the principal component N. Hence obtaining all the target weight coefficients $w = \{w_1, w_2 \ldots w_m\}$. The population evolution is sorted by PCA dominant mechanism, and the nondominated solutions are selected by difference selection operator. The specific formula is as follows:

$$f_i(a) - f_i(b) + \sum_{j=1, j \neq i}^{m} w_j(f_j(a) - f_j(b)) < 0,$$

a and b are two different individuals, f_i is any target, $f_i(a)$ is the value of target f_i of individual a. If this inequality is true, it means that individual a dominates individual b.

6.4 Algorithm Flow

See Table 2.

Table 2. Algorithm flow

Step1	Initial population
Step2	Population iteration yields the nondominated solution set P1
Step3	Perform PCA analysis on the target matrix of the nondominated solution set P1, and calculate the weight ratio w_i of the eigenvalue i
Step4	According to the dimensionality reduction principle of PCA, the ith principal component corresponding to the eigenvalue i can be obtained, and the nonredundant target f_i can be further obtained. Then, the weight of this target is w_i
Step5	Reinitialize the population and merge P1 and new population into one population Pt
	Determine whether the number of iterations reaches the upper limit. If yes, the algorithm ends. Otherwise, execute Step7
Step7	Selecting individuals from the population Pt with a certain probability for crossover and mutation operation, and combining the generated new individual and the original population into a new population
Step8	Put the optimal solution gbest into the population $P_{t+1,}$, which determined by nondominated sorting with the PCA dominant mechanism and execute Step6

7 Case Study

7.1 Case Description

The production data of a batch of castings from a foundry is analyzed. The relevant data includes 10 castings, each of which has 7 processes. The optional processing machine for each process is shown in Table 3. Table 4 shows the specific production data. In Table 5, the Power is the rated power of processing machines, the Power is the rated power of dust removal equipment corresponding to each processing machine, and the Cost is the processing cost per hour of machine processing. When calculating carbon emissions according to the standard, the carbon emission coefficient of electric energy is 4.035 $kgCO_2e/kgce$, and the power standard conversion coefficient is 0.1229 kgce/(kW·h).

Table 3. Process machine set of each process

	Operations						
	Pouring(I_1)	Shake out(I_2)	Initial shot blasting(I_3)	Polishing(I_4)	Rough cast(I_5)	Finishing(I_6)	Fine buffing(I_7)
M	[M_1–M_2]	[M_3–M_5]	[M_6–M_9]	[M_{10}–M_{13}]	[M_6–M_9]	[M_{14}, M_{16}]	[M_6–M_9]

Table 4. The specific production data

The casting	Process	Machine	Process time (h)	Buffer time interval (h)
Job1	I_1	[1,2]	[0.1,0.1]	15
	I_2	[3,4,5]	[1,1,1.5]	3
	I_3	[6,7,8,9]	[0.8,0.6,0.5,1]	1
	I_4	[10,11,12,13]	[1,1,1,0.5]	1.5
	I_5	[6,7,8,9]	[2,1,1,3]	1.5
	I_6	[14,15,16]	[4,4,3.5]	1
	I_7	[6,7,8,9]	[3,2,2,4]	15
Job2	I_1	[1,2]	[0.15,0.2]	20
	I_2	[3,4,5]	[1.8,1.5,2]	2.5
	I_3	[6,7,8,9]	[0.5,0.6,0.6,0.4]	1
	I_4	[10,11,12,13]	[0.5,0.5,0,5,1]	2
	I_5	[6,7,8,9]	[1,1,2,2.5]	1.5
	I_6	[14,15,16]	[3,3,3.5]	1.5
	I_7	[6,7,8,9]	[2,2,3,2.5]	20
Job3	I_1	[1,2]	[0.05,0.05]	10
	I_2	[3,4,5]	[0.5,0.5,0.8]	2.5
	I_3	[6,7,8,9]	[0.5,0.4,0.4,0.6]	0.5
	I_4	[10,11,12,13]	[0.5,0.5,0.5,0.5]	1
	I_5	[6,7,8,9]	[1.5,1.1,2]	1
	I_6	[14,15,16]	[3.5,4,3.5]	1
	I_7	[6,7,8,9]	[2.5,2,2,2.5]	10
Job4	I_1	[1,2]	[0.1,0.1]	16
	I_2	[3,4,5]	[1,1,0.8]	2
	I_3	[6,7,8,9]	[0.5,0.6,0.6,0.4]	0.5
	I_4	[10,11,12,13]	[1,2,1,1]	2
	I_5	[6,7,8,9]	[1.5,2,2,1]	1
	I_6	[14,15,16]	[3,3.5,4]	2
	I_7	[6,7,8,9]	[2,3,3,2]	16
Job5	I_1	[1,2]	[0.1,0.15]	17
	I_2	[3,4,5]	[1,1,1]	1.5
	I_3	[6,7,8,9]	[0.8,0.7,0.6,0.9]	0.5
	I_4	[10,11,12,13]	[1,2,2,2]	5
	I_5	[6,7,8,9]	[2.5,1.5,1,2]	1
	I_6	[14,15,16]	[2,2,2.5]	2
	I_7	[6,7,8,9]	[3.5,2.5,2,4]	17

Table 5. The Specific Production Data

M	M1	M2	M3	M4	M5	M6	M7	M8
Power	8	9	60	48	54	56	46	64
Power0	0	0	40	40	10	10	35	30
Cost	80	70	75	80	75	70	80	35
Power	42	5	6	4	4.5	0	0	0
Power0	0	0	0	0	0	0	0	0
Cost	40	40	30	35	45	50	30	35

8 Analysis of Results

An actual case is solved using this algorithm to verify the feasibility of the algorithm. And NSGAII DPSO PESAII as the comparison algorithm, the results of the four algorithms are compared as shown in Fig. 3. Set the same population size 50, maximum number of iterations 100, crossover probability 0.8, mutation probability 0.3, RPX parameter pfmax = 0.9, RPX parameter pfmin = 0.2. Considering the distribution of each target value, PCA-NSGAII algorithm is better than other algorithms. The result obtained by processing the data independently run by the four algorithms is shown in Table 6.

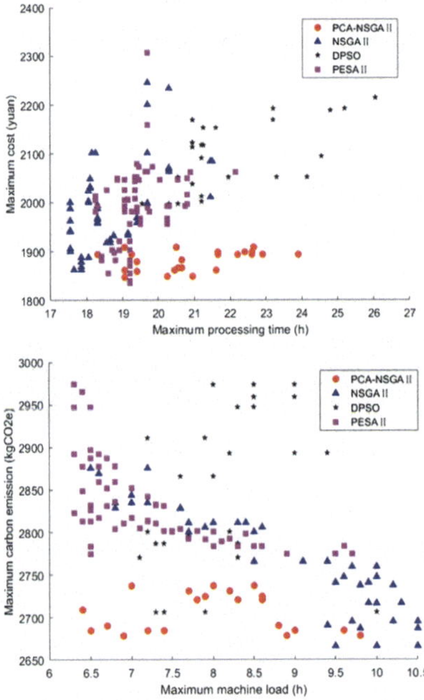

Fig. 3. Magnetization as a function of applied field.

Table 6. Four algorithms comparison results

Algorithms		PCA-NSGAII	NSGAII	DPSO	PESAII
Maximum machine load	Δmin	6.4	6.5	6.5	6.3
	Δmax	9.8	10.5	10	9.7
	Δavg	8.164	9.121	7.845	7.375
Maximum carbon emission	Δmin	2678.987	2666.68	2706.615	2774.831
	Δmax	2737.585	2875.972	2973.528	2974.506
	Δavg	2698.64	2752.615	2832.649	2825.996
Maximum processing time	Δmin	18.3	17.55	19.55	18.25
	Δmax	23.9	21.5	26.05	22.15
	Δavg	20.675	18.339	21.23233	19.351
Maximum cost	Δmin	1848	1863	1997.5	1882
	Δmax	1909	2246	2214.5	2306.5
	Δavg	1876.633	1971.66	2069.227	2007.32

It can be seen from the minimum Δmin and maximum Δmax and the mean value Δavg of the four targets that PCA-NSGAII significantly optimizes the carbon emission and processing cost while maintaining the processing time and the maximum machine load at a general level. Considering four goals in general, the PCA-NSGAII algorithm is better than other algorithms.

9 Conclusion

Under the background of sand casting for high-dimensional objective flexible job-shop scheduling problem, adopting the PCA algorithm instead of the traditional nondominated sorting algorithm effectively solves the problem that the number of the non-dominated solutions increases sharply, when the selection pressure decreases due to the increase of targets. However, at the same time, the running time also increases. In subsequent studies, the algorithm of taboo search and simulated annealing can be considered to improve the local search ability of the algorithm and the selection ability of the algorithm. In the actual production process, there are often multiple objectives to meet the needs of various aspects. For high-dimensional target optimization problems, more effective fitness value calculation methods can be explored. For the concept of green development, constraints such as noise and dust can be added to the model.

References

1. Singh, A., Philip, D., Ramkumar, J., Das, M.: A simulation based approach to realize green factory from unit green manufacturing processes. J. Clean. Prod. **182**, 67–81 (2018)
2. Jeswiet, J., Kara, S.: Carbon emissions and CES™ in manufacturing. CIRP Ann. Manuf. Technol. **57**(1), 17–20 (2008)

3. Zeng, Z., Hong, M., Man, Y., Li, J., Zhang, Y., Liu, H.: Multi-object optimization of flexible flow shop scheduling with batch process—consideration total electricity consumption and material wastage. J. Clean. Prod. **183**, 925–939 (2018). (In Chinese)
4. Blazewicz, J., Domschke, W., Pesch, E.: The job shop scheduling problem: conventional and new solution techniques. Eur. J. Oper. Res. **93**(1), 2 (1996)
5. Kato, E.R.R., Aranha, G.D.D.A., Tsunaki, R.H.: A new approach to solve the flexible job shop problem based on a hybrid particle swarm optimization and Random-Restart Hill Climbing. Comput. Ind. Eng. **125**, 178–189 (2018)
6. Kim, Y., Worrel, E.: International comparison of CO2 emission trends in the iron and steel industry. Energy Policy **30**, 927–929 (2002)
7. Yin, R.X., Cao, J.H., Li, H.C., Du, Y.B.: Carbon emission quantification method of sand casting process and its application. Comput. Integ. Manuf. Syst. **18**(05), 1071–1076 (2012). (In Chinese)
8. Junjie, B., Ningsheng, W., Dunbing, T.: Improved PSO algorithm for multi-objective optimization flexible job shop scheduling problems. J. Nanjing Aerosp. Univ. **42**(04), 447–453 (2010). (in Chinese)
9. Balasubbareddy, M.: Multi-objective optimization in the presence of ramp-rate limits using non-dominated sorting hybrid fruit fly algorithm. Ain Shams Engineering Journal **7**(2), 895–905 (2016)
10. Liu, Q., Xiong, S., Zhan, M.: Many-objectives scheduling optimization based on PCA-NSGAII withan improved elite strategy. Comput. Integ. Manuf. Syst., 1–13 (2019). (In Chinese)
11. L Y.Z, W Y.H.: PCA and algorithms analysis. Journal Of Suzhou University (Natural Science Edition), (01), 32–36 (2005). (In Chinese)
12. Zhang, G., Gao, L., Shi, Y.: An effective genetic algorithm for the flexible job-shop scheduling problem. Expert Syst. Appl. **38**(4), 3563–3573 (2011). (in Chinese)
13. Zhang, J., Wang, W., Xu, X.: A hybrid discrete particle swarm optimization for dual-resource constrained job shop scheduling with resource flexibility. J. Intell. Manuf. **28**(8), 1–12 (2015). (in Chinese)
14. Zhang, C.Y., Rao, Y.Q., Liu, X.J.: Solving job-shop scheduling problem based on genetic algorithm based on pox crossing. China Mech. Eng. **15**(23), 2149–2153 (2004). (In Chinese)

Evaluation Index System of Production Planning in Manufacturing Enterprise

Jing-ying Zhao(✉)

College of Mechanical and Electrical Engineering, Guangdong University of Petrochemical Technology, Maoming 525000, Guangdong, China
guptzjy@126.com

Abstract. This paper expounds the definition of production plan from two aspects of broad sense and narrow sense. What's more, we classify the production plan from time, space, object and resource, and divide the production plan into layers. According to give the function and purpose of production plan, the paper summarizes the influencing factors of production plan and the common reasons why production plan could not be realized, and expounds the process of making production plan, and points out the principles and ideas that establishes the evaluation index system of production plan. On this basis, the evaluation index system of manufacturing enterprise production planning is established.

Keywords: Evaluation index system · Manufacturing enterprise · Principles · Ideas

1 Introduction

With the development of science and technology, customers' demands on manufacturing products are becoming higher and higher. They only rely on improving the quality of their own products and reducing production costs. The upgrade of enterprise's overall production management mode becomes particularly important. In order to upgrade the overall production management model of manufacturing enterprise, we must first make a production plan that conforms to the actual production situation of the enterprise.

2 Definition, Classification and Hierarchy of Production Planning

2.1 The Generalized Definition of Production Planning

Production planning refers to the preparation of activities related to production. Time planning: how to arrange the schedule; space planning: how to plan the site; resource planning: such as the allocation of materials.

C.-F. Chien et al. (Eds.): *IE&EM 2019*, pp. 261–268, 2020.
https://doi.org/10.1007/978-981-15-4530-6_26

2.2 The Narrow Definition of Production Planning

Production planning refers to the production planning of all production workshops centered on the production planning department, which is generally expressed in the form of a form. For example, monthly production plan, weekly production plan, daily production plan, etc. [1].

2.3 Classification of Production Plans

(1) *Time classification*: The annual production plan, the quarterly production plan, the monthly production plan, the weekly production plan, the daily production plan, the production plan per hour, the production plan per minute and the production plan per second.
(2) *Organization classification*: Company production plan; department production plan; section production plan; group production plan; class production plan; line production plan; single machine production plan; single person production plan.
(3) *Object classification*: Order production plan, finished product production plan, component production plan, part production plan, production process plan.
(4) Resource classification: Personnel production plan; equipment production plan; material production plan; industrial production plan; environmental production plan; energy production plan.

2.4 Hierarchical Division of Production Planning

The production plan of manufacturing enterprise can be divided into three levels: strategic plan, tactical plan and job plan. The strategic plan is related to the future development of the whole enterprise. The formulation of the tactical plan defines the objectives to be achieved by the enterprise in production, and the job plan clarifies the working arrangements in the enterprise's production process: schedule, quantity, time, etc. See Table 1.

Table 1. Hierarchical division of production planning

Content	Strategy	Tactic
Time of plan	$\geq$5 years	1 year
Range	Enterprise	Factory
Uncertainty	High	Medium
Management	Top	Middle
Trait	Acquirement of resource	The use of resource

3 Function and Purpose of Production Planning

3.1 Function of Production Planning

(1) *Planning function*: Production activities can be organized, organized and organized.
(2) *Prediction function*: The essence of production planning is to predict the production activities and to study and analyze the production activities [2].
(3) *Management function*: Production planning is also a kind of management, a clear objective management of the operation of the enterprise.
(4) *Supervisory function*: The production plan can exert the function of the production plan only by strictly supervising the actual production of the enterprise.
(5) *Coordination function*: The production plan studies and analyzes the problems that appear in the enterprise and resolves the problems.

3.2 Purpose of Production Planning

Improve production, work efficiency; properly reduce inventory; achieve balanced production.

To realize the production planning system with "monthly production plan as the leader and weekly production plan as the guide", and to realize the production plan management mode with "production planning, supervision during production, summary after production"; To realize the production mode of "production department as the center, each workshop comprehensive production", to prepare well before production, to study the external factors affecting production, and to enhance the balance of production.

4 The Influencing Factors and the Causes of Difficulty

4.1 Factors Affecting Production Planning

There are a number of factors that affect production planning, such as markets, resources, targets, warehouses, equipment, production methods, etc., as shown in Table 2.

Table 2. Factors affecting production planning

Factor	Content
Market	Demand, variety, price, etc.
Resource	Raw materials, equipment, manpower, etc.
Target	Cost minimization, production maximization, etc.
Depot	Maximize storage capacity, etc.
Equipment	Equipment maintenance cost minimization, etc.
Manufacturing	Single type mass production, multi-type production, etc.

(1) *Consideration of the ability*: According to the production capacity, the production planning problem can be divided into three kinds of production planning problems: no ability constraint, ability constraint and variable capacity. It can be solved by changing labor force, changing overtime time, changing equipment resources, etc. Make the production plan with different production capacity, different characteristics of production costs [3].
(2) *Consideration of energy conservation and environmental protection*: With the continuous development of economy, environmental problems have become increasingly prominent, manufacturing enterprises must abandon the previous economic benefit only the idea, and to consider sustainable development. To achieve the sustainable development of enterprises, mainly to the performance of equipment and production process improvement, reduce waste emissions increase to the same enterprise benefit goal the establishment of a position. The coordination of economy and environment production planning will take waste emissions into consideration.
(3) *Consideration of uncertainty*: When making production plans, enterprises must take into account many uncertain factors, such as market demand, price, production time, raw material supply, equipment problems, operational problems, etc. These uncertainties can affect the normal operation of the enterprise. To deal with these uncertainties, we need to do the following: for uncertain information, use probability distribution to express it. The production plan should be supported by the probability theory, the fuzzy information should be transformed into the quantitative probability theory, the quantitative probability theory should be solved, and the fuzzy information should be expressed by the way of fuzzy number. Fuzzy production planning problem should be solved by fuzzy theory and fuzzy technology.

4.2 Reasons Why the Production Plan Is Difficult to Realize

When the production plan is difficult to realize, there are usually the following reasons: the high level of the enterprise pays less attention to the production plan and thinks that the production plan is not necessary; the enterprise lacks the professional skills to make the production plan, The resulting production plan cannot be implemented; the actual situation is not fully taken into account when the production plan is drawn up, and the production plan does not conform to the actual production of the enterprise. There is not enough coordination and cooperation between the various departments, resulting in production planning could not be carried out.

5 Establishment of Evaluation Index System for Manufacturing Enterprise Production Planning

5.1 The Principle of Setting up an Index System

When an enterprise establishes an index system for production planning, it should consider the actual situation of the enterprise itself, and only establish an index system that conforms to the conditions of the enterprise itself. The following principles should be followed when establishing the evaluation index system of production plan.

(1) *The principle of qualitative and quantitative interaction*: The implementation of the enterprise production plan will change the original state of the enterprise, and the impact of the implementation of the production plan on the enterprise should be fully considered in the establishment of the index system. Therefore, when establishing the index system, we should not only consider quantitative indicators. Also consider qualitative indicators.
(2) *The level and quantity of indicators should be appropriate*: If there are more index levels and the number of indicators, the evaluation will be more complex, which may lead to the decline of the correct rate of evaluation; if the index level and the number of indicators are too small, the enterprise production plan can't be evaluated completely.
(3) *The establishment of evaluation index should be in accordance with the conditions of the enterprise itself*: To establish an index system of production planning, we should consider the actual situation of the enterprise itself. Only by establishing an index system that conforms to the conditions of the enterprise itself, can the enterprise better realize its strategic objectives [4].
(4) *Combined with production planning*: The establishment of the evaluation index of production planning should not only reflect the current production planning situation of the enterprise, but also reflect the possible changes of the future production plan of the enterprise.
(5) *Operational principle*: The establishment of evaluation index should accord with objective facts, be simple to calculate, and the index data should be defined, otherwise, it may lead to poor maneuverability.
(6) *Having the characteristics of comparison*: The evaluation index should not only show the actual situation of its own enterprise, but also compare the situation of other enterprises and the same industry, and then, through comparison, improve its own enterprise's shortcomings and enhance its competitive power in the industry.
(7) *Principle of integrity*: Because the factors affecting the development of the enterprise include many aspects, the evaluation index system should be comprehensive, complete, and should be considered in many aspects at the same time.

5.2 The Idea of Establishing Index System

(1) *Understanding the necessity of establishing Evaluation Indexes for production Planning*: There are quite obvious differences between production plans in different industries, and the establishment of evaluation indicators will vary widely. The weights of the indicators should be adjusted according to the actual conditions of their own enterprises. Finally, set up a specific evaluation index of production plan in line with their own enterprise.
(2) *The expectations of the evaluation indicators for each aspect of the production plan vary*: When selecting the evaluation index, we should consider and analyze all aspects, and finally choose an evaluation index system which can fully reflect the expectations of the production plan.

(3) *To analyze the degree of difficulty and ease of Evaluation Index*: The more specific the evaluation indicators, the more accurate and reliable the results will be, but the higher the cost of establishing such evaluation indicators, the more effective and cost should be taken into account in the establishment of evaluation indicators, and to establish a reasonable evaluation index system [5].
(4) *Establishment of Evaluation Index should be practical*: The established evaluation index should play an important role in the enterprise's production plan, and the evaluation index should be combined with practice and theory in order to play its greatest role.

5.3 Establishment of Index System

From the aspects of financial management, operation management, customer, product quality, growth and learning, the following evaluation index system is established. See Table 3.

Table 3. Evaluation index system of production planning

First class index	Second class index	Third class index
Financial management	Enterprise manager	Return on assets
		Asset turnover rate
		Profit growth rate
		Sales growth rate
	Enterprise owner	Return on assets
Operation management	Production flexibility	Quantitative flexibility Time flexibility
	Operation process	Rate of sale of marketed goods
		Stock turnover
	Operating cost	Human cost
Customer	Product quality	Bounce rate
		Qualified rate
	Reliability	Rate of on time delivery
		Accurate delivery rate
		Customer complaint rate
		Effective trading rate
		Completion of order rate
Product quality	Product quality	Scrap rate Product rework rate
		One-time qualification rate
Growth learning	Information sharing	Effective rate of information
		Rate of change of information
	Innovation ability	Ratio of new technologies
		R & D investment rate

(1) Consideration of financial management: As far as enterprise managers are concerned, they are usually most concerned about the economic benefits, assets operation, development of enterprises, etc. The following indicators are commonly used to express.

Return on assets = Total profit/Value of assets

Asset turnover = Value of business income/Assets

Profit growth rate = (Current profit − Prior period profit)/Prior period profit

Sales growth rate = (Current sales volume − Prior period sales volume)/Prior period sales volume

Return on assets = 2 * Assets income/(Initial owner's income + Final owner's income)

(2) Consideration of operational management: Quantitative flexibility and time flexibility are utilized to express the manufacturing flexibility.

Quantitative flexibility = Enterprise product amount/Demanded amount from customer

Production and marketing rate = Sales volume/Manufactured volume

Human cost = Direct human cost + Indirect human cost

(3) Consideration of customer: Bounce rate and qualified rate are often taken into consideration when it refers to the product quality.

Bounce rate = Return quantity/Total sales

Qualified rate = Qualified amount/All the product amount

Rate of on time delivery = Times of on time delivery/Times of all the delivery

Accurate delivery rate = Times of accurate delivery/Times of trade

Completion of order rate = Times of completion from customers/Times of trade

Effective trading rate = Times of effective trading/Times of trade

Completed order rate = Completed order amount/All the order amount

(4) Consideration of the quality of the product: Scrap rate, product rework rate and one-time qualification rate are utilized to define the quality of the product.

Scrap rate = Scrap amount/Product amount

Product rework rate = Product rework amount/Product amount

One-time qualification rate = The amount of one-time qualification product/Product amount

(5) Consideration of growth and learning: In the aspect of information sharing, the effective information rate and the information change rate are commonly used.

Effective information rate = Converted information/All the information

The information change rate = (Current information − Prior period information)/All the information

Ratio of new technologies = Benefit from using new technologies/Benefit of the whole enterprise

R & D investment rate = Cost on R & D/Cost of the whole enterprise

Acknowledgment. Project supported by the Science and Technology Panning Project of Guangdong Province of China (2015A030401102).

References

1. Ming, D., Wu, Y., Wu, C.: Simulation optimization for semiconductor manufacturing logistics systems under stochastic demand. J. Shanghai Jiao tong Univ. **44**(3), 306–310 (2010)
2. Drira, A., Pierreval, H., Hajri-Gabouj, S.: Facility layout problems: a survey. Annual Review in Control **31**(6), 255–267 (2007)
3. Xia, J., Fang, R., Lan, X.: Hot research areas of simulation technique in the country—review of journal of system simulation in recent years. J. Syst. Simul. **16**(9), 1910–1913 (2004)
4. Cui, Y.D., Huang, L.: Dynamic programming algorithms for generating optimal strip layouts. Comput. Optim. Appl. **33**(23), 287–301 (2006)
5. Fung, S.H.: A virtual warehouse system for production logistics. Prod. Plan. Control **6**, 597–607 (2005)

Online 3D Printing Remote Monitoring and Control System Based on Internet of Things and Cloud Platform

Hong-fei Guo[1], Zhen-hao Huo[1], Ru Zhang[1,2(✉)], Ting Qu[1], Cong-dong Li[1], Bao Chao[1,4], Xiao-feng Luo[5], and Zhi-hui He[3]

[1] Institute of Physical Internet, Jinan University, Zhuhai 519070, China
zhangru@stu2016.jnu.edu.cn
[2] Institute of Management Science and Engineering, Jinan University, Zhuhai 519070, China
[3] Zhuhai Hengqin Building & Construction Quality Inspection Centre Co., Ltd., Zhuhai 519031, China
[4] School of Mechanical Engineering, Inner Mongolia University of Technology, Hohhot 010051, China
[5] College of Information Science and Technology, Jinan University, Guangzhou 511436, China

Abstract. For the situation where 3D printers have not been widely popularized, and there is a high demand for 3D printing, a visual and controllable online 3D printing cloud platform technology based on the Internet of Things is proposed. The main components of the system consist of a remote terminal module (based on the Arduino mega2560 main control board), a remote communication module (using the ESP8266 wireless module) and a web-side web page. The system uses the Arduino UNO controller as the control core to coordinate the stepper motor drive of the 3D printer, real-time temperature detection of the nozzle body, real-time image return and heating system. The required data is transmitted to the server and finally realized on the webpage end. The user can observe the real-time data on the webpage side, and the printing progress, etc., and can also control the printer on the webpage side, thereby realizing the purpose of 3D printing remote control.

Keywords: 3D printing · Internet of Things · Remote control · Real-time monitoring

1 Introduction

The concept of 3D printing had emerged as early as 19th century, and was subsequently promoted and developed globally in the twentieth century. 3D printing, compared with the original traditional production process, uses additive manufacturing as brand-new technology and combines the characteristics of many fields such as hardware control, molten materials, machinery, etc. Through selective bonding layer by layer stacking method, 3D printer manufactures parts using bonding materials such as wire plastic or photocurable resin. According to the path witch automatically generated

C.-F. Chien et al. (Eds.): *IE&EM 2019*, pp. 269–275, 2020.
https://doi.org/10.1007/978-981-15-4530-6_27

by machine, it could print out the pre-designed parts quickly layer by layer [1–3]. Because of the unique advantages of 3D printing, it has become a development direction in the industrial field. With the rapid development of 3D printing technology, it has been widely used in many fields such as Aerospace, Medical Equipment, Automotive Molds, Bioreforestation, Electronic Devices, and Architectural Design [4–6].

The emergence of 3D printing has had a huge impact on the manufacturing industry. The arrival of the 3D printing era provides a new solution which combines the Internet technology and 3D printing technology. Through the terminal wireless network technology, 3D printing has transformed from the "simple manufacturing" of the original single device to the "intelligent manufacturing" of the network informationization [4]. We could remotely control 3D printers by changing the format of limited transmission data to wireless transmission data.

2 Research Status and Development Trend

As early as 1988, a company in the United States introduced the first commercial-purpose model for the masses based on the first proposed 3D printer principle using liquid photosensitive resin curing technology, which enabled the public to have a preliminary understanding of 3D printing technology [7]. In 1992, the concept of "3D printing technology" was proposed by two professors at the Massachusetts Institute of Technology in the United States, brought the 3D printing to the public and began to be noticed by everyone [8]. Along with the rapid development of RP, a serious of 3D printing technology like Three-dimensional Printing, FDM, Laser Selective Melting

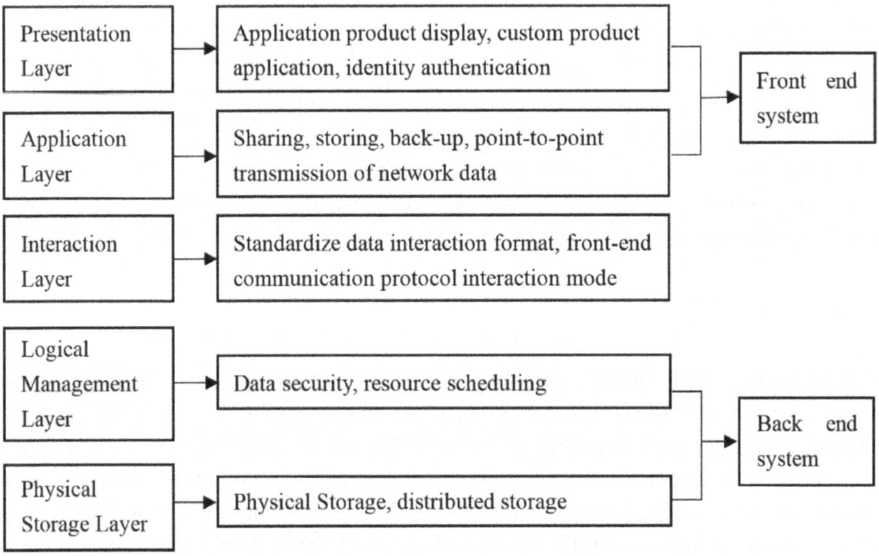

Fig. 1. Cloud platform overall application architecture diagram

Molding, Selective Laser Sintering, Electron Beam Melting Deposition Molding, Layered Solid Modeling etc. have been developed.

In recent years, with the development of emerging industries such as Information, Internet and Electronic Technology, the subject of the Internet of Things has been developed. It is based on computer and Internet-based technologies. Through the combination of the Internet of Things and 3D printers, it is basically possible to control the 3D printer inside the LAN (Local Area Network). The current development direction of 3D printing is the combination of monitoring technology and new technologies of the Internet of Things to do remote multi-machine joint printing.

3 3D Printing Cloud Platform

According to the overall needs of the website, the overall structure of the customized printing platform is as follows:

3.1 Presentation Layer

The part that the user directly touches is the presentation layer part of the application, which can have a general display of the product. It can be a web interface, and the user interacts with the website through operations.

3.2 Application Layer

By storing the data to implement the functions of the application layer, data storage, modification, backup and peer-to-peer transmission of the late data can be performed. The overall functionality is greatly enhanced by packaging the data and providing a connection interface.

3.3 Interaction Layer

The browser connects to the server through the interaction layer. The data interaction layer transmits the data of the browser to the server, but the premise is that the data format must be unified, and the same communication protocol must be observed. We choose to use Http communication, it ensures the stability and reliability of communication and prevents data loss.

3.4 Logical Management Layer

Through the logical management layer to interact with the data. Coordinating multiple servers to work together through algorithms, managing these devices collaboratively, and then performing resource allocation and data scheduling through distributed management methods.

3.5 Physical Storage Layer

As the lowest layer of the entire platform architecture, physical devices can be used to store data information, and redundant resources can be integrated for reasonable allocation, saving space and improving efficiency. As in Fig. 1.

4 Overall Design of System

The entire system framework is based on web pages, cloud platforms, local LAN gateways, and 3D printers. Information exchange through a LAN connection.

The user sends a connection request to the 3D printer through a prescribed instruction which is used to connect web side with the cloud platform. After parsing this request through the 3D printing cloud platform, the required request is sent to the 3D printer, waiting for the 3D printer to confirm. When the 3D printer receives the request, it feeds back the status to the cloud platform according to its own situation and then feeds back to the user terminal. As in Fig. 2.

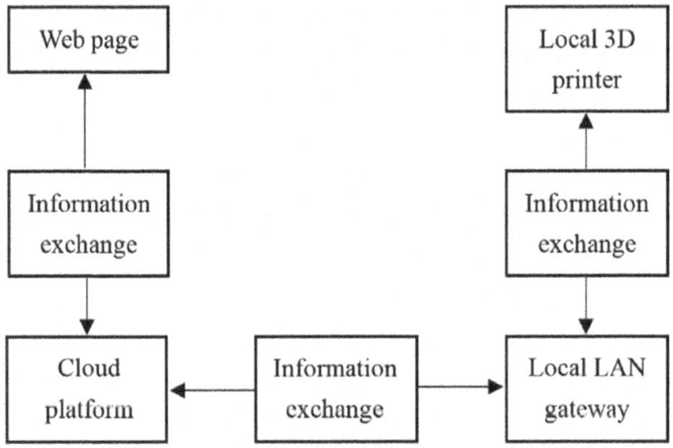

Fig. 2. System overall design

5 Data Storage System

First, the user on the web side uploads the model file to the web page, and then the web client sends the model file to the server. The server receives the model file and stores it in the cloud storage space. Finally, the URL (Uniform Resource Locator) address of the model file is written into the data table. We can find the corresponding model file by URL address. As in Fig. 3.

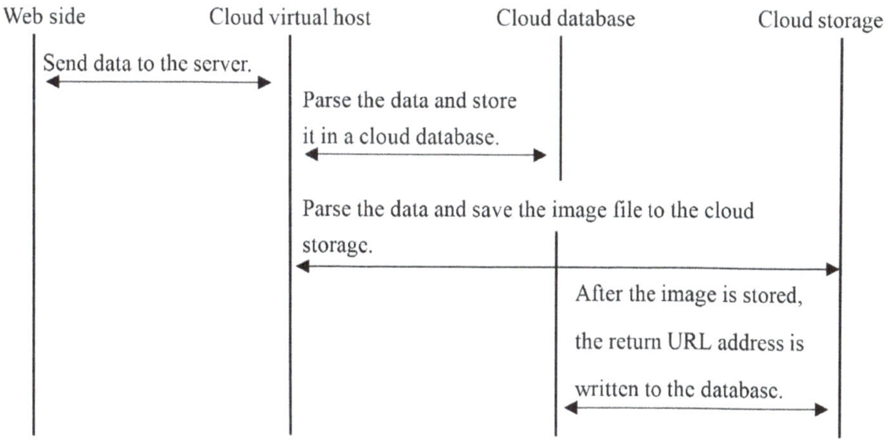

Fig. 3. Model storage implementation process

6 System Development

6.1 Log in System

The login interface serves as the website facade. In order to verify the user's login rights, the user name and password are verified, and the user behavior is recorded. This site uses HTML CSS (Hyper Text Markup Language, Cascading Style Sheets) styles to design the site. As in Fig. 4.

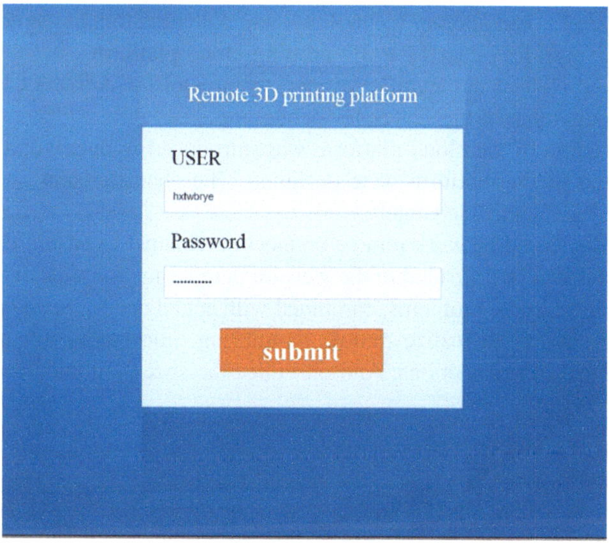

Fig. 4. Login interface diagram

6.2 System Main Interface

The main interface of the system is shown in the figure, which can display device information and member management options. It allows simple operations for websites and 3D printers. As in Fig. 5.

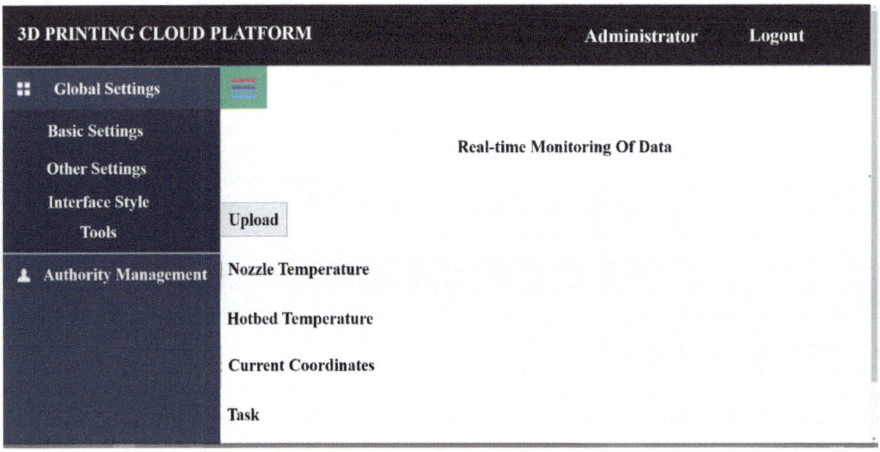

Fig. 5. System main interface

7 Conclusion

1. Analyzed the necessity of connecting 3D printers into the network under the development of contemporary technology, and paid attention to the development of 3D printing. And introduced the 3D printing cloud platform.
2. Analyzed and studied the architecture and features of the Alibaba Cloud platform, and actually purchased the server for debugging.
3. The organization of the cloud platform was introduced in detail, and the front end and back end of the platform were designed. The characteristics and structure of each part are explained separately.
4. Establish a local and network remote connection channel to ensure that the data of the 3D printer can be uploaded to the website in real time for real-time data display.
5. Starting from practical problems, combined with actual needs, the 3D printing cloud platform is realized to realize remote monitoring, information transmission, data monitoring and other functions. And introduce the specific implementation of each function.

Acknowledgment. The research and publication of their article was supported by the Fundamental Research Funds for the Central Universities [grant number 21618412; grant number 21618804]; the National Natural Science Foundation of China [grant number 51475095]; Project of Guangdong; Natural Science Foundation [grant number 2016A030311041]; 2015 Guangdong

Special Support Scheme [grant number 2014TQ01X706]; High-level Talent Scheme of Guangdong Education Department [grant number 2014–2016]; the Guangdong Natural Science Foundation [grant number 2017A030313401]; Inner Mongolia Autonomous Region Science and Technology Innovation Guide Award Fund Project [grant number 103-413193]; Key Research Projects of Henan Higher Education Institutions [grant number 19A630037].

References

1. Sun, Z.Q.: Development status and prospects of China's 3D printing industry. Jiangsu Sci. Technol. Inf. (6), 19–20 (2014). (Chinese)
2. Chapiro, M.: Current achievements and future outlook for composites in 3D printing. Reinf. Plast. **60**(6), 372–375 (2016)
3. Duda, T., Raghavan, L.V.: 3D metal printing technology. IFAC-PapersOnLine **49**(29), 103–110 (2016)
4. Jin, T., Ma, J., Yang, L.C.: Design of remote control system of 3D printer based on WiFi. Transducer Microsyst. Technol. **37**(11), 109–112 (2018). (Chinese)
5. Hwa, L.C., Rajoo, S., Noor, A.M., Ahmad, N., Uday, M.B.: Recent advances in 3D printing of porous ceramics: a review. Curr. Opin. Solid State Mater. Sci. **21**(6), 323–347 (2017)
6. Zhang, S., Xu, Y.S., Sun, S.S., Zang, W.H., Sun, J., Gu, X.Y.: A review on the progress of 3D printing materials. China Plast. **30**(1), 7–14 (2016). (Chinese)
7. Li, D.F., Chen, J.M., Yuan, Y.P., Huang, K., Fang, H.B.: Development and application of stereo lithography apparatus. Beijing Univ. Technol. **41**(12), 1769–1774 (2015). (Chinese)
8. Walker, J.M., et al.: Real-time process monitoring of core shifts during metal casting with wireless sensing and 3D sand printing. Addit. Manuf. **27**, 54–60 (2019)

Bibliometric Analysis of Internet of Things Based on CiteSpace

Hong-fei Guo[1], Chun-yu Xu[1], Ru Zhang[1,2], Jin-cheng Shi[5(✉)], Ting Qu[1], Cong-dong Li[1], Yun-gong Cai[1], Xiao-feng Luo[4], and Zhi-hui He[3]

[1] Institute of Physical Internet, Jinan University, Zhuhai 519070, China
[2] Institute of Management Science and Engineering, Jinan University, Zhuhai 519070, China
[3] Zhuhai Hengqin Building & Construction Quality Inspection Centre Co., Ltd., Zhuhai 519031, China
[4] College of Information Science and Technology, Jinan University, Guangzhou 511436, China
[5] Office of Scientific R&D, Jinan University, Guangzhou 510632, China
tshijc@jnu.edu.cn

Abstract. The Internet of Things (IoT) has become the central issue and research focus of the global academic community and industry, in order to more clearly understand the research status and future development direction of the IoT. This paper uses the Dewent-Innovation-Index patent database as a data source, using the relevant theories and methods of bibliometrics and content analysis, and using the data visualization tool CiteSpace to conduct 25,775 related patents in the global IoT field between 2006 and 2018. Classification statistics, content analysis and knowledge map mapping, focusing on the analysis and comparison of the IoT technology patent annual analysis, ownership analysis, research fields, research hotspots, and evolution trends, provide reference for China's IoT development model and technological innovation.

Keywords: Bibliometrics · CiteSpace · Internet of Things · IoT · Knowledge map

1 Introduction

The IoT is a network that can be interconnected by ordinary information objects such as traditional telecommunication networks and the Internet, so that all ordinary objects that can independently perform their functions can be interconnected. The concept first appeared in the definition of the global standard system for RFID and sensor combinations in Ashton in 1999 [1] and officially defined in the 2005 "ITU Internet Report2005: The Internet of Things" [2] report. Based on the sensor network, it is further expanded into a new dimension of modern Internet extension. RFID, sensor network technology, intelligent devices and devices, nanotechnology and miniaturization technology are recommended technologies for the development of the IoT.

C.-F. Chien et al. (Eds.): *IE&EM 2019*, pp. 276–283, 2020.
https://doi.org/10.1007/978-981-15-4530-6_28

Therefore, the emerging field of the IoT has received wide attention from political, academic and industrial circles around the world.

In order to better study the development prospects of the IoT technology, this paper uses the visualization tool CiteSpace V5.3.R4 software, combined with the relevant theories and methods of bibliometric analysis, the Dewent-Innovation-Index database between 2006 and 2018. The relevant research and comparative analysis of 25,775 valid patent data in the field of global IoT are aimed at comparing and analyzing three aspects of the IoT technology patent research field, research hotspots, and evolution trends. Technical research and practical application in the field of IoT provide direction guidance and experience.

2 Research Methods and Data Collection

2.1 Research Methods

This study uses a combination of statistical analysis, content analysis and knowledge mapping analysis to measure and analyze patent data in the field of IoT. Statistical analysis refers to the use of comprehensive evaluation index, the number of patent documents and other literature measurement indicators to quantify the influence of the subject of the research field [3]; Content analysis is generally used to identify key feature words in document data, and subject frequency statistics are used for subject analysis to avoid subjectivity and uncertainty of pure qualitative analysis [4]; Knowledge map analysis refers to the mapping of common words network maps, distribution timing diagrams and other knowledge maps through the visualization tool Citespace [5], and further analysis of these maps can be used to know the latest research hotspots, knowledge bases and evolution trends in a certain field.

2.2 Data Source

This paper searches all patents filed in the IoT field in the Dewent-Innovation-Index database from 2006 to 2018. In the database, a total of 26,371 patent data in the IoT field with the words "The IoT" and "Sensor Networks" were retrieved. Manual identification and screening were used to eliminate irrelevant patent data and integrate duplicate patent data. After the operation, 25 775 pieces of valid analysis data were finally obtained. The 25,775 patent data was used as the basic data of this analysis, and it was researched, discussed and analyzed with data visualization software such as CiteSpace and Excel.

3 Analysis of Patent Research Fields of IoT Technology

According to the different fields of each technology, the International Patent Classification (IPC) has created a patent technology classification system that is different from the specific language symbols. Because of its effectiveness in classifying patent technology, it is compared with other the patent classification system [6] has a higher impact. When a patent is applied to multiple fields, correspondingly, it will have different patent numbers in different fields, and thus a patent has multiple classification numbers. Most scholars believe that the higher the frequency of co-occurrence for the two classification numbers, the higher the degree of association between the two technical fields, and the flow and integration of related technologies between these technical fields. More frequent and fast, this is conducive to the promotion of technological innovation and technological progress [7–10]. This paper adopts the international patent classification number system, combined with the social network analysis method, to visually analyze the multi-domain linkages and innovations in the global IoT technology field, and finally obtain the global network diagram of the global IoT patent technology field, as shown in Fig. 1.

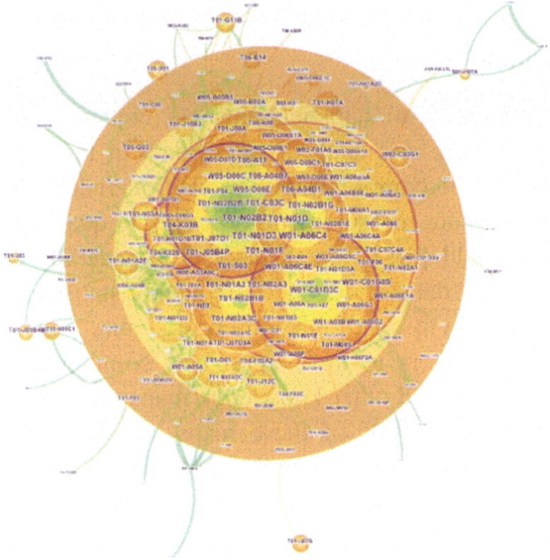

Fig. 1. Global network of IoT patent technology

As can be seen from Fig. 1, the links between different patented technical fields are very close, and the same technology is fully utilized in different application fields, thus forming clustering circles of different sizes on the knowledge map. In order to better analyze the relationship between IoT technologies, select the top ten IoT technology fields for detailed analysis.

As can be seen from Table 1, the most frequently occurring discipline is T01-N01F (IoT), which is the study of interconnection/interconnection between computers, devices and systems used in applications such as home automation and smart grid. From a realistic point of view, in recent years, scientists and engineers in various countries have become more and more enthusiastic about research on smart homes and intelligent robots. For this reason, the T01-N01F (IoT) discipline will have such a high frequency. For the short-range system radio link studied by W01-A06C4 (radio link), including IEEE 802.15 radio link, including Bluetooth, Dect-based radio link, IEEE

802.11 radio link, etc., since the launch of the smartphone in 2009 At the beginning, it attracted a large number of researchers to study it.

Table 1. Top 10 IoT technology fields

Ranking	Frequency	Centrality degree	Year	Subject code
1	4 611	0.06	2015	T01-N01F
2	2 735	0.16	2009	W01-A06C4
3	2 045	0.12	2009	T01-N02B2
4	2 028	0.08	2007	T01-C03C
5	2 023	0.08	2007	T01-N01D3
6	1 969	0.04	2010	T01-N01D
7	1 533	0.13	2009	T01-N02B2B
8	1 516	0.06	2006	W01-C01D3C
9	1 415	0.07	2007	T01-S03
10	1 233	0.11	2010	T06-A11

Different IoT technologies have different research priorities. For the research of intelligent hardware, there are mainly T01-N02B2 (monitoring) research including monitoring computer/network communication and hardware monitoring, and its subsystem research T01-N02B2B (system and fault monitoring) including monitoring system for monitoring computer hardware operations, log events, reporting failures, online (Internet-based) monitoring and online diagnostics of any electronic system. For the research of mobile communication and its interface, there are mainly T01-C03C (radio link) including satellite, radio, infrared and other interfaces for accessing the network; T01-N01D3 (Getting services from a remote site or server) Includes a network running applications on the server under the control of the client system; T01-N01D (data transmission) includes downloading files from a remote site (FTP) and T06-A11 (control system related (data) communication arrangement). For product research, there are mainly W01-C01D3C (portable; handheld), that is, the invention related to "portable terminal" is the development and design of mobile devices and T01-S03 (claimed software products), that is, a software-based product is claimed and stored, for example, in a CD-ROM, where a computer program or software component is used in the independent claims.

4 Analysis of Hotspots in Patent Research of IoT Technology

The word frequency analysis method is a kind of bibliometric method, which counts the frequency of occurrence of keywords in a research field to determine the research hotspots and development trends in this field [11]. Using CiteSpace V5.3.R4 software, after a series of operations such as clustering and visualization, the knowledge map of global Internet of Things technology research hotspots in this time period is obtained, as shown in Fig. 2.

According to statistics, the map has 141 nodes and 715 connections. The network density is 0.0724. This means that the research hotspots between different IoT patent technologies are very close, and the patent hotspots form an interactive overlap complex network structure. Among them, can be clearly seen from Fig. 2, t01 (digital computer), w01 (telephone and data transmission system), w05 (alarm; signal; telemetry and remote control), t06 (process and machine control), w02 (broadcast, radio and line transmission systems), t04 (computer peripherals), w06 (aerospace and radar systems), t05 (counting, checking vending machines and POS systems), w04 (audio/video recording and systems), s03 (science Instruments) These 10 IoT technology patent hotspot words are obviously larger than other words, which means that these 10 words are the hot high-frequency words of research. These hot words cover the IoT technologies in the fields of digital information transmission, electrical signal transmission, alarm signal transmission, telemetry and remote control, and aerospace radar system control.

That is to say, the application of IoT technology in these fields is the most extensive. Other technologies such as q63 (coupling; clutch; brake; damper), p61 (grinding; polishing), q43 (general building structure), m24 (steel metallurgy, including manufacturing and processing, steel processing melting and changing steel physics Properties, control/test methods, lubricating oil furnaces and converter metallurgical coking processes), p42 (spraying; atomization), etc. Although their frequency does not appear high, but since their application year is in the near future (2018). Therefore, it can be seen that the latest development direction of the IoT technology is in the optimization of industrial assembly lines, building structure, machine processing, etc., that is, intelligent technology fields such as smart factories and smart homes.

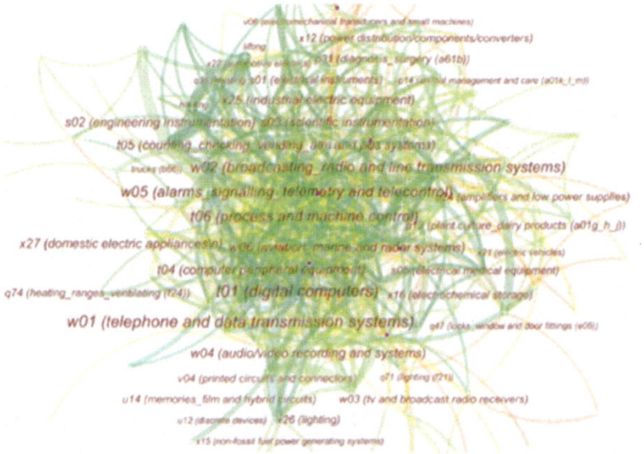

Fig. 2. Global IoT patent research hotspot knowledge map

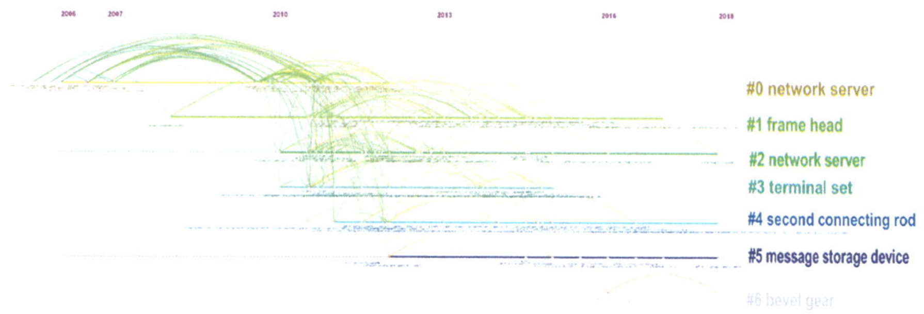

Fig. 3. Global IoT patent research evolution trends

5 Analysis of the Evolution Trend of IoT Technology Patents

The evolution of the discipline refers to the distribution of research hotspots and research topics in different years by acquiring subject areas. The knowledge map of the time series diagram is used to further analyze the evolution trend of the field of IoT research. According to the patent data collected in the Dewent-Innovation-Index database, the knowledge map of the evolution trend of the IoT research field is drawn, as shown in Fig. 3.

It can be seen from Fig. 3 that the global IoT patent data is merged and classified by Cite Space's clustering algorithm, and finally divided into seven research topics or application directions. They are: network server, frame head, network server2, terminal set, second connecting rod, message storage device, and bevel gear. This means that the process of development and evolution of the IoT continues to infiltrate and integrate with other disciplines, and forms patents based on this. At the beginning of the IoT, the International IoT is in the market cultivation period of "concept exploration, government-led, application demonstration" [12]. Most of the IoT patents are about Internet network services, including network service interfaces and their communication, definition of network file format, design of network service equipment, etc. With the development of technology, the international IoT has entered the industrial outbreak period of "cross-border integration, integrated innovation, scale application, and ecological acceleration" [12], when the IoT patents began to more and various disciplines, including materials, Electronic communication, mechanical manufacturing, etc., and integration, so new technologies such as smart home, cloud computing, and car networking are constantly appearing and being applied.

6 Conclusion

Looking at the full text, the analysis of global IoT patent data from 2006 to 2018 shows that: In the current research on the IoT, T01-N01F (IoT), T01-N02B2 (monitoring), T01-C03C (radio link), W01-C01D3C (portable; handheld) and other technologies are very popular. This is because in recent years, with the continuous development of the IoT technology, intelligent devices such as smart phones and smart robots are increasingly popular among consumers. Therefore, making these devices more convenient, more complete functions, and easier and faster human-computer interaction will become the research focus of researchers. The research hotspots of the IoT technology have shifted from the research on traditional communication methods such as digital transmission, information transmission and signal control to the research on materials and transmission media that can better realize the "Internet of Everything". That is to say, the IoT technology has evolved from its own definition to a new stage of integration with various disciplines. Based on the above conclusions, two suggestions are made: (1) Research must be developed on interdisciplinary research on the IoT, especially for new media for information transmission. (2) We must pay attention to the R&D investment of IoT technology and strengthen the research and development of IoT technology. Only continue to innovate on the basis of the cutting-edge IoT technology. Therefore, it is possible to apply for its own basic patents, peripheral patents, etc. in key areas and core technologies. In order to enable China's IoT technology to achieve rapid development, its scientific research results can also achieve rapid industrial development.

Acknowledgement. 2018 the Ministry of Education's Industry-University Cooperation Collaborative Education Project "Research and Reform of the IoT Security Course System Based on New Network Security Technology" [grant number 201802007015]; 2019 the teaching reform research project of Jinan University "innovated and practiced the teaching model of the 'Intelligent Engineering Group' based on the double-drive mode of production and education" [grant number JC2019168]; 2019 the Research and Practice of the Teaching Reform Research Project of Jinan University "Based on Scientific Research and Cultivation of Innovative Talents" [grant number JC2019181].

References

1. Ashton, K.: That 'Internet of Things' thing. RFID J. **22**(7), 97–114 (2009)
2. International Telecommunication Union (ITU): ITU Internet Reports 2005: The Internet of Things. ITU, Geneva, Switzerland (2005)
3. Zhang, X.P., Zhang, F.F., Xu, Y., Wang, D.X., Wu, Z.H.: Comparison of mapping knowledge domain of ecological footprint researches between Chinese and foreign literatures based on CiteSpace. Resour. Dev. Market **33**(11), 1347–1353+1339 (2017). (in Chinese)
4. Xiao, L.V., Niu, S.D., Zhang, Q.J., Zhong, T.Y., Chen, C.L.: Evolution of policies on rural collective-owned construction land circulation in China based on content analysis method. China Land Sci. **29**(4), 25–33 (2015). (in Chinese)
5. Chen, M., Zhu, J.W.: Focus and trend of project management research in China based on knowledge graph. J. Eng. Manag. **30**(3), 105–109 (2016). (in Chinese)
6. Xiao, M., Kong, C.G.: Big data: where to go——based on bibliometric perspective. J. Library Sci. **36**(11), 110–117 (2014). (in Chinese)
7. Romm, N.R.A.: Interdisciplinary practice as reflexivity. Syst. Pract. Action Res. **11**(1), 63–77 (1998)
8. Roper, A., Brookes, M.: Theory and reality of interdisciplinary research. Int. J. Contemp. Hosp. Manag. **11**(4), 174–180 (1999)

9. Alves, J., Saur, I., Marques, M.J.: Envisioning the House of the Future: a multisectorial and interdisciplinary approach to innovation. In: E-Core Conference, ECCREDI, Maastricht, Holland (2004)
10. Saur, I., Marques, M.J., Alves, J.: Multisectoral cooperation networks as stimuli of knowledge processes leading to innovation. In: ISPIM 2005, Porto, Portugal (2005)
11. Xu, Z.: Mapping knowledge domain analysis of library and information science knowledge service. J. Mod. Inf. **33**(2), 166–170 (2013). (in Chinese)
12. Luo, S., Xu, H.Y.: Overview of the development of the IoT industry and the trend of technological innovation. Inf. Commun. Technol. **12**(4), 4–8 (2018). (in Chinese)

Research on the Development Situation of Industrial Internet of Things Based on Mapping Knowledge Domain

Hong-fei Guo[1], Ru Zhang[1,2(✉)], Ting Qu[1], Cong-dong Li[1], Zheng-wei Zou[5], Yi Zhou[1], Qiu-fan Chen[1], Hao Jiang[1], Bao Chao[1,4], Xiao-feng Luo[5], and Zhi-hui He[3]

[1] Institute of Physical Internet, Jinan University, Zhuhai 519070, China
zhangru@stu2016.jnu.edu.cn
[2] Institute of Management Science and Engineering, Jinan University, Zhuhai 519070, China
[3] Zhuhai Hengqin Building & Construction Quality Inspection Centre Co., Ltd., Zhuhai 519031, China
[4] School of Mechanical Engineering, Inner Mongolia University of Technology, Hohhot 010051, China
[5] School of Statistics, Jiangxi University of Finance and Economics, Nanchang 330013, China

Abstract. The rapid development of Industrial Internet of Things has attracted extensive attention of scholars in related fields. In order to fully understand the research progress in the field of Industrial Internet of Things, the WOS (Web of Science) database is used as the data source, the qualitative research and quantitative research are combined, the CiteSpace III software is used as the data visualization tool, and the literature published in the field of Industrial Internet of Things from 2011 to 2019 is used as the research basis to draw the map of scientific knowledge. From three aspects of research hotspot, knowledge base and development trend, it is concluded that the integrated application of information technology such as Internet, big data, 5G and cloud computing will be the direction of development in the field of Industrial Internet of Things.

Keywords: Industrial Internet of Things · Bibliometrics · Mapping knowledge domain · CiteSpace · Research progress

1 Introduction

The Industrial Internet of Things integrates various acquisition and control sensors or controllers with sensing and monitoring capabilities, as well as mobile communication and intelligent analysis technologies into all aspects of the industrial production process, thereby greatly improving manufacturing efficiency, improving product quality, and reducing products cost and resource consumption, eventually bringing the traditional industry to a new stage of intelligence. From the application form, the Industrial Internet of Things application has the characteristics of real-time, automation, embedded (software), security and information interconnection.

C.-F. Chien et al. (Eds.): *IE&EM 2019*, pp. 284–293, 2020.
https://doi.org/10.1007/978-981-15-4530-6_29

The academician of the Chinese Academy of Engineering, Hequan Wu, expands the application of the Internet of Things in the field of production and life on the basis of systematically introducing the composition, network relationship, technology and functions of the Internet of Things [1]. Qibo Sun et al. used the literature analysis method to analyze the basic concepts, features, architecture, key technologies and standardization of the Internet of Things, and sorted them out in different categories [2]. Jorge E. Ibarra-Esquer et al. analyzed the evolution process of various application fields from the perspective of the evolution of Internet of Things theory and technology [3]. Yang et al. classified and systematically reviewed the commercial literature of the Internet of Things from the perspective of individual users and organizations. Some researchers have conducted more in-depth quantitative analysis of information science from the field of Internet of Things research [4]. Yangping He based on the core collection of WOS, through the CiteSpace, HistCite and Pajek tools to analyze and summarize the Internet of Things literature, to obtain visual results [5]. Xinli Zhou et al. applied scientific measurement and visualization method to statistically analyzed the data of the Internet of Things research papers of CNKI (china national knowledge infrastructure) literature database from 2003 to 2012 [6]. Ruiying Sun et al. mapped the mapping knowledge domain and analysis of the Internet of Things related literature in the CNKI literature database from 2005 to 2015 [7]. Juan et al. used the ScientoPy tool developed in Python language to conduct quantitative statistical analysis of the Internet of Things research in WOS and Scopus literature database between 2002 and 2016 [8]. Erfanmanesh further carried out bibliometric analysis, altmetric analysis and network analysis on the literature data related to the Internet of Things research of the Scopus database from 2006 to 2015 [9].

The current Industry 4.0 has an important connection with the Industrial Internet of Things. From the development of industry and the use of intelligent machines related to the Internet to achieve intelligent mechanical production, it can be seen that the Industrial Internet of Things is based on the development of 5G technology. Industrial Internet of Things will be more likely to grow with the support of 5G networks. However, because the Industrial Internet of Things has not been produced for a long time, the topics, contents and methods involved are also extensive. Therefore, it is necessary to fully understand the research hotspots, knowledge bases and development trends in the field of Industrial INTERNET OF THINGS in recent years. Therefore, based on bibliometrics theory and CiteSpace software, this paper sorts out and analyzes the related research in SCI, core journals and Industrial Internet of Things from 2011 to 2019, in order to provide a certain degree of guidance and reference for the theoretical research and development direction of the Industrial Internet of Things.

2 Research Method and Data Source

2.1 Research Method

The research method in this paper is a combination of qualitative analysis and quantitative analysis. Qualitative analysis is to sort out the main content of the literature in the field of Industrial Internet of things, especially the classic literature, and analyze its role in the development of the discipline. Quantitative analysis is a statistical analysis of high-frequency vocabulary and frequency in the field of Industrial Internet of Things through bibliometrics, using CiteSpace V as a visualization tool for research, to draw the keyword network, literature indexing network, and distribution time series map, to analyze the research hotspots, knowledge base and evolution trend in the field of Industrial Internet of Things.

CiteSpace is a data visualization analysis software developed by Professor Chaomei Chen from Drexel University. It combines the disciplines of bibliometrics, statistics, information science, etc., and uses the visual form of spatial form to visually display the research hotspots, knowledge structure, development history and distribution of the Industrial Internet of Things researched in this paper.

2.2 Data Source

The research object of this paper is related literature in the field of international Industrial Internet of Things. The data comes from the WOS database and the retrieval time is July 24, 2019. In order to cover the research situation in the field of Industrial Internet of Things as much as possible, this paper selects the topic retrieval method, the time range is defined as 2011–2019, and the keyword is "Industrial Internet of Things". In addition, through manual identification and screening, the invalid records such as call notice, journal statement, conference news and unrelated documents were excluded. This paper finally determined 2055 related documents of WOS database.

3 Analysis of Research Hotspots in Industrial Internet of Things

High-frequency keywords can highlight research hotspots in the subject area. The higher the frequency of similar keywords, the more attention is paid to the research content of the keyword. The more concerned research content usually indicates the future development trend of the field, which is beneficial for researchers to understand the development of the field as a whole. Therefore, in order to understand the latest developments in the field of Internet of Things, using the literature data of WOS data, draw a mapping knowledge domain of research hotspots in the field of knowledge management, as in Fig. 1. There are 413 nodes in the formed map, and the size of the nodes represents the frequency of occurrence of keywords. The top ten research hotspots are listed in the order of the frequency of occurrence of keywords, as shown in Table I.

Analysis of Fig. 1 and Table I shows that in the mapping knowledge domain-co-word network diagram, the co-word network uses "internet" as the network center, and the keywords "thing", "system", "internet of thing" and so on closely surround the network center. The vast majority of keywords at or near the core of Internet of Things research are highly representative of the latest research in this field. In general, the mapping knowledge domain -co-word network has many nodes with similar sizes, and the keyword frequency distribution is balanced. The difference in network center degree between different high-frequency words is small, indicating that there are many

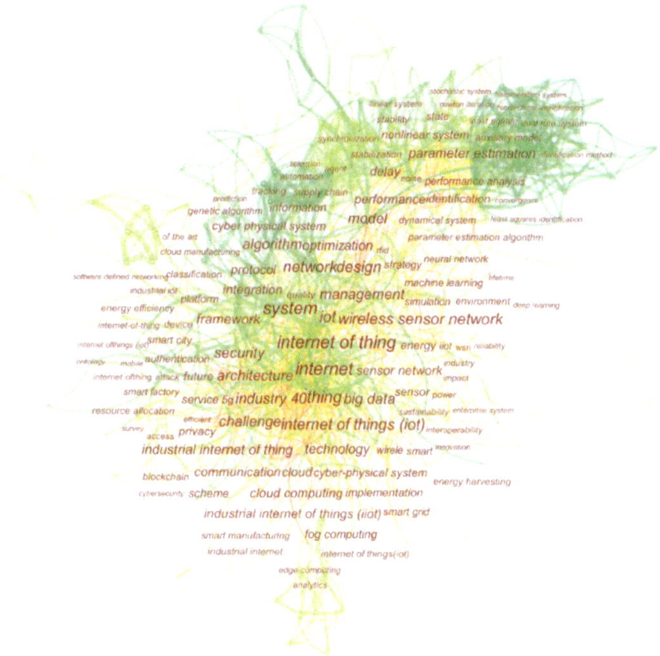

Fig. 1. Mapping knowledge domain – co-word network

Table 1. Top ten research hotspot statistics

Ranking	High frequency word	Frequency	Centrality
1	Internet	331	0.03
2	Internet of Thing	305	0.04
3	System	265	0.03
4	Thing	225	0.03
5	Design	170	0.12
6	IoT	154	0.06
7	Network	145	0.05
8	Internet of Things	143	0.02
9	Wireless sensor network	142	0.02
10	Model	123	0.18

research directions in the Industrial Internet of Things field. The focus of research hotspots is relatively close. Moreover, the overall density of the common word network is high, the network knowledge points are clustered, and there are many cross-disciplinary studies in the field, indicating that the interdisciplinary cooperation in the field of Industrial Internet of Things research is both normal and close. The specific analysis of research hotspots from 2011 to 2019 is as follows:

(1) Communication and data processing area: The frontier technologies of the Internet of Things and computer science, such as big data, artificial intelligence, and cloud computing, are closely combined. Through literature analysis, many scholars have used the relevant theories and technologies of big data, artificial intelligence and cloud computing to innovate and expand the Industrial Internet of Things function.
(2) wireless sensor network area: Industrial Internet of Things development requires more accurate, smarter, more efficient and more compatible sensor technology, wireless sensor network as the basic technology for the development of Industrial Internet of Things technology. With the ubiquity of information, Industrial sensors and sensing devices are required to be more miniaturized, intelligent, and low-power.
(3) Smart factory: Smart factory construction is one of the most widely used applications for Industrial Internet of Things, it mainly includes resource allocation optimization, production process optimization, and management visualization in intelligent industrial production.

4 Evolution Trend of International Industrial Internet of Things

When analyzing the evolution trend, we can use Burst Detection in CiteSpace software to get the research topic of sudden increase in frequency or obvious increase in frequency from keywords. According to the display of the burst terms in the map at different time periods, this method can clearly analyze the evolutionary trend of the Industrial Internet of Things field. Using "keyword" as a node, draw a high-frequency burst terms sequence diagram in the field of Industrial Internet of Things. The "keyword" is used as the analysis node, and the "Timezone" method is used to display the time distribution sequence diagram of the burst terms in international Industrial Internet of Things field, as in Fig. 2.

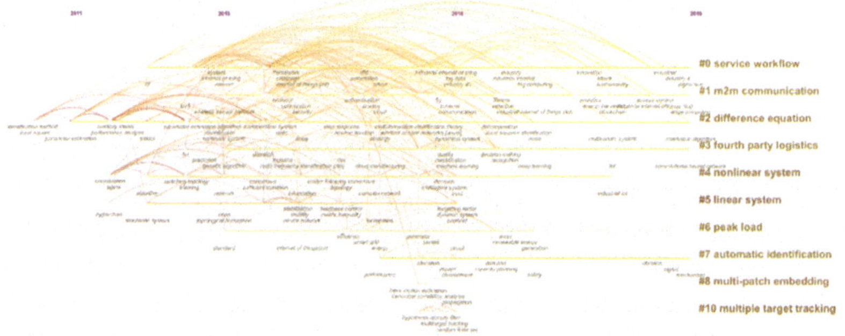

Fig. 2. Burst terms sequence diagram of international Industrial Internet of Things

In-depth analysis of the evolution trend of the 2011–2019 Industrial Internet of Things field, as follows:

During 2011 to 2013, the key words with high international mutation intensity are "IoT", "parameter estimation" and "model", indicating that during this period, the international focus on the integration process of industry and the Internet of Things. It is expected that by analyzing the Internet of Things itself and parameter estimation and experiments applied to the industrial field to build a better model to optimally combine industrial production and Internet of Things technology,

During 2013 to 2016, keywords such as "network", "machine learning" and "wireless sensor network" emerged, indicating that, internationally, further application of network technology and processing of information in a more advanced manner were valued in the field of Industrial Internet of Things. The network is one of the core components of the Industrial Internet of Things, and data is transmitted over the network between different levels of the system. The network is divided into a wired network and a wireless network. The wired network is generally applied to cluster server of a data processing center, local area network of factory and some of the field bus control network, it can provide a high-speed and high-bandwidth data transmission channel. Industrial wireless sensor networks are an emerging technology that uses wireless technology for sensor networking and data transmission. The application of wireless network technology can greatly reduce the wiring cost of industrial sensors, which is conducive to the expansion of sensor functions, thus attracting the attention of many enterprises and scientific research institutions at home and abroad.

During 2016 to 2019, themes such as "Industrial Internet of Things" and "fog computing" were emerged, this indicates that the international Industrial Internet of Things system has been initially completed and is trying to expand it into new areas. With the popularization of information technology and the strengthening of global information trends, the integration of Industrial Internet of Things with other advanced information technologies such as cloud computing and big data will become an area worthy of further study.

5 Knowledge Base Analysis in the Field of International Industrial Internet of Things

The knowledge base is the citation and co-introduction trajectory in the scientific literature at the forefront of research in a certain field. Select "cited reference" as the network node, find out the authoritative classic literature in the field of international Industrial Internet of Things, and conduct knowledge base analysis, so as to further clarify its development context and knowledge research foundation. Through the "Generate a narrative" operation in CiteSpace V, you can get list information of important literatures, as in Fig. 3. After many search and screening, it has identified 10 international articles on the international Industrial Internet of Things with high frequency. Analysis shows: Based on the system performance index, Xu Ling uses the delay of the first-order Taylor expansion approximation system, and proposes an algorithm to determine the controller parameters, so that the system achieves the expected dynamic performance [10].

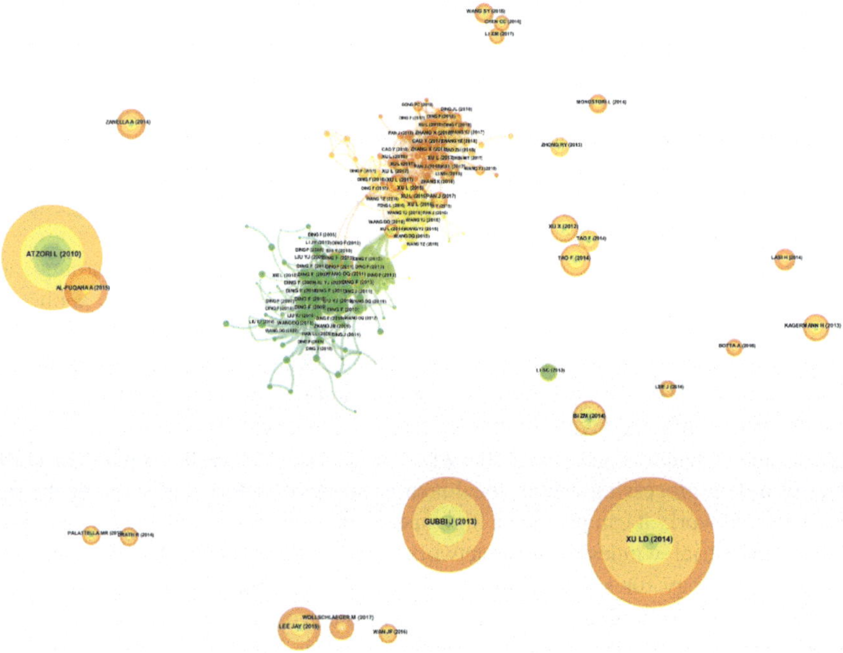

Fig. 3. Mapping knowledge domain - citation network

Ding Feng et al. used a decomposition-based hierarchical recognition principle to derive a Hammerstein system iterative algorithm based on least squares and a gradient-based iterative algorithm for Hammerstein systems [11]. Based on the step response analysis, Xu Ling proposed the Newton iterative algorithm. In addition, in order to verify the accuracy of the estimated parameters, the frequency and step response experiments were applied to the dynamical system between the estimated model and the real model [12]. Xu Ling et al. proposed a Newton iterative identification method for estimating the parameters of second-order dynamic systems using step response data. In addition, in order to obtain the ideal dynamic performance, a controller design method based on root locus is proposed to meet the requirements of overshoot dynamic performance [13]. Dongqing Wang et al. derived a Wiener nonlinear system iterative identification algorithm based on least squares and gradients. This method decomposes a bilinear cost function into two linear cost functions and directly estimates the parameters of the Wiener system. There is no need to re-parameterize, resulting in redundant estimates [14]. Zhang Xiao et al. studied the identification problem of bilinear systems that can measure noise in the form of moving average model, and proposed an unmeasurable state based on hierarchical identification principle and interactive estimation algorithm based on parameters [15]. Yuan Cao et al. According to CTCS-3 (China's train control system level 3) GSM wireless system standard—the control data transmission delay should be no more than 500 ms, the probability is greater than 99%, considering the coverage of non-redundant networks and cross-redundant networks, and in the case of single-row MTs (mobile terminals) and redundant MTs, the corresponding vehicle-to-ground communication model, delay model and fault model are established [16]. Chen Jing et al. studied the identification problem of multi-input and multi-output nonlinear systems. The difficulty in identifying such system parameters is that the information vector in the identification model contains unknown variables. The solution of Jing Chen et al. is to overcome this difficulty by using auxiliary model identification ideas, and to extend the innovation vector into an innovation matrix, and propose a multi-innovation extended stochastic gradient algorithm based on the auxiliary model [17]. Xiangkui Wan et al. introduced a method combining MMF and WT to overcome the shortcomings of existing methods in suppressing BW (baseline wander). In order to verify the effectiveness of the proposed method, Xiangkui Wan et al. used artificial ECG signals containing clinical BW for numerical simulation and established a realistic baseline drift model. The results show that the BW effect of this method is better than other methods [18]. Yanjun Liu et al. applying the idea of auxiliary model identification and the theory of multi-innovation identification, proposed a multi-innovation stochastic gradient algorithm based on the auxiliary model, and the simulation example shows that the algorithm works well [19].

6 Conclusion

Research uses the CiteSpace tool to draw a visual mapping knowledge domain, the relevant literatures in the field of Industrial Internet at home and abroad in the WOS database from 2011 to 2019 are used as research data samples to scientifically analyze the research hotspots, knowledge structure, development process and distribution of sample literatures, obtained the following research conclusions and corresponding recommendations:

(1) From the research hotspot level, there are close links between high-frequency keywords in the Industrial Internet field worldwide, forming a wide range of cross-research fields, and the research direction is relatively scattered. On the basis of maintaining in-depth exploration and discussion of basic knowledge theory, the related literature pays close attention to the practical application and method research of communication data processing, wireless sensor network, smart factory construction, etc. Most of them focus on a single subject research field, and relatively lack scientific research on each crossing field.
(2) From the knowledge base, the main knowledge base of Industrial Internet research at home and abroad is Industrial Internet system design, parameter control, model construction, algorithm deduction, basic theoretical research is more in-depth, and the connection between research scholars is relatively close, the technology application research needs continuous exploration, lacking the emerging power of more fruitful results.
(3) From the perspective of evolutionary trends, relevant research scholars at home and abroad have paid close attention to the development history and application prospect for Industrial Internet. In recent years, research directions have gradually turned to network information processing, wireless sensing control, cloud

computing and big data. The development status of the Industrial Internet field in the new era is changing with each passing day, but the comprehensive research in related fields has not yet attracted widespread attention, the integrated application research in the aspects of modern new information technology also needs to be explored.

In response to the above research conclusions, the corresponding scientific recommendations for promoting the development of China's Industrial Internet: On the one hand, the academic and industrial circles should keep abreast of and grasp the first dynamics in the field of international Industrial Internet, strengthen the informationization, technicalization and integration in the field of Industrial Internet, and continuously develop and improve the basic theoretical system of the Industrial Internet field in our country. On the other hand, industrial enterprises should be the mainstay and guided by national policies, leading the relevant industries to develop towards modernization and interconnection. Focus on promoting technological innovation and application innovation simultaneously, and promote the diversification, phase, generalization, collaboration and integration of the Industrial Internet.

Acknowledgment. The research and publication of their article was supported by the Fundamental Research Funds for the Central Universities [grant number 21618412; grant number 21618804]; the National Natural Science Foundation of China [grant number 51475095]; Project of Guangdong; Natural Science Foundation [grant number 2016A030311041]; 2015 Guangdong Special Support Scheme [grant number 2014TQ01X706]; High-level Talent Scheme of Guangdong Education Department [grant number 2014–2016]; the Guangdong Natural Science Foundation [grant number 2017A030313401]; Inner Mongolia Autonomous Region Science and Technology Innovation Guide Award Fund Project [grant number 103-413193]; Key Research Projects of Henan Higher Education Institutions [grant number 19A630037].

References

1. Wu, H.Q.: Review on Internet of Things: application and challenges. J. Chongqing Univ Posts Telecommun. **22**(5), 526–531 (2010). (Chinese)
2. Sun, Q.B., Liu, J., Li, W., Fan, C.X., Sun, J.J.: Internet of Things: summarize on concepts, architecture and key technology problem. J. Beijing Univ. Posts Telecommun. **33**(3), 1–9 (2010). (Chinese)
3. Ibarra-Esquer, J.E., Gonzalez-Navarro, F.F., Flores-Rios, B.L., Burtseva, L., Astorga-Vargas, M.A.: Tracking the evolution of the Internet of Things concept across different application domains. Sensors **17**(6), 2017 (2017). (Basel, Switzerland)
4. Lu, Y., Papagiannidis, S., Alamanos, E.: Internet of Things: a systematic review of the business literature from the user and organisational perspectives. Technol. Forecast. Soc. Chang. **136**, 285–297 (2018)
5. He, Y.P., He, X.P.: Visualization analysis of Internet of Things research based on knowledge mapping. Inf. Res. **6**, 116–123 (2017). (Chinese)
6. Zhou, X.L., Deng, L., Yang, L.B.: Mapping knowledge domain-based visualized analysis of studies on Internet of Things. Internet of Things Technol. **5**(8), 83–87 (2015). (Chinese)
7. Sun, R.Y., Wang, X.: Analysis on the status of the Internet of Things in China based on the bibliometrics statistical methods. J. Mod. Inf. **36**(1), 153–159 (2016). (Chinese)

8. Ruiz-Rosero, J., Ramirez-Gonzalez, G., Williams, J.M., Liu, H., Khanna, R., Pisharody, G.: Internet of Things: a scientometric review. Symmetry **9**(12), 301 (2017). Basel
9. Erfanmanesh, M., Abrizah, A.: Mapping worldwide research on the Internet of Things during 2011–2016. Electron. Libr. **36**(6), 979–992 (2018)
10. Xu, L.: A proportional differential control method for a time-delay system using the Taylor expansion approximation. Appl. Math. Comput. **236**, 391–399 (2014)
11. Ding, F., Liu, X., Chu, J.: Gradient-based and least-squares-based iterative algorithms for Hammerstein systems using the hierarchical identification principle. IET Control Theory Appl. **7**(2), 176–184 (2013)
12. Xu, L.: Application of the Newton iteration algorithm to the parameter estimation for dynamical systems. J. Comput. Appl. Math. **288**, 33–43 (2015)
13. Xu, L., Chen, L., Xiong, W.: Parameter estimation and controller design for dynamic systems from the step responses based on the Newton iteration. Nonlinear Dyn. **79**(3), 2155–2163 (2015)
14. Wang, D.Q., Feng, D.: Least squares based and gradient based iterative identification for Wiener nonlinear systems. Signal Process. **91**(5), 1182–1189 (2011)
15. Zhang, X., Xu, L., Ding, F., Hayat, T.: Combined state and parameter estimation for a bilinear state space system with moving average noise. J. Franklin Inst. Eng. Appl. Math. **355**(6), 3079–3103 (2018)
16. Cao, Y., Ma, L., Xiao, S., Zhang, X., Xu, W.: Standard analysis for transfer delay in CTCS-3. Chin. J. Electron. **26**(5), 1057–1063 (2017)
17. Jing, C., Yan, Z., Ding, R.: Auxiliary model based multi-innovation algorithms for multivariable nonlinear systems. Math.Comput. Model. **52**(9–10), 1428–1434 (2010)
18. Wan, X.-K., et al.: Electrocardiogram baseline wander suppression based on the combination of morphological and wavelet transformation based filtering. Comput. Math. Methods Med. **2019**, 7 (2019). Article no. 7196156
19. Liu, Y.J., Xiao, Y.S., Zhao, X.L.: Multi-innovation stochastic gradient algorithm for multiple-input single-output systems using the auxiliary model. Appl. Math. Comput. **215**(4), 1477–1483 (2009)

Author Index

C.-F. Chien et al. (Eds.): *IE&EM 2019*, pp. 295–296, 2020.
https://doi.org/10.1007/978-981-15-4530-6

The manufacturer's authorised representative in the EU is Springer Nature Customer Service Centre GmbH, Europaplatz 3, 69115 Heidelberg, Germany. If you have any concerns regarding our products, please contact ProductSafety@springernature.com

Printed and bound by CPI Group (UK) Ltd, Croydon, CR0 4YY
05/10/2026
02228120-0007